Research and Development in Intelligent Systems XXXII

Incorporating Applications and Innovations in Intelligent Systems XXIII

Max Bramer · Miltos Petridis
Editors

Research and Development in Intelligent Systems XXXII

Incorporating Applications and Innovations in Intelligent Systems XXIII

Proceedings of AI-2015, The Thirty-Fifth SGAI International Conference on Innovative Techniques and Applications of Artificial Intelligence

Editors
Max Bramer
School of Computing
University of Portsmouth
Portsmouth
UK

Miltos Petridis
School of Computing, Engineering and Mathematics
University of Brighton
Brighton
UK

ISBN 978-3-319-25030-4 ISBN 978-3-319-25032-8 (eBook)
DOI 10.1007/978-3-319-25032-8

Library of Congress Control Number: 2015952537

Springer Cham Heidelberg New York Dordrecht London

Printed on acid-free paper

Springer International Publishing AG Switzerland is part of Springer Science+Business Media (www.springer.com)

Programme Chairs' Introduction

This volume comprises the refereed papers presented at AI-2015, the Thirty-fifth SGAI International Conference on Innovative Techniques and Applications of Artificial Intelligence, held in Cambridge in December 2015 in both the technical and the application streams. The conference was organised by SGAI, the British Computer Society Specialist Group on Artificial Intelligence.

The technical papers included present new and innovative developments in the field, divided into sections on Knowledge Discovery and Data Mining, Machine Learning and Knowledge Acquisition, and AI in Action. This year's Donald Michie Memorial Award for the best refereed technical paper was won by a paper entitled "Sparse Covariance Matrix Adaptation Techniques for Evolution Strategies" by S. Meyer-Nieberg and E. Kropat (Universität der Bundeswehr, München, Germany).

The application papers included present innovative applications of AI techniques in a number of subject domains. This year, the papers are divided into sections on Applications of Genetic Algorithms, Applications of Intelligent Agents and Evolutionary Techniques, and AI Applications. This year's Rob Milne Memorial Award for the best refereed application paper was won by a paper entitled "Development of Temporal Logic-Based Fuzzy Decision Support System for Diagnosis of Acute Rheumatic Fever/Rheumatic Heart Disease" by Sanjib Raj Pandey, Jixin Ma and Chong Hong Lai (University of Greenwich, UK).

The volume also includes the text of short papers presented as posters at the conference.

On behalf of the conference organising committee we would like to thank all those who contributed to the organisation of this year's programme, in particular the programme committee members, the executive programme committees and our administrators Mandy Bauer and Bryony Bramer.

Max Bramer, Technical Programme Chair, AI-2015.

Miltos Petridis, Application Programme Chair, AI-2015.

Acknowledgements/Committees

AI-2015 Conference Committee

Prof. Max Bramer, University of Portsmouth (Conference Chair)
Prof. Max Bramer, University of Portsmouth (Technical Programme Chair)
Prof. Miltos Petridis, University of Brighton (Application Programme Chair)
Dr. Jixin Ma, University of Greenwich (Deputy Application Programme Chair)
Prof. Adrian Hopgood, Sheffield Hallam University (Workshop Organiser)
Rosemary Gilligan, University of Hertfordshire (Treasurer)
Dr. Nirmalie Wiratunga, The Robert Gordon University (Poster Session Organiser)
Andrew Lea, Primary Key Associates Ltd. (AI Open Mic and Panel Session Organiser)
Giovanna Martinez, Nottingham Trent University and Christo Fogelberg, Palantir Technologies (FAIRS 2015)
Prof. Miltos Petridis, University of Brighton and Prof. Thomas Roth-Berghofer, University of West London (UK CBR Organisers)
Mandy Bauer, BCS (Conference Administrator)
Bryony Bramer (Paper Administrator)

Technical Executive Programme Committee

Prof. Max Bramer, University of Portsmouth (Chair)
Dr. John Kingston, Tribal Group
Prof. Thomas Roth-Berghofer, University of West London
Dr. Nirmalie Wiratunga, Robert Gordon University, Aberdeen

Applications Executive Programme Committee

Technical Programme Committee

Application Programme Committee

Contents

Applications of Genetic Algorithms

Applications of Intelligent Agents and Evolutionary Techniques

AI Applications

Short Papers

Research and Development in Intelligent Systems XXXII

Best Technical Paper

Sparse Covariance Matrix Adaptation Techniques for Evolution Strategies

Silja Meyer-Nieberg and Erik Kropat

Abstract Evolution strategies are variants of evolutionary algorithms. In contrast to genetic algorithms, their search process depends strongly on mutation. Since the search space is often continuous, evolution strategies use a multivariate normal distribution as search distribution. This necessitates the tuning and adaptation of the covariance matrix. Modern evolution strategies apply covariance matrix adaptation mechanisms in order to achieve this end. However, the covariance estimation is conducted with small sample sizes compared to the search space dimensionality. Regarding the agreement of sample estimate and true covariance, this represents a potential problem. This paper introduces a new approach by changing the coordinate systems and implements several sparse covariance matrix techniques. The results are evaluated in experiments.

1 Introduction

Evolution strategies (ESs) are a variant of evolutionary algorithms often used for continuous black-box optimization. They differ from many other evolutionary algorithms in the role of mutation: While it is only a background operator in genetic algorithms, it represents the main search operator here. Evolution strategies operate with a multivariate normal distribution which is used to generate a population of new search points. Its parameters, the mean and the covariance matrix, must be updated during a run in order for the strategy to reach the vicinity of optimal points fast and reliably. The adaptation of the parameters takes the search history and the present population into account. Due to its importance, research focussed and focusses on the covariance matrix. The main techniques introduced are based on the sample covariance matrix [1]. The usage of this estimator may bear potential improvement points

S. Meyer-Nieberg (✉) · E. Kropat
Department of Computer Science, Universität der Bundeswehr München,
Werner-Heisenberg Weg 39, 85577 Neubiberg, Germany
e-mail: silja.meyer-nieberg@unibw.de

E. Kropat
e-mail: erik.kropat@unibw.de

M. Bramer and M. Petridis (eds.), *Research and Development in Intelligent Systems XXXII*, DOI 10.1007/978-3-319-25032-8_1

within itself: Evolution strategies typically operate with small population or sample sizes. The size of the population does not exceed the search space dimensionality. Estimating the $N \times N$ dimensional covariance matrix with a sample size of $\mu < N$ or $\mu \approx N$ leads to unreliable estimates. All adaptation techniques introduced so far consider correction terms. However, the question remains whether an ES may benefit if techniques developed for and tailored to the task at hand were introduced.

Literature concerning attempts of combining evolutionary algorithms or related approaches with statistical estimation methods of high-dimensional covariance matrices is scarce. So far, we were only able to identify two approaches aside from our own research: In the first [6], the authors investigated estimation of distribution algorithms (EDAs) for continuous spaces. The EDA applied a Gaussian search distortion similar to evolution strategies. The estimation of the covariance matrix resulted however in matrices that were not positive definite. To circumvent the problem, a shrinkage procedure was introduced, see e.g. [13]. Very recently, a shrinkage estimator was integrated into an evolution strategy variant with a single search point [12].

The research presented here is part of an ongoing investigation into alternative estimation techniques for high-dimensional covariances [15, 16]. In [15, 16] Ledoit-Wolf shrinkage estimators were analyzed. While the results were promising, finding the appropriate shrinkage intensity represented a challenge. Therefore, in [14] another computational simple estimation method was introduced: thresholding. Here the work begun in [14] is continued by addressing two of open problems remaining: The first concerns the choice of the thresholding function, the latter the influence of an important parameter of the thresholding.

The paper is structured as follows. First, the evolution strategy variant considered in this paper is introduced. Afterwards, we argue why high-dimensional estimation techniques might improve the performance. The next section introduces the sparse covariance estimation evolution strategy developed. An experimental analysis of the approaches follows, before the paper closes with an outlook on potential future research.

2 Evolution Strategies

Evolution strategies (ESs) [18, 19] are used for continuous black-box optimization $f : \mathbb{R}^N \rightarrow \mathbb{R}$. Several variants have been introduced (see e.g. [1, 3]). In many cases, a population of μ parents is used to create a set of λ offspring, with $\mu \leq \lambda$. Like all evolutionary algorithms, evolution strategies operate in a sequence of generations. In each generation, the same cycle of processes is carried out. In general, these are parent selection, recombination, mutation, and survivor selection. In the following, the processes are described based on the ES variant considered. Here, all μ parents contribute to create the offspring. First recombination is performed, that is, the centroid of the parents is computed [3]. All offspring are based on the same origin and differ only in their mutation vector, a normally distributed random variable with zero mean and covariance matrix $\sigma^2\mathbf{C}$ which is added to the mean. After the λ offspring $\mathbf{y}_1, \ldots, \mathbf{y}_\lambda$ have been created, the individuals are evaluated. In most cases, the function to be optimized is used directly. In that case, the function is also called fitness. Selection in evolution strategies takes often only the λ offspring into account of which the μ best $\mathbf{y}_{1:\lambda}, \ldots, \mathbf{y}_{\mu:\lambda}$ are chosen.

The most important factor concerning the mutation is the covariance matrix. It must be adapted during the run and fitted to the landscape. Otherwise, the performance may be low. Therefore, research on controlling the mutation has a long tradition in ESs. First approaches were already considered in [18]. The next section describes the variant considered in this paper.

2.1 Covariance Matrix Adaptation: The Population Covariance

To our knowledge, covariance matrix adaptation comprises two main classes: one applied in the *covariance matrix adaptation evolution strategy* (CMA-ES) [11] and an alternative used in the *covariance matrix self-adaptation evolution strategy* (CMSA-ES) [4]. Both are based on a variant of the sample covariance, correcting the estimate with information from the search history. The present paper focuses on the CMSA-ES leaving the CMA-ES for future research. One of the reasons is that the CMSA-ES does only include one additional correction term making it easier to assess the effects of the thresholding operator. The CMSA-ES considers the covariance matrix $(\sigma^{(g)})^2\mathbf{C}^{(g)}$ with $\sigma^{(g)}$ denoting general scaling factor (or step-size or mutation strength) and with $\mathbf{C}^{(g)}$ a rotation matrix. Following the usual practice in literature on evolution strategies the latter matrix $\mathbf{C}^{(g)}$ is referred to as *covariance matrix* in the remainder of the paper. The CMSA uses covariance matrix adaptation for the matrix $\mathbf{C}^{(g)}$ and self-adaptation for the mutation strength.

The covariance matrix update is based upon the common estimate of the covariance matrix using the newly created population. However, the sample consists of the selected parents and not of the complete set. Restricting the sample, shall induce a bias towards promising parts of the search space. Since the adaptation of the mutation strength happens separately, the sample is normalized with $\mathbf{z}_{m:\lambda}^{(g+1)} := \left(\mathbf{x}_{m:\lambda}^{(g+1)} - \mathbf{m}^{(g)}\right)/\sigma^{(g)}$ before estimating the covariance, see also [11]. Since the centroid used for the mutation is known, the covariance matrix estimation does not need to re-estimate the mean. The rank-μ update then obtains the covariance matrix as

$$\mathbf{C}_\mu^{(g+1)} := \sum_{m=1}^{\mu} w_m \mathbf{z}_{m:\lambda}^{(g+1)} (\mathbf{z}_{m:\lambda}^{(g+1)})^{\mathrm{T}} \tag{1}$$

which is usually a positive semi-definite matrix since $\mu \ll N$. The weights w_m should fulfill $w_1 \geq w_2 \geq \ldots \geq w_m$ with $\sum_{m=1}^{\mu} w_i = 1$. While it is possible to consider unequal weights, the CMSA-ES usually operates with $w_m = 1/\mu$. To derive reliable estimates larger population sizes are required which would lower the algorithm's speed. Therefore, past covariance matrices are taken into account via the convex combination of (1) with the sample covariance and the old covariance

$$\mathbf{C}^{(g+1)} := (1 - \frac{1}{c_\tau})\mathbf{C}^{(g)} + \frac{1}{c_\tau}\mathbf{C}_\mu^{(g+1)} \quad (2)$$

with the weights usually set to $w_m = 1/\mu$ and following [4]

$$c_\tau = 1 + \frac{N(N+1)}{2\mu}. \quad (3)$$

2.2 *Step-Size Adaptation*

The CMSA implements the step-size using *self-adaptation* first introduced in [18] and developed further in [19]. Here, evolution is used to tune the strategy parameters of the mutation process. In other words, these parameters undergo recombination, mutation, and indirect selection processes. The working principle is based on an indirect stochastic linkage between good individuals and appropriate parameters: Well-adapted parameters should result more often in better offspring than too large or too small values or misleading directions. Although self-adaptation has been developed to adapt the whole covariance matrix, it is applied today mainly to adapt the step-size or a diagonal covariance matrix. In the case of the mutation strength, usually a log-normal distribution $\sigma_l^{(g)} = \sigma^{(g)}\exp(\tau\mathcal{N}(0,1))$ is used for the mutation of the mutation strength. The parameter τ, the *learning rate*, should scale with $1/\sqrt{2N}$. The CMSA-ES often uses recombination. Among others, self-adaptation with recombination improves the performance in the presence of noise [2]. While the recombination of the mutation strength could be realized in several ways, it normally follows the recombination of the objective values in computing the mean of the mutation strengths of the parents. The newly created mutation strength $\sigma_l^{(g)}$ is then used for mutating the objective values of the offspring. If the resulting offspring is sufficiently good, the scale factor is passed to the next generation.

3 A Sparse Covariance Matrix Adaptation

This section introduces the new covariance adaptation technique which uses thresholding to transform the population covariance matrix. The decision for thresholding is based upon the comparatively computational efficiency of the approach.

The sample covariance (1) has a strong influence on the adaptation. However, the good properties of the maximum likelihood estimator hold for the case $\mu \gg N$ and $\mu \gg 1$. In evolution strategies, the sample size seldom exceeds the search space dimension with $\mu < N$. For example, [9] recommends to use $\lambda = \lfloor \log(3N) \rfloor + 4$ offspring and to set the size of the parent population to $\mu = \lfloor \lambda/2 \rfloor$. Thus, a potential problem arises in high-dimensional settings. For $N \to \infty$, we have $\mu/N \to 0$ contradicting the assumptions on which the estimator was based.

In order to assess the problem in evolution strategies, we take a closer look at the eigenvalues of the covariance matrix for some selected functions. Figure 1 shows the development of the ratio of the largest to the smallest eigenvalue of the covariance matrix on the sphere $f(\mathbf{x}) = \|\mathbf{x}\|^2$ and on the discus $f(\mathbf{x}) = 10^6 x_1^2 + \sum_{i=2}^{N} x_i^2$. In the latter case it can be argued that the behavior observed is beneficial. For the sphere, the figures hint at a potential problem: The gap between largest and smallest eigenvalue widens for all runs with the problem being more pronounced for the smaller search space dimensionalities. Furthermore, the extremely small sample size for $N = 10$ causes a large variation between the runs. It is interestingly less distinct in the case of the higher dimensional search spaces. This is probably an effect of the parameter c_τ which follows $\lim_{N \to \infty} c_\tau(N) = \infty$ as long as $\mu \propto \log(N)$ or $\mu \propto N$. Thus, the influence of the population covariance lessens. In statistics the problem is well-known [20, 21] with a long research tradition concerning approaches to counteract the problematic properties, see e.g. [17] for an overview. Among others, it has been shown that the eigenstructures of the estimate and the covariance do not agree well.

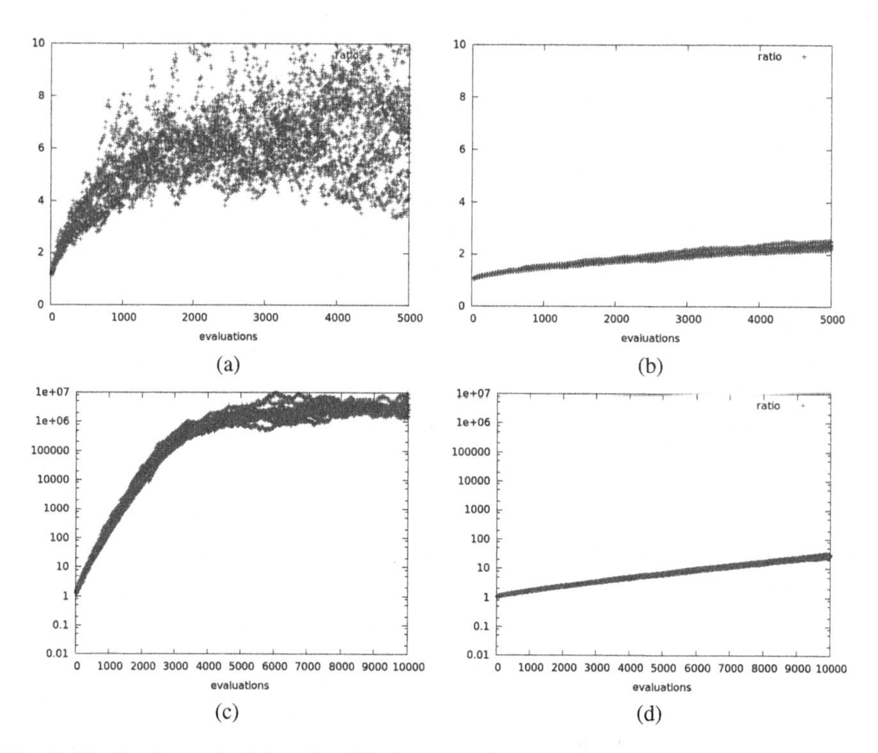

Fig. 1 The development of the ratio of the largest to the smallest eigenvalue of the covariance for the CMSA-ES on the sphere and the discus. Shown are the results from 15 runs for each dimensionality. **a** Sphere, N = 10. **b** Sphere, N = 40. **c** Discus, N = 10. **d** Discus, N = 40

3.1 Space Transformation

Several types of estimators assume a sparse structure of the covariance matrix. Shortly stated, these approaches work well if many entries are small or even zero. Then, computationally simple estimation techniques can be applied. In the case of evolution strategies, a sparseness assumption may not hold in every situation. The form of the covariance matrix depends strongly on the function landscape and may vary widely in practice. Furthermore, there may not be any information available concerning the type of the function itself. Therefore, the covariance matrix is not considered in the original space but in the eigenspace of the previous covariance matrix $\mathbf{C}^{(g)}$.

Let the covariance matrix $\mathbf{C}^{(g)}$ be a symmetric, positive definite $N \times N$ matrix. The condition holds for the original adaptation since (2) combines a positive definite with a positive semi-definite matrix. As we will see below, in the case of thresholding the assumption may not always be fulfilled. Let $\mathbf{v}_1, \ldots, \mathbf{v}_N$ denote the N eigenvectors with the eigenvalues $\lambda_1, \ldots, \lambda_N$, $\lambda_j > 0$. The definiteness of $\mathbf{C}^{(g)}$ guarantees their existence. The eigenvectors form a orthonormal basis of $\mathbb{R}^N$, i.e., $\mathbf{v}_i^{\mathrm{T}}\mathbf{v}_i = 1$ and $\mathbf{v}_i^{\mathrm{T}}\mathbf{v}_j = 0$, if $i \neq j$. Define $\mathbf{V} := (\mathbf{v}_1, \ldots, \mathbf{v}_N)$. It then holds that $\mathbf{V}^{-1} = \mathbf{V}^{\mathrm{T}}$. Switching to the eigenspace of $\mathbf{C}^{(g)}$ results in the representation of the covariance matrix $\Lambda^{(g)} = \mathbf{V}\mathbf{C}^{(g)}\mathbf{V}^{\mathrm{T}}$ with $\Lambda^{(g)}$ a diagonal matrix containing the eigenvalues. Diagonal matrices are sparse, thus for the estimation of the covariance matrix the more efficient procedures for sparse structures could be used. However, it is not the goal to re-estimate $\mathbf{C}^{(g)}$ but to estimate the true covariance matrix of the distribution indicated by the sample $\mathbf{z}_{1;\lambda}, \ldots, \mathbf{z}_{\mu;\lambda}$.

Before continuing, it should be noted that several definitions of sparseness have been introduced. For instance, the number of non-zero elements in a row may not exceed a predefined limit $s_0(N) > 0$, i.e., $\max_i \sum_{j=1}^{N} \delta(|a_{ij}| > 0) \leq s_0(N)$, which should grow only slowly with N. This definition can, however, be relaxed to a more general definition of sparseness, also referred to as approximate sparseness [5] on which the adaptive thresholding considered is based. Applying thresholding in our case requires that the true covariance matrix of the selected set has an approximately sparse structure in the eigenspace of $\mathbf{C}^{(g)}$. Assuming the validity of the assumption, we change the coordinate system in order to perform the covariance matrix estimation. Reconsider the normalized (aside from the covariance matrix) mutation vectors $\mathbf{z}_{1;\lambda}, \ldots, \mathbf{z}_{\mu;\lambda}$ that were associated with the μ best offspring. Denoting their representation in the eigenspace as $\hat{\mathbf{z}}_{m;\lambda} = \mathbf{V}^{\mathrm{T}}\mathbf{z}_{m;\lambda}$ for $m = 1, \ldots, \mu$ leads to the new population covariance

$$\hat{\mathbf{C}}_\mu = \sum_{i=1}^{\mu} w_i \hat{\mathbf{z}}_{m;\lambda} \hat{\mathbf{z}}_{m;\lambda}^{\mathrm{T}} \tag{4}$$

which is used to derive the final estimate. In the next section, potential techniques for sparse covariance matrices are discussed.

3.2 Sparse Covariance Matrix Estimation

Several methods have been developed for estimating sparse covariance matrices: Among others banding, tapering, and thresholding can be applied, see e.g. [17]. While all three are based on the assumption that many entries of the true covariance are zero, banding and tapering assume an ordering of the variables which is not present in the case of evolution strategies.

Therefore, only thresholding remains. Thresholding discards entries which are smaller than a given threshold $\varepsilon > 0$. For a matrix $\mathbf{A} \in \mathbb{R}^{N \times N}$, the thresholding operator $T_\varepsilon(\mathbf{A})$ is defined as

$$T_\varepsilon(\mathbf{A}) := (a_{ij}\delta(|a_{ij}| \geq \varepsilon))_{N \times N} \tag{5}$$

with $\delta(\cdot) = 1$ if the condition is fulfilled and zero otherwise. The choice of the threshold is critical for the quality of the resulting estimate. Equation (5) represents an example of universal thresholding with a hard thresholding function. Soft thresholding is also common, examples of this function class comprise e.g.

$$s_\lambda(x) = \text{sign}(x)(|x| - \lambda)_+ \quad \text{(soft-thresholding)} \tag{6}$$

$$s_\lambda(x) = |x|(1 - |\frac{\lambda}{x}|^\eta)_+ \quad \text{(Lasso)} \tag{7}$$

with $(x)_+ := \max(0, x)$. Adaptive thresholding which considers the current data for determining the threshold λ_{ij} appears as more appropriate for evolution strategies than using constant thresholds. Following [5], we use

$$\lambda_{ij} := \lambda_{ij}(\delta) = \delta\sqrt{\frac{\hat{\theta}_{ij}\log N}{\mu}} \tag{8}$$

where $\delta > 0$ can be either chosen as a constant or be obtained using cross-validation. The variable $\hat{\theta}_{ij}$ in (8) is determined as $\hat{\theta}_{ij} = \frac{1}{\mu}\sum_{m=1}^{\mu}[(\hat{z}_{mi} - \overline{Z^i})(\hat{z}_{mj} - \overline{Z^j}) - \hat{c}_{ij}^{\mu}]^2$ with $\hat{c}_{ij}^{\mu}$ denoting the (i, j)-entry of $\hat{\mathbf{C}}_\mu^{(g+1)}$, $\hat{z}_{mi}$ the ith component of $\hat{\mathbf{z}}_{m:\lambda}$, and $\overline{Z^i} := (1/\mu)\sum_{m=1}^{\mu}\hat{z}_{mi}$.

While thresholding respects symmetry and non-negativeness properties, it results only in asymptotically positive definite matrices. Thus, for finite sample sizes, it does neither preserve nor induce positive definiteness in general. Due to this potential problem, future research will investigate repair mechanisms as well as alternative thresholding functions, see e.g. [7]. Here, the soft-thresholding (6) and the Lasso thresholding function (7) are considered. While it is common to exclude the diagonal entries of the covariance from thresholding, this may not be always appropriate for optimization since the nature of the functions may vary widely. Our previous experiments did not show a clear advantage for either method. Therefore, both versions are taken into account. In combination with the thresholding function, the following four

ES types are investigated: (1) CMSA-Thres-ES (abbreviated to Thres): an evolution strategy with CMSA which applies thresholding in the eigenspace of the covariance with soft-thresholding, (2) CMSA-ThresL-ES (abbreviated to ThresL): the same as above but using the Lasso thresholding, (3) CMSA-Diag-ES (abbreviated to Diag): an ES with covariance matrix adaptation with thresholding in the eigenspace of the covariance, preserving the diagonal elements, and using soft-thresholding, (4) CMSA-DiagL-ES (abbreviated to DiagL): the variation with the Lasso function.

4 Experiments

Two series of experiments were conducted: The first with the aim to gain more insight regarding the choice of the parameter δ. Our first approach was to make this parameter data dependent by setting it to $\delta = 2\max(\hat{\mathbf{C}}_\mu)$. Since [5] recommends to use either $\delta = 2$ or to conduct cross-validation, we performed a short experimental analysis and took a closer look at the development of the eigenvalues on the sphere and on the discus. We considered the $\delta = 2, 3$, and 4 for the CMSA-ThresL-ES with the search space dimensionalities set to $N = 10, 20, 40$, and 100.

The second series of experiments compares the different shrinkage variants with the original CMSA-ES. Two thresholding operators, soft thresholding and Lasso thresholding (with $\eta = 4$), are taken into account. The comparison is based on the search space dimensions $N = 10$ and 20. The second series of experiments uses a maximal number of fitness evaluations of $\text{FE}_{\text{max}} = 2 \times 10^5 N$. While the experiments revealed that longer experiments are necessary in order to derive meaningful findings for the difficult multimodal functions, the task was delegated to future research because of the computing time required.

All strategies start from randomly chosen positions, sampled uniformly from the interval $[-4, 4]^N$. The ESs used $\lambda = \lfloor \log(3N) + 8 \rfloor$ offspring and $\mu = \lceil \lambda/4 \rceil$ parents. An equal setting of weights w_m was used with $w_m = 1/\mu$. A run terminates before reaching the maximal number of evaluations, if the difference between the best value obtained so far and the optimal fitness value $|f_{\text{best}} - f_{\text{opt}}|$ is below a predefined target precision set to 10^{-8}. For each fitness function and dimension, 15 runs are conducted. In order not to waste resources, a run is restarted when a search stagnation is observed. The latter is characterized by observing changes of the best values below 10^{-8} over the last $10 + \lceil 30N/\lambda \rceil$ generations.

4.1 Test Suite Und Performance Measure

The algorithms were implemented in MATLAB. The paper uses the black box optimization benchmarking (BBOB) software framework and test suite, see [10]. The framework[1] can be used to benchmark and compare continuous black-box optimiz-

[1]Current software and tutorials under http://coco.gforge.inria.fr.

ers and provides easy means to generate tables and figures. This paper considers the 24 noiseless functions of the test suite [8]. In order to lower the possibility that an algorithm benefits from initialisation effects, the position of the optimum is changed from run to run. The test suite comprises four function classes which differ in the degree of difficulty they pose for the optimization: separable functions (function ids 1–5), functions with low/moderate conditioning (ids 6–9), functions with high conditioning (ids 10–14), and two groups of multimodal functions (ids 15–24), with the last comprising functions with a weak global structure.

Following [10], the expected running time (ERT) is used as the performance measure. It is defined as the expected value of the function evaluations (f-evaluations) required to reach the target value with the required precision for the first time, see [10]. In this paper, ERT $= \frac{\#(FEs(f_{\text{best}} \geq f_{\text{target}}))}{\#succ}$ is used as an estimate. It is obtained by summing up the evaluations $FEs(f_{\text{best}} \geq f_{\text{target}})$ in each run until the fitness of the best individual is smaller than the target value, divided by the number of all successful runs.

4.2 Results and Discussion

First, we describe the results from the parameter dependency experiments. The thresholding should on the one hand "stabilize" the covariance matrix in the sense that the eigenvalues do not diverge unless of course it is required to optimize the function. On the other hand, it should not delay or prohibit the adaptation of the covariance matrix to the function space. Summarizing the effects of operating with a data independent δ from $\{2, 3, 4\}$, this detrimental behavior can be observed. Thus, Fig. 2 only shows the development of the ratio of the largest and the smallest eigenvalue for the sphere and the discus for two exemplary search space dimensions using the data dependent δ. Comparing Figs. 2 to 1 reveals that for the sphere the variation between the runs is reduced even for the smaller search space. In the case of the discus, the increase of the ratio is slower, which could result in slower convergence.

The findings for the BBOB test suite indicate advantages for thresholding in many cases. The outcome of the comparison depends on the function class. In the case of the separable functions with ids 1–5, the strategies behave on the whole very similar for 10D and 20D. Concerning the particular functions, differences are revealed as Tables 1 and 2 show for the expected running time (ERT) provided for several precision targets. In the case of the sphere (function with id 1) and the separable ellipsoid (id 2), all strategies reach the final precision goal in all runs. For both functions, ESs with thresholding are the fastest. In the case of the sphere, preserving the diagonal elements appears slightly advantageous, however, all variants are close together. For the ellipsoid, the gap widens. Interestingly, two variants remain close together: the CMSA-Thres-ES and the CMSA-DiagL-ES which differ in the thresholding function as well as in the decision whether to subject the diagonal entries to thresholding or not. No strategy reaches the required target precision in the case of the separable

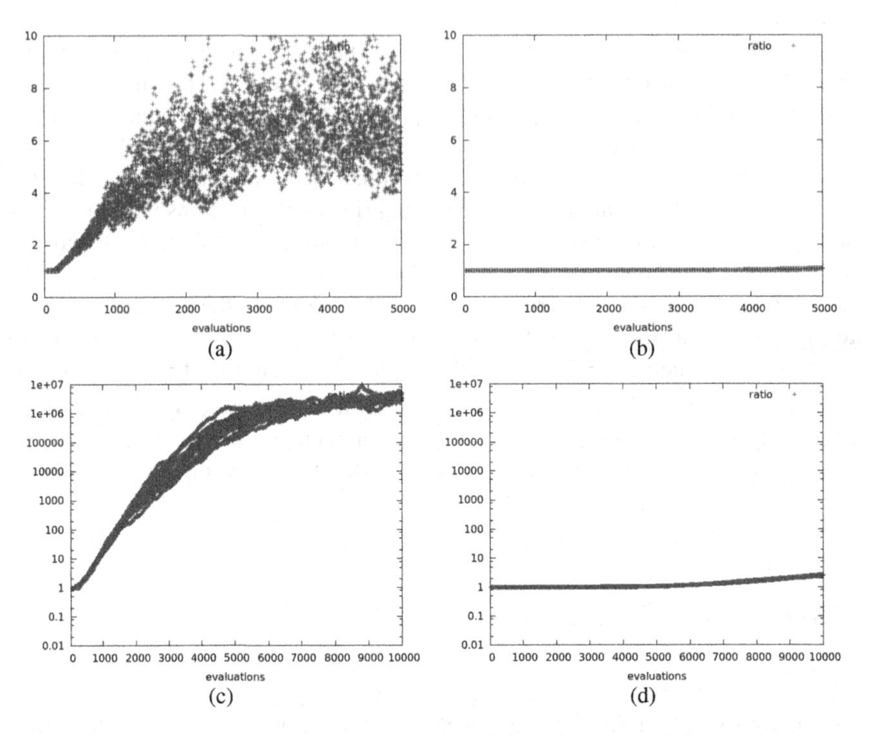

Fig. 2 Development of the ratio of the largest and the smallest eigenvalue of the covariance matrix of the CMSA-ThresL-ES on the sphere. Shown are the results from 15 repeats for each dimensionality. **a** Sphere, N = 10. **b** Sphere, N = 40. **c** Discus, N = 10. **d** Discus, N = 40

Rastrigin (id 3) and the separable Rastrigin-Bueche (id 4). Since all strategies only achieve the lowest target precision of 10^1, a comparison is not performed and due to page restrictions the data are removed from the tables. In the case of the linear slope (id 5) all strategies are successful. While the thresholding variants perform better for the smaller search space probably due to the more stable behavior of the covariance adaptation, the advantage is lost for $N = 20$ as Table 2 shows.

In the case of the functions with low to moderate conditioning (id 6–9), the step ellipsoid with id 7 is the most difficult function to optimize for the ESs. Experiments with a larger number of maximal function evaluations will be performed in future research. In the case of the remaining functions, we see a separation between the attractive sector (id 6) and the Rosenbrock variants (ids 8 and 9). In the case of the attractive sector and $N = 10$, see Table 1, the original CMSA-ES could only reach the required target precision in eight of the 15 runs, whereas the thresholding variants resulted only in one or two unsuccessful runs. Increasing the search space dimensionality, causes all runs of the CMSA-ES to be unsuccessful while the thresholding variants still achieve two or three successful runs. On the original Rosenbrock function (id 8), the CMSA-Thres-ES with the soft-thresholding function is the worst performing strategy. For $N = 20$, Table 2, the CMSA-DiagL-ES which uses the Lasso also

Table 1 Expected running time (ERT in number of function evaluations) divided by the respective best ERT measured during BBOB-2009 in dimension 10

Δf_{opt}	1e1	1e0	1e-1	1e-2	1e-3	1e-5	1e-7	#succ	Δf_{opt}	1e1	1e0	1e-1	1e-2	1e-3	1e-5	1e-7	#succ
f1	22	23	23	23	23	23	23	15/15	**f11**	266	1041	2602	2954	3338	4092	4843	15/15
CMSA	4.4(2)	10(5)	15(6)	21(7)	26(6)	38(9)	51(3)	15/15	CMSA	17(12)	6.7(3)	3.3(2)	3.7(2)	3.8(1)	3.6(1)	3.4(0.8)	15/15
Thres	3.7(1)	8.3(1)	14(2)	19(2)	24(3)	35(6)	46(4)	15/15	Thres	108(22)	36(10)	16(4)	15(4)	13(3)	11(3)	10(5)	15/15
Diag	3.9(1)	8.5(4)	13(2)	**18**(3)	23(5)	**33**(6)	45(8)	15/15	Diag	19(6)	7.2(2)	3.6(1.0)	3.7(1)	3.7(0.8)	3.4(0.7)	3.0(0.7)	15/15
DiagL	**3.2**(2)	**7.5**(1)	**12**(2)	18(3)	**23**(3)	34(3)	**44**(3)	15/15	DiagL	84(28)	30(8)	13(6)	12(2)	11(2)	10(2)	9.0(3)	15/15
ThresL	4.2(3)	9.2(3)	14(3)	18(3)	24(5)	35(5)	46(5)	15/15	ThresL	**17**(2)	**6.1**(2)	**3.0**(1)	**3.0**(0.7)	**3.0**(0.6)	**2.9**(1)	**2.7**(1)	15/15
Δf_{opt}	1e1	1e0	1e-1	1e-2	1e-3	1e-5	1e-7	#succ	Δf_{opt}	1e1	1e0	1e-1	1e-2	1e-3	1e-5	1e-7	#succ
f2	187	190	191	191	193	194	195	15/15	**f12**	515	896	1240	1390	1569	3660	5154	15/15
CMSA	70(33)	85(38)	95(30)	104(54)	109(40)	117(53)	120(54)	15/15	CMSA	**4.7**(8)	**13**(9)	**15**(8)	**17**(7)	**17**(7)	**10**(4)	**9.2**(4)	15/15
Thres	**41**(32)	**64**(56)	**78**(63)	**87**(46)	**91**(59)	**97**(62)	**101**(62)	15/15	Thres	30(50)	36(30)	37(22)	39(16)	37(16)	18(3)	14(5)	15/15
Diag	67(43)	103(20)	124(48)	133(44)	141(24)	149(53)	153(46)	15/15	Diag	7.4(16)	16(19)	18(15)	19(16)	20(14)	11(5)	9.2(4)	15/15
DiagL	55(31)	73(49)	88(61)	97(56)	101(62)	107(56)	111(63)	15/15	DiagL	14(20)	29(27)	33(10)	35(11)	34(8)	17(5)	15(10)	15/15
ThresL	71(23)	88(23)	100(21)	109(25)	113(19)	120(19)	125(20)	15/15	ThresL	13(15)	24(16)	24(14)	24(8)	24(12)	12(5)	12(7)	15/15
Δf_{opt}	1e1	1e0	1e-1	1e-2	1e-3	1e-5	1e-7	#succ	Δf_{opt}	1e1	1e0	1e-1	1e-2	1e-3	1e-5	1e-7	#succ
f5	20	20	20	20	20	20	20	15/15	**f13**	387	596	797	1014	4587	6208	7779	15/15
CMSA	12(4)	17(4)	18(13)	18(11)	18(10)	18(9)	18(10)	15/15	CMSA	8.2(20)	**22**(19)	**43**(74)	**65**(101)	**27**(31)	**106**(53)	**377**(559)	1/15
Thres	**11**(2)	**16**(8)	**16**(4)	**16**(4)	**17**(8)	**17**(3)	**17**(8)	15/15	Thres	**4.8**(7)	30(19)	73(124)	219(320)	135(171)	∞	∞ *2e5*	0/15
Diag	15(3)	23(9)	24(14)	24(11)	24(10)	24(13)	24(13)	15/15	Diag	7.4(7)	45(65)	67(48)	124(235)	74(63)	236(129)	382(154)	1/15
DiagL	13(4)	17(6)	18(4)	18(6)	18(11)	18(11)	18(8)	15/15	DiagL	12(23)	52(168)	71(53)	164(104)	117(155)	∞	∞ *2e5*	0/15
ThresL	14(4)	19(6)	21(6)	21(9)	21(11)	21(7)	21(10)	15/15	ThresL	6.3(13)	35(27)	56(83)	107(117)	72(88)	461(540)	∞ *2e5*	0/15
Δf_{opt}	1e1	1e0	1e-1	1e-2	1e-3	1e-5	1e-7	#succ	Δf_{opt}	1e1	1e0	1e-1	1e-2	1e-3	1e-5	1e-7	#succ
f6	412	623	826	1039	1292	1841	2370	15/15	**f14**	37	98	133	205	392	687	4305	15/15
CMSA	2.0(0.9)	2.9(5)	5.3(17)	7.7(9)	13(28)	21(17)	112(104)	8/15	CMSA	1.2(0.8)	2.2(0.9)	2.9(1.0)	3.2(1)	**4.1**(1)	10(5)	5.5(8)	15/15
Thres	**1.3**(0.5)	**2.7**(3)	5.1(1)	8.0(10)	13(6)	24(26)	35(91)	13/15	Thres	1.1(0.9)	2.1(0.8)	2.7(0.5)	**3.2**(0.7)	10(6)	30(16)	14(8)	15/15
Diag	2.2(2)	3.8(1)	**4.2**(5)	6.0(8)	13(13)	23(40)	43(21)	12/15	Diag	1.3(1)	2.8(0.6)	3.1(1.0)	3.4(1)	5.2(1)	**8.5**(2)	5.4(2)	15/15
DiagL	1.6(0.6)	2.8(1)	4.3(2)	**4.4**(4)	**4.7**(3)	**13**(31)	**21**(45)	13/15	DiagL	**0.64**(0.2)	**1.8**(0.7)	2.7(1)	3.6(1.0)	9.1(3)	23(8)	9.1(3)	15/15
ThresL	1.8(0.6)	5.4(22)	7.0(4)	6.9(11)	10(25)	20(2)	30(27)	14/15	ThresL	0.90(0.7)	2.2(0.9)	**2.6**(1.0)	3.2(0.9)	4.3(2)	8.7(3)	**4.4**(1)	15/15

(continued)

Table 1 (continued)

Δf_{opt}	1e1	1e0	1e-1	1e-2	1e-3	1e-5	1e-7	#succ	Δf_{opt}	1e1	1e0	1e-1	1e-2	1e-3	1e-5	1e-7	#succ
f8	326	921	1114	1217	1267	1315	1343	15/15	**f18**	238	836	7012	15928	27536	37234	42708	15/15
CMSA	**3.5**(4)	18(13)	19(10)	18(5)	18(4)	18(11)	19(7)	15/15	CMSA	5.8(34)	289(1098)	399(606)	∞	∞	∞	∞ *2e5*	0/15
Thres	7.5(7)	35(46)	33(13)	32(31)	31(33)	31(31)	31(8)	15/15	Thres	31(0.4)	193(177)	399(278)	∞	∞	∞	∞ *2e5*	0/15
Diag	6.4(7)	**14**(10)	**15**(8)	**15**(8)	**15**(3)	**15**(8)	**16**(6)	15/15	Diag	7.9(0.7)	**182**(161)	196(128)	∞	∞	∞	∞ *2e5*	0/15
DiagL	6.6(0.3)	17(4)	18(3)	17(7)	17(4)	18(2)	18(4)	15/15	DiagL	**5.1**(0.5)	189(299)	192(268)	∞	∞	∞	∞ *2e5*	0/15
ThresL	8.5(10)	18(11)	18(10)	18(9)	18(7)	18(7)	18(7)	15/15	ThresL	88(241)	313(509)	**130**(110)	**184**(270)	∞	∞	∞ *2e5*	0/15
Δf_{opt}	1e1	1e0	1e-1	1e-2	1e-3	1e-5	1e-7	#succ	Δf_{opt}	1e1	1e0	1e-1	1e-2	1e-3	1e-5	1e-7	#succ
f9	200	648	857	993	1065	1138	1185	15/15	**f21**	130	2236	4392	4487	4618	5074	11329	8/15
CMSA	**2.2**(0.7)	31(53)	28(13)	26(36)	25(32)	24(13)	24(9)	15/15	CMSA	7.6(12)	20(38)	17(10)	17(8)	16(13)	15(20)	6.7(8)	13/15
Thres	3.3(1)	34(22)	30(16)	28(7)	27(2)	26(12)	25(12)	15/15	Thres	10(34)	**8.8**(12)	**12**(22)	**12**(39)	**12**(5)	**11**(6)	**4.8**(2)	13/15
Diag	7.8(14)	24(15)	22(13)	20(3)	20(3)	20(3)	20(9)	15/15	Diag	8.0(0.8)	19(22)	19(29)	18(36)	18(19)	16(25)	7.3(8)	13/15
DiagL	4.9(10)	35(29)	31(13)	29(18)	27(9)	27(15)	26(12)	15/15	DiagL	21(60)	19(37)	20(20)	20(14)	19(57)	18(39)	8.0(9)	12/15
ThresL	4.4(5)	**18**(19)	**19**(8)	**18**(12)	**17**(12)	**17**(3)	**17**(7)	15/15	ThresL	**5.9**(10)	17(8)	15(10)	15(15)	14(26)	13(23)	5.9(12)	13/15
Δf_{opt}	1e1	1e0	1e-1	1e-2	1e-3	1e-5	1e-7	#succ	Δf_{opt}	1e1	1e0	1e-1	1e-2	1e-3	1e-5	1e-7	#succ
f10	1835	2172	2455	2728	2802	4543	4739	15/15	**f22**	98	2839	6353	6620	6798	8296	10351	6/15
CMSA	6.8(3)	**7.9**(2)	**7.9**(2)	**7.5**(3)	**7.8**(2)	5.2(0.9)	**5.1**(1)	15/15	CMSA	26(60)	**4.1**(11)	**4.6**(17)	**4.6**(12)	**4.5**(5)	**3.9**(3)	**3.2**(7)	14/15
Thres	16(7)	16(3)	16(3)	14(3)	14(3)	9.1(2)	9.0(2)	15/15	Thres	26(26)	10(17)	12(15)	12(11)	12(7)	10(8)	8.4(8)	13/15
Diag	8.7(3)	10(3)	9.4(2)	9.0(2)	9.4(2)	6.1(1)	6.1(1)	15/15	Diag	**18**(12)	8.1(14)	8.6(21)	8.4(10)	8.3(7)	6.9(10)	5.7(7)	14/15
DiagL	14(4)	14(3)	14(2)	13(2)	13(2)	8.4(1)	8.3(1)	15/15	DiagL	60(15)	8.9(29)	8.8(13)	8.8(20)	8.8(14)	7.8(13)	6.5(6)	14/15
ThresL	**6.6**(2)	8.3(2)	8.2(2)	7.8(1)	8.0(2)	**5.2**(0.8)	5.3(1)	15/15	ThresL	30(27)	8.8(8)	13(53)	12(19)	12(10)	10(12)	8.2(16)	13/15

The ERT and in braces, as dispersion measure, the half difference between 90 and 10 %-tile of bootstrapped run lengths appear for each algorithm and target, the corresponding best ERT in the first row. The different target Δf-values are shown in the top row. #succ is the number of trials that reached the (final) target $f_{opt} + 10^{-8}$. The median number of conducted function evaluations is additionally given in *italics*, if the target in the last column was never reached. Entries, succeeded by a star, are statistically significantly better (according to the rank-sum test) when compared to all other algorithms of the table, with $p = 0.05$ or $p = 10^{-k}$ when the number k following the star is larger than 1, with Bonferroni correction by the number of instances

Table 2 Expected running time (ERT in number of function evaluations) divided by the respective best ERT measured during BBOB-2009 in dimension 20

Δf_{opt}	1e1	1e0	1e-1	1e-2	1e-3	1e-5	1e-7	#succ	Δf_{opt}	1e1	1e0	1e-1	1e-2	1e-3	1e-5	1e-7	#succ
f1	43	43	43	43	43	43	43	15/15	**f9**	1716	3102	3277	3379	3455	3594	3727	15/15
CMSA	**4.7**(0.7)	10(2)	14(2)	19(3)	25(2)	35(3)	45(2)	15/15	CMSA	**14**(7)	51(33)	52(62)	51(64)	51(91)	49(29)	48(135)	12/15
Thres	4.9(1)	9.3(1)	14(2)	19(2)	23(2)	33(1)	42(1)	15/15	Thres	34(19)	51(3)	55(32)	57(32)	59(31)	60(29)	60(30)	14/15
Diag	5.3(3)	9.5(3)	14(3)	**18**(1)	**22**(3)	**31**(4)	**40**(4)	15/15	Diag	18(6)	53(101)	53(70)	53(62)	52(61)	51(4)	50(29)	12/15
DiagL	4.9(1)	**9.2**(2)	**13**(2)	18(2)	23(3)	33(2)	42(3)	15/15	DiagL	36(11)	52(35)	54(6)	56(5)	57(32)	57(57)	57(27)	14/15
ThresL	4.9(1)	9.5(1)	14(1)	18(3)	23(2)	33(4)	42(4)	15/15	ThresL	15(3)	**20**(6)	**22**(7)	**23**(7)	**23**(7)	**23**(2)	**23**(5)	15/15
Δf_{opt}	1e1	1e0	1e-1	1e-2	1e-3	1e-5	1e-7	#succ	Δf_{opt}	1e1	1e0	1e-1	1e-2	1e-3	1e-5	1e-7	#succ
f2	385	386	387	388	390	391	393	15/15	**f10**	7413	8661	10735	13641	14920	17073	17476	15/15
CMSA	167(44)	221(53)	245(51)	268(51)	279(35)	290(40)	298(39)	15/15	CMSA	8.6(0.8)	**10**(1)	**9.0**(1)	**7.3**(2)	**6.9**(0.8)	**6.3**(0.4)	**6.4**(1.0)	15/15
Thres	**95**(15)	**109**(12)	**117**(4)	**120**(8)	**124**(7)	**127**(10)	**127**(7)	15/15	Thres	35(3)	33(2)	27(2)	22(2)	21(1)	18(2)	18(1)	15/15
Diag	169(46)	231(41)	246(67)	261(41)	270(37)	282(32)	291(30)	15/15	Diag	9.1(4)	10(2)	9.1(2)	7.6(0.9)	7.1(0.5)	6.5(0.5)	6.6(0.4)	15/15
DiagL	96(20)	113(11)	122(12)	126(5)	128(14)	130(10)	130(9)	15/15	DiagL	33(2)	31(3)	26(2)	20(2)	19(2)	17(1)	17(1)	15/15
ThresL	154(35)	212(65)	245(26)	259(12)	265(34)	273(31)	282(38)	15/15	ThresL	**8.4**(3)	10(2)	9.1(2)	7.8(1)	7.4(1.0)	6.8(0.5)	6.8(0.7)	15/15
Δf_{opt}	1e1	1e0	1e-1	1e-2	1e-3	1e-5	1e-7	#succ	Δf_{opt}	1e1	1e0	1e-1	1e-2	1e-3	1e-5	1e-7	#succ
f5	41	41	41	41	41	41	41	15/15	**f11**	1002	2228	6278	8586	9762	12285	14831	15/15
CMSA	**11**(4)	**13**(6)	**14**(3)	**14**(4)	**14**(6)	**14**(2)	**14**(5)	15/15	CMSA	**12**(2)$^{\star 2}$	**7.1**(0.9)	**3.0**(0.6)	**2.6**(0.6)	2.7(0.6)	2.7(0.7)	2.7(0.7)	15/15
Thres	12(4)	15(4)	15(6)	15(6)	15(6)	15(8)	15(5)	15/15	Thres	260(34)	130(7)	48(5)	36(3)	32(7)	26(4)	22(3)	15/15
Diag	11(4)	15(6)	16(6)	16(4)	16(6)	16(6)	16(5)	15/15	Diag	16(3)	9.1(0.9)	3.6(0.5)	3.0(0.5)	2.9(0.5)	2.7(0.3)	2.5(0.5)	15/15
DiagL	14(5)	17(7)	18(8)	18(8)	18(13)	18(7)	18(6)	15/15	DiagL	223(54)	115(18)	43(8)	33(7)	30(3)	24(4)	20(3)	15/15
ThresL	13(6)	17(6)	18(13)	18(14)	18(5)	18(10)	18(11)	15/15	ThresL	15(2)	8.1(0.8)	3.2(0.3)	2.6(0.3)	**2.6**(0.3)	**2.3**(0.4)	**2.2**(0.3)	15/15

(continued)

Table 2 (continued)

Δf_{opt}	1e1	1e0	1e-1	1e-2	1e-3	1e-5	1e-7	#succ	Δf_{opt}	1e1	1e0	1e-1	1e-2	1e-3	1e-5	1e-7	#succ
f6	1296	2343	3413	4255	5220	6728	8409	15/15	**f12**	1042	1938	2740	3156	4140	12407	13827	15/15
CMSA	**2.1**(0.8)	**3.4**(6)	5.4(9)	18(12)	51(89)	255(307)	∞ *4e5*	0/15	CMSA	**3.2**(7)	**7.9**(14)	**13**(6)	**14**(6)	**13**(9)	**5.6**(2)	**5.8**(1)	15/15
Thres	5.2(15)	16(32)	29(28)	37(25)	47(33)	121(158)	317(416)	2/15	Thres	114(160)	150(148)	279(254)	309(127)	480(291)	∞	∞ *4e5*	0/15
Diag	3.4(5)	4.2(2)	6.1(9)	14(29)	34(38)	**64**(137)	211(184)	3/15	Diag	6.4(0.2)	15(15)	17(14)	18(11)	17(7)	6.9(3)	7.0(2)	15/15
DiagL	17(34)	26(35)	32(27)	51(40)	64(69)	135(127)	**204**(279)	3/15	DiagL	18(64)	96(83)	103(65)	123(40)	113(6)	59(33)	86(37)	5/15
ThresL	3.2(2)	4.4(6)	**5.0**(6)	**7.9**(12)	**21**(22)	83(75)	324(290)	2/15	ThresL	14(27)	21(20)	23(13)	25(15)	22(8)	8.6(2)	8.5(3)	15/15
Δf_{opt}	1e1	1e0	1e-1	1e-2	1e-3	1e-5	1e-7	#succ	Δf_{opt}	1e1	1e0	1e-1	1e-2	1e-3	1e-5	1e-7	#succ
f8	2039	3871	4040	4148	4219	4371	4484	15/15	**f14**	75	239	304	451	932	1648	15661	15/15
CMSA	**12**(8)	**31**(80)	**32**(34)	**32**(53)	**32**(71)	**32**(53)	**32**(93)	13/15	CMSA	2.0(2)	2.1(0.5)	2.6(0.4)	3.3(1)	**4.9**(0.9)*3	**12**(2)	4.1(1)	15/15
Thres	28(9)	70(86)	73(77)	74(27)	75(26)	75(27)	75(49)	11/15	Thres	**1.7**(1)	1.9(0.8)	**2.3**(0.5)	3.2(0.5)	15(3)	117(15)	18(3)	11/12
Diag	14(7)	31(4)	32(6)	33(28)	33(4)	32(25)	32(4)	13/15	Diag	1.8(1)	2.0(0.9)	2.5(0.8)	3.2(1)	7.0(1)	13(0.9)	4.0(0.5)	15/15
DiagL	28(14)	82(132)	84(82)	85(75)	86(53)	86(49)	85(68)	10/15	DiagL	1.8(1)	1.9(0.5)	2.4(0.6)	3.2(0.4)	15(4)	101(32)	16(2)	15/15
ThresL	15(6)	33(78)	34(27)	34(27)	34(3)	34(25)	34(68)	13/15	ThresL	1.9(1)	**1.8**(0.7)	2.3(0.3)	**3.0**(0.6)	6.8(0.6)	14(1)	**3.7**(0.9)	15/15

The ERT and in braces, as dispersion measure, the half difference between 90 and 10 %-tile of bootstrapped run lengths appear for each algorithm and target, the corresponding best ERT in the first row. The different target Δf-values are shown in the top row. #succ is the number of trials that reached the (final) target $f_{opt} + 10^{-8}$. The median number of conducted function evaluations is additionally given in *italics*, if the target in the last column was never reached. Entries, succeeded by a star, are statistically significantly better (according to the rank-sum test) when compared to all other algorithms of the table, with $p = 0.05$ or $p = 10^{-k}$ when the number k following the star is larger than 1, with Bonferroni correction by the number of instances exhibits slower convergence. Here, the CMSA-ES is marked as the best strategy for many intermediate precision targets. However, the CMSA-Diag-ES and the CMSA-ThresL-ES achieve very similar results. Interesting is the mixture of thresholding target and thresholding function. The interactions will be investigated more closely in future work. In the case of the rotated Rosenbrock (id 9), the CMSA-ThresL-ES shows the best results.

For the ill-conditioned functions (id 10–14), the findings are mixed. On some functions, especially on the ellipsoid (id 10) and the bent cigar (id 12), the original CMSA-ES has the lowest ERT values for the precision targets. For $N = 10$, Table 1, all strategies are successful for the ellipsoid (id 10), the discus (id 11), the bent cigar (id 12), and the sum of different powers (id 14). For the higher-dimensional search space, the bent cigar leads to problems for the CMSA-Thres-ES and the CMSA-DiagL-ES. Again, there appears to be an interaction between thresholding target and function. Only the CMSA-ES and the CMSA-Diag-ES are able to reach the final precision target on the sharp ridge (id 13) for $N = 10$. Since this occurs just once in both cases, more experiments are clearly necessary. Interestingly, differences between the group consisting of the CMSA-ES, the CMSA-ThresL-ES, and the CMSA-Diag-ES and the remaining strategies can be observed. The latter group is unable to achieve comparable performance on f11, f12, and f13 with more unsuccessful runs and larger expected numbers of function evaluations especially for the lower targets.

The group of multi-modal functions represents challenges for all ESs under consideration: The functions Rastrigin (id 15), Weierstrass (id 16), Schaffer F7 with condition number 10 (id 17), Schaffer F7 with condition 1000 (id 18), and Griewank-Rosenbrock F8F2 (id 19) cannot be solved with the final target precision required. Partly, this may be due to the maximal number of fitness evaluations. Even the best performing methods of the 2009 BBOB workshop required more evaluations than we allowed in total. Thus, longer experiments should be conducted in future research. Concerning the preliminary targets with lower precision, thresholding variants often achieve the best results. However, more experiments are required. In the case of $N = 20$, the number of function evaluations necessary for the best algorithms of 2009 to reach even the lower precision target of 10^{-1} exceeds our total budget. Therefore, no analysis is attempted and the results are not shown in Table 2.

The last group, the multi-modal functions with weak global structures, are also difficult to solve and struck from Table 2. Only for function 21, Gallagher 101 peaks, and function 22, Gallagher 21 peaks, successful runs are observed for $N = 10$, see Table 1. In the case of the first, the CMSA-Thres-ES achieves the best results, whereas the original CMSA-ES is the best strategy to tackle function 22.

To summarize the findings, thresholding appears as a means to improve the performance. However, we observe an interaction between thresholding function and threshold target that should be analyzed further.

5 Conclusions and Outlook

The focus of the paper lay on the covariance matrix adaptation in evolution strategies. In many cases, the sample covariance is used which gives cause for concern regarding that its quality may be poor in situations where the estimation is only based on a small sample. Alternative approaches have been developed in the field of statistics. Evolution strategies require, however, methods that do not increase the computational effort considerably. Therefore, the paper investigated and compared several thresholding techniques which originate from estimation theory for high-dimensional spaces. The performance of the resulting new evolution strategies were compared to the original variant on the black-box optimization benchmarking test suite. The results were promising with the new variants performing better for several function classes. Concerning the variants of thresholding, more experiments and analyses are required in order to identify the best solution and to shed more light on the interaction between thresholding function and thresholding target.

References

1. Bäck, T., Foussette, C., Krause, P.: Contemporary Evolution Strategies. Natural Computing. Springer, Berlin (2013)
2. Beyer, H.-G., Meyer-Nieberg, S.: Self-adaptation of evolution strategies under noisy fitness evaluations. Genetic Program. Evolvable Mach. **7**(4), 295–328 (2006)
3. Beyer, H.-G., Schwefel, H.-P.: Evolution strategies: a comprehensive introduction. Nat. Comput. **1**(1), 3–52 (2002)
4. Beyer, H.-G., Sendhoff, B.: Covariance matrix adaptation revisited—the CMSA evolution strategy. In: Rudolph, G. et al. (eds.), *PPSN*, volume 5199 of *Lecture Notes in Computer Science*, pp. 123–132. Springer (2008)
5. Cai, T., Liu, W.: Adaptive thresholding for sparse covariance matrix estimation. J. Am. Stat. Assoc. **106**(494), 672–684 (2011)
6. Dong, W., Yao, X.: Covariance matrix repairing in Gaussian based EDAs. In: IEEE Congress on Evolutionary Computation, CEC 2007, pp. 415–422 (2007)
7. Fan, J., Liao, Y., Liu, H.: An overview on the estimation of large covariance and precision matrices. arXiv:1504.02995
8. Finck, S., Hansen, N., Ros, R., Auger, A.: Real-parameter black-box optimization benchmarking 2010: presentation of the noiseless functions. Technical report, Institute National de Recherche en Informatique et Automatique, 2009/22 (2010)
9. Hansen, N.: The CMA evolution strategy: a comparing review. In: Lozano, J. et al. (eds.) Towards a New Evolutionary Computation. Advances in Estimation of Distribution Algorithms, pp. 75–102. Springer (2006)
10. Hansen, N., Auger, A., Finck, S., Ros, R.: Real-parameter black-box optimization benchmarking 2012: experimental setup. Technical report, INRIA (2012)
11. Hansen, N., Ostermeier, A.: Completely derandomized self-adaptation in evolution strategies. Evolut. Comput. **9**(2), 159–195 (2001)
12. Kramer, O.: Evolution strategies with Ledoit-Wolf covariance matrix estimation. In: 2015 IEEE Congress on Evolutionary Computation (IEEE CEC) (2015)
13. Ledoit, O., Wolf, M.: A well-conditioned estimator for large dimensional covariance matrices. J. Multivar. Anal. Arch. **88**(2), 265–411 (2004)

14. Meyer-Nieberg, S., Kropat, E.: Small populations, high-dimensional search spaces: sparse covariance matrix adaptation. submitted
15. Meyer-Nieberg, S., Kropat, E.: Adapting the covariance in evolution strategies. In: Proceedings of ICORES 2014, pp. 89–99. SCITEPRESS (2014)
16. Meyer-Nieberg, S., Kropat, E.: A new look at the covariance matrix estimation in evolution strategies. In: Pinson, E., Valente, F., Vitoriano, B. (eds.) Operations Research and Enterprise Systems. Communications in Computer and Information Science, vol. 509, pp. 157–172. Springer International Publishing (2015)
17. Pourahmadi, M.: High-Dimensional Covariance Estimation: With High-Dimensional Data. Wiley, New York (2013)
18. Rechenberg, I.: Evolutionsstrategie: Optimierung technischer Systeme nach Prinzipien der biologischen Evolution. Frommann-Holzboog Verlag, Stuttgart (1973)
19. Schwefel, H.-P.: Numerical Optimization of Computer Models. Wiley, Chichester (1981)
20. Stein, C.: Inadmissibility of the usual estimator for the mean of a multivariate distribution. In: Proceedings of the 3rd Berkeley Symposium on Mathematical Statistics and Probability 1, pp. 197–206. Berkeley (1956)
21. Stein, C.: Estimation of a covariance matrix. In: Rietz Lecture, 39th Annual Meeting. IMS, Atlanta (1975)

Knowledge Discovery and Data Mining

A Replicator Dynamics Approach to Collective Feature Engineering in Random Forests

Khaled Fawgreh, Mohamed Medhat Gaber and Eyad Elyan

Abstract It has been demonstrated how random subspaces can be used to create a Diversified Random Forest, which in turn can lead to better performance in terms of predictive accuracy. Motivated by the fact that each subsforest is built using a set of features that can overlap with those sets of features in other subforests, we hypothesise that using *Replicator Dynamics* can perform a collective feature engineering, by allowing subforests with better performance to grow and those with lower performance to shrink. In this paper, we propose a new method to further improve the performance of Diversified Random Forest using Replicator Dynamics which has been used extensively in evolutionary game dynamics. A thorough experimental study on 15 real datasets showed favourable results, demonstrating the potential of the proposed method. Some experiments reported a boost in predictive accuracy of over 10 % consistently, evidencing the effectiveness of the iterative feature engineering achieved through the *Replicator Dynamics* procedure.

1 Introduction

Diversified Random Forest (DRF) [8] is an extension of Random Forest (RF) which is an ensemble learning technique used for classification and regression. Ensemble learning is a supervised machine learning paradigm where multiple models are used to solve the same problem [23]. Since single classifier systems have limited predictive performance [20, 23, 25, 29], ensemble classification was developed to overcome this limitation [20, 23, 25], and thus boosting the accuracy of classification. In such an ensemble, multiple classifiers are used. In its basic mechanism, majority

K. Fawgreh · M.M. Gaber (✉) · E. Elyan
School of Computing Science and Digital Medial, Robert Gordon University,
Garthdee Road, Aberdeen, AB10 7GJ, UK
e-mail: m.gaber1@rgu.ac.uk

K. Fawgreh
e-mail: k.fawagreh@rgu.ac.uk

E. Elyan
e-mail: e.elyan@rgu.ac.uk

M. Bramer and M. Petridis (eds.), *Research and Development in Intelligent Systems XXXII*, DOI 10.1007/978-3-319-25032-8_2

voting is then used to determine the class label for unlabelled instances where each classifier in the ensemble is asked to predict the class label of the instance being considered. Once all the classifiers have been queried, the class that receives the greatest number of votes is returned as the final decision of the ensemble.

Boosting, bagging, and stacking are the three widely used and adopted ensemble methods. Boosting is an incremental process of building a sequence of classifiers, where each classifier has the incorrectly classified instances of the previous one in the sequence emphasised. AdaBoost [11] is the representative of this class of techniques. However, AdaBoost is prone to overfitting, due to the nature of the process. The other class of ensemble approaches is the bootstrap aggregating (bagging) [4]. Bagging involves building each classifier in the ensemble using a randomly drawn bootstrap sample of the data, having each classifier giving an equal vote when classifying unlabelled instances. Bagging is known to be more robust than boosting against model overfitting. Random Forest (RF) is the main representative of bagging [5]. Stacking (sometimes called stacked generalisation) extends the cross-validation technique that partitions the dataset into a held-in data set and a held-out data set; training the models on the held-in data; and then choosing whichever of those trained models performs best on the held-out data. Instead of choosing among the models, stacking combines them, thereby typically getting performance better than any single one of the trained models [28]. It is worth noting that each of these three methods diversify among the classifiers of the ensemble. Diversity plays an important role in the success of the ensemble.

Since RF has been proved to be the state-of-the-art ensemble classification technique, and since it has been proven empirically that ensembles tend to yield better results when there is a significant diversity among the models [1, 6, 18, 26], DRF was developed as an extension of RF by injecting a new level of diversity [8]. An observation in DRF has motivated the work proposed in this paper; that subforests created in DRF can exhibit varying discriminative power according to the randomly drawn subspace (more details about DRF is given later in the paper). We hypothesise that if suforests with relatively better accuracy are allowed to grow, and those with less accuracy can shrink, we can perform inherit feature engineering of the ensemble in a collective manner. This is done by means of subforest growth and shrinking that can be looked at as emphasising a set of features collectively over other sets. This unique way of feature engineering allows for feature interactivity to take place, thus providing a new method that favours feature co-existence in some subforests. As such, this paper investigates how to further boost the performance of DRF by using *Replicator Dynamics* that provides the mechanism for growing and shrinking subforests.

This paper is organised as follows. First, an overview of DRF is presented in Sect. 2. This is followed by Sect. 3 that presents a brief introduction to Replicator Dynamics. Section 4 demonstrates how Replicator Dynamics can be utilised to boost the performance of DRF. Experimental results demonstrating the superiority of the proposed method is detailed in Sect. 5. In Sect. 6, we describe related work. The paper is then concluded with a summary and pointers to future directions in Sect. 7.

2 Diversified Random Forests: An Overview

As detailed in [8], creating an DRF proceeds as follows. First, from the training set, we create a number of random subspaces. The number of subspaces is determined by the following equation:

$$Subspaces = \alpha \times Z \tag{1}$$

where α denotes the subspace factor such that $0 < \alpha \leq 1$, and Z is the size of DRF to be created. Each subspace will contain a fixed randomized subset of the total number of features and will correspond to a sub-forest. A projected training dataset will be then created for each subspace and will be used to create the trees in the corresponding sub-forest. The following equation calculates the number of trees in each sub-forest:

$$Trees = \frac{Z}{Subspaces} \tag{2}$$

Next, a weight is assigned to each projected training dataset using the *Absolute Predictive Power (APP)* given by Cuzzocrea et al. [7]. Given a dataset S, the APP is defined by the following equation:

$$APP(S) = \frac{1}{|Att(S)|} \times \sum_{A \in Att(S)} \frac{I(S,A)}{E(S)} \tag{3}$$

where $E(S)$ is the entropy of a given dataset S having K instances and $I(S,A)$ is the information gain of a given attribute A in a dataset S. $E(S)$ is a measure of the uncertainty in a random variable and is given by the following equation:

$$E(S) = \sum_{i=1}^{K} -p_i(x_i) \log_2 p_i(x_i) \tag{4}$$

where x_i refers to a generic instance of S and $p_i(x_i)$ denotes the probability that the instance x_i occurs in S. $I(S,A)$ is given by the following equation:

$$I(S,A) = E(S) - \sum_{v \in Val(A)} \left(\frac{|S_v|}{|S|} \right) E(S_v) \tag{5}$$

where $E(S)$ denotes the entropy of S, $Val(A)$ denotes the set of possible values for A, S_v refers to the subset of S for which A has the value v, and $E(S_v)$ denotes the entropy of S_v.

The weight given in Eq. 3 above will be inherited by the corresponding sub-forest and will be used in the voting process. This means that the standard voting technique, currently used in the standard RF, is going to be replaced by a weighted voting technique, in order to classify the instances in the testing dataset.

3 Replicator Dynamics

Replicator Dynamics (RD) is a deterministic monotone non-linear and also non-innovative game dynamic (older solutions are obsolete) used in evolutionary game theory [16, 27]. It provides a convenient way to represent selection among a population of diverse types. To illustrate how it works, assume that selection occurs between periods after dividing time into discrete periods. The proportion of each type in the next period is given by the replicator equation as a function of the type's payoffs and its current proportion in the population. Types that score above the average payoffs increase in proportion, while types that score below the average payoffs decrease in proportion. The amount of increase or decrease depends on a type's proportion in the current population and on it's relative payoffs.

The most general continuous form is given by the differential equation

$$\dot{x}_i = x_i[f_i(x) - \phi(x)] \tag{6}$$

such that

$$\phi(x) = \sum_{j=1}^{n} x_j f_j(x) \tag{7}$$

where x_i is the proportion of type i in the population, $x = (x_1, \dots, x_n)$ is the vector of the distribution of types in the population, $f_i(x)$ is the fitness of type i (which is dependent on the population), and $\phi(x)$ is the average population fitness (given by the weighted average of the fitness of the n types in the population).

In the next section, we shall see how to apply these equations in order to boost the performance of an DRF. To the best of our knowledge, RD has never been used before in ensemble learning.

4 Applying Replicator Dynamics to an DRF

As was detailed in Sect. 2, the population is n sub-forests where each sub-forest represents a type. The discrete periods mentioned in the previous section correspond to loop iterations. At each iteration, the accuracy of the sub-forest being processed is compared with the average accuracy of the entire DRF which is calculated as the average accuracy of the sub-forests. If it is greater, then the size of the sub-forest grows and if it less, the size shrinks.

For growing and shrinking the sub-forest, two variations will be used. In the first one, the size grows/shrinks by a fixed number as shown in the following equations:

$$\text{treesToAdd} = \beta \tag{8}$$

$$\text{treesToRemove} = \gamma \tag{9}$$

In the second variation, the sub-forest grows/shrinks by adding/removing a variable number of trees according to the following equations:

$$treesToAdd = \lfloor ((subforestAccuracy(i) - DRFAccuracy) \times numTrees) \rfloor \quad (10)$$

$$treesToRemove = \lfloor ((DRFAccuracy - subforestAccuracy(i)) \times numTrees) \rfloor \quad (11)$$

where *subforestAccuracy(i)* refers to the accuracy of subforest(i) being processed, and *numTrees* refers to the initial number of trees that was used to construct the subforest (given by Eq. 2 above). The *floor* function maps a real number to the largest previous integer. The *DRFAccuracy* refers to the average accuracy of the entire DRF which can be calculated as follows:

$$DRFAccuracy = \frac{1}{Subspaces} \sum_{i=1}^{Subspaces} subforestAccuracy(i) \quad (12)$$

where the constant *Subspaces* is given by Eq. 1.

Figure 1 outlines the main steps involved in applying RD to a DRF to further optimise its performance. As shown in the figure, RD is performed over DRF in an iterative procedure. Thus, we can identify two main stages to the method proposed in this paper: (1) building a DRF; and (2) applying RD a preset number of iterations. The outcome is a number of subforests, each with potentially a different number of trees than the number initially set when building the DRF in the first step.

Algorithm 1 illustrates how RD can be applied to a DRF to grow or shrink the subs-forests. At each iteration, the average accuracy of all subforests is calculated as well as the accuracy of each subforest. This process is used to determine those subforests that have an above average accuracy to grow them further. On the other hand, subforests that have accuracy below the calculated average are forced to shrink.

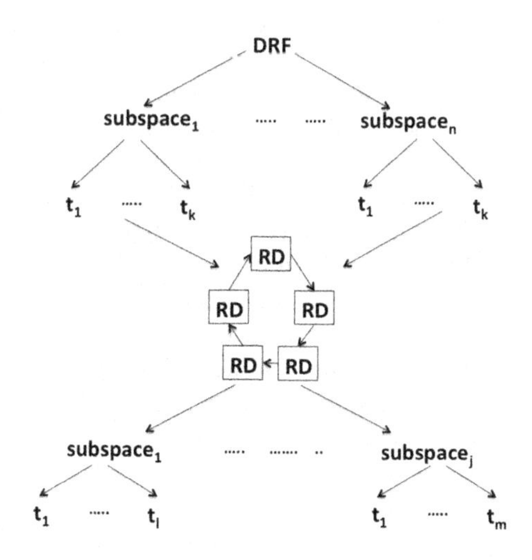

Fig. 1 Applying RD to an DRF

The process performs inherit collective feature engineering, as subforets that grow have the features that collectively produce accurate trees.

Algorithm 1 Applying Replicator Dynamics to a DRF

```
{User Settings}
input DRF
input numberofIterations
{Process}
for i = 1 → numberofIterations do
   for j = 1 → DRF.numberofSubforests() do
      DRFAccuracy ← CalculateDRFAccuracy(DRF)
      subforestAccuracy ← CalculateSubforestAccuracy(DRF.subforest(j))
      if (subforestAccuracy > DRFAccuracy) then
         Apply Eq. 8 or 10 to determine treestoAdd
         AddTrees(DRF.subforest(j), treestoAdd)
      else if (subforestAccuracy < DRFAccuracy) then
         Apply Eq. 9 or 11 to determine treesToRemove
         RemoveTrees(DRF.subforest(j), treestoRemove)
      end if
   end for
end for
{Output}
A new DRF with grown or shrunk sub-forests
```

5 Experimental Study

Several experiments were performed on 15 real datasets from the UCI repository [2]. All the DRFs created had an initial size of 500 trees.

We used 2 subspace factors of 2 and 4 %. According to Eq. 1, these factors produced DRFs with 10 and 20 sub-forests respectively. We used a random 70 % of the features for each subspace, which has proved empirically to lead good performance in the traditional version of DRF. By Eq. 2, each sub-forest contained 50 trees for the DRF with 10 sub-forests, and 25 trees for the DRF with 20 sub-forests. For the number of iterations (refer to *Number of Iterations* in Algorithm 1 above), we used 25, 50, 100, 150, and 1000 iterations. Table 1 lists the different experimental scenarios that will be discussed in details later on in the paper. It is worth pointing out that results reported in this section captured all the experiments that have been carried out. However, due to space limitation, the results of each scenario are summarized in a separate table, followed by a detailed table of winning solutions that represent the significant difference in accuracy that has been achieved frequently.

In the first set of experiments, we used the first variation given by Eqs. 8 and 9 above. Here we chose $\beta = \gamma = 1$, therefore, only 1 tree may be added/removed in each iteration. A new experiment was performed on each of the 10 scenarios listed

Table 1 Scenario reference table

Scenario#	Number of sub-forests	Number of trees per sub-forest	Number of iterations
1	10	50	25
2	10	50	50
3	10	50	100
4	10	50	150
5	10	50	1000
6	20	25	25
7	20	25	50
8	20	25	100
9	20	25	150
10	20	25	1000

in Table 1. We then compared the performance of DRF before and after applying RD to see if there is a performance gain and recorded the number of wins, losses, and ties sorted by number of wins in descending order as shown in Table 2. As shown in the table, the number of wins of the new DRF (after applying RD) exceeded the number of losses in the majority of the scenarios. By wins we refer to the number of datasets where the new DRF outperformed the initial DRF. In fact, with the exception of scenario 2, the adoption of RD has consistently shown superiority. It is worth noting that the number of iterations in this scenario is only 50. This suggests that the larger number of iterations can results in a better performance. We argue that with the larger number of iterations, the feature engineering process is stabilised.

Table 2 First experiments set results

Scenario	Wins	Losses	Ties
8	6	4	5
9	6	4	5
4	6	2	7
3	6	3	6
6	6	3	6
1	5	1	9
7	5	4	6
5	5	4	6
10	4	4	7
2	3	4	8

Accuracy of the winning datasets is shown in Table 3. As demonstrated in the table, the improvement in the accuracy after applying RD can be very significant (e.g. in scenario 5, for the squash-unstored dataset, the accuracy was boosted from 61.1 to 77.7 %). This provides the evidence of the positive effect of the feature engineering provided by the RD process.

In the second set of experiments and using again the first variation, performance was compared with traditional RFs of identical size (500 trees). Table 4 shows the results. Once again, the number of wins exceeded the number of losses in the majority of the scenarios. In this experiment, scenario 10 that was experimented with 1000 iterations of RD has shown the best performance in terms of predictive accuracy. This also confirms that feature engineering through RD requires a large number of iterations to stabilise, as we argued in the first set of experiments.

Table 5 shows the accuracy of the winning datasets. As DRF has already improved the performance of the ensemble, the results in this table show more datasets that have had their accuracy boosted. It is important to note that the boost in the accuracy is due to two factors: (1) the diversification achieved through random subspacing; and (2) the feature engineering provided through RD.

The third and fourth set of experiments were performed using the second variation given by Eqs. 10 and 11 above where a fraction of the initial number of trees in the sub-forest may be added or removed. Similar to Table 2, Table 6 compares the performance of DRF with itself before and after applying RD. This represents the typical implementation of the RD, where the subforest with a considerable over average accuracy can grow with more than one tree in each iteration.

Table 7 shows the accuracy of the winning datasets. As demonstrated in the previously discussed experiments, the increase in accuracy is significant in some of the datasets, suggesting the effectiveness of RD in the feature engineering process.

Likewise, Table 8 is similar to Table 4 where the DRFs (after applying RD) are compared with traditional RFs of identical size. As was the case with the first and second set of experiments, favourable results were achieved since, as demonstrated in these tables, the number of wins exceeded the number of losses in the majority of the scenarios. We also note that the two scenarios with the highest number of iterations (1000 iterations) have demonstrated the best performance. This confirms that a large number of iterations is required for the feature engineering to be effective, as argued earlier in this experimental study.

Accuracy of the winning datasets is shown in Table 9. Significant boost in the performance has been demonstrated as shown in the table.

The extensive experimental study discussed in this section has demonstrated the effectiveness of the *Replicator Dynamics* in collective feature engineering. As subforests that contain subspaces with high discriminative power features can grow–so that their contributions to the vote also grow. On the other hand, subforests that contain subspaces with low discriminative power features can shrink, and potentially disappear.

Table 3 First experiments set winning datasets accuracy

Scenario	Dataset	Accuracy before (%)	Accuracy after (%)
8	Squash-unstored	61.11	77.78
	Vote	96.62	97.30
	Vehicle	72.92	74.65
	Eucalyptus	23.60	29.60
	Audit	96.76	96.91
	Soybean	77.59	78.88
9	Squash-unstored	61.11	77.78
	Vehicle	72.57	73.96
	Eucalyptus	280	28.80
	Sonar	00.00	01.41
	Audit	96.62	97.06
	Soybean	73.71	85.78
4	Squash-unstored	61.11	72.22
	Squash-stored	44.44	50.00
	White clover	61.90	66.67
	Vehicle	72.57	72.92
	Eucalyptus	20.80	26.40
	Credit	77.94	78.82
3	Squash-unstored	61.11	66.67
	Glass	10.96	12.33
	Diabetes	74.33	74.71
	Vehicle	70.83	72.57
	Eucalyptus	18.00	24.00
	Sonar	00.00	01.41
6	Squash-unstored	61.11	66.67
	Diabetes	73.56	75.10
	Vehicle	72.22	73.26
	Eucalyptus	22.00	26.40
	Credit	75.29	76.18
	Audit	96.47	96.76
1	Diabetes	72.41	73.18
	Vehicle	73.61	75.00
	Eucalyptus	21.20	26.40
	Credit	75.29	76.18
	Soybean	73.71	76.29
7	Squash-unstored	61.11	66.66
	Diabetes	73.56	73.95
	Vehicle	73.26	73.61
	Eucalyptus	28.00	29.20
	Audit	97.06	97.21

(continued)

Table 3 (continued)

Scenario	Dataset	Accuracy before (%)	Accuracy after (%)
5	Squash-unstored	50.00	61.11
	White clover	61.90	66.67
	Vehicle	72.92	73.26
	Eucalyptus	17.20	24.00
	Soybean	73.28	75.43
10	Pasture	33.33	41.67
	Squash-unstored	61.11	72.22
	Eucalyptus	23.60	30.40
	Soybean	77.15	79.74
2	Squash-unstored	55.56	61.11
	Eucalyptus	21.60	24.00
	Credit	74.12	76.18

Table 4 Second experiments set results

Scenario	Wins	Losses	Ties
10	9	4	2
1	8	4	3
6	8	4	3
7	8	4	3
3	8	5	2
9	8	4	3
4	7	4	4
2	6	4	5
8	6	6	3
5	5	6	4

6 Related Work

Random Forest has proved to be the state-of-the-art classification technique when compared with other methods in a recent large scale experimental study [10]. Previous work to enhance its performance has been also reported recently in [9]. However, feature engineering through an evolutionary method like *Replicator Dynamics* has not been attempted.

As aforementioned, RD has been used extensively in the domain of evolutionary game theory [13–16, 21, 24, 27]. To a lesser extent, however, it was also used in other domains. In medicine, [19] used RD to analyse fMRI data of the human brain. In mathematical ecology, [3] used RD to describe the interaction of two populations over time. In mathematical biology, [17] used RD in the study of permanence, i.e., the study of the long-term survival of each species in a set of populations. Multi-agent

Table 5 Second experiments set winning datasets accuracy

Scenario	Dataset	Accuracy classic RF (%)	Accuracy DRF (%)
10	Pasture	41.67	41.67
	Squash-unstored	61.11	72.22
	Diabetes	73.56	74.71
	Vehicle	72.92	73.96
	Eucalyptus	21.20	30.40
	Sonar	00.00	01.41
	Credit	75.88	76.76
	Audit	96.32	96.91
	Soybean	75.86	79.74
1	Pasture	41.67	41.67
	Breast-cancer	72.16	75.26
	Vehicle	72.92	75.00
	Eucalyptus	21.20	26.40
	Sonar	00.00	01.41
	Credit	75.88	76.18
	Audit	96.32	96.76
	Soybean	75.86	76.29
6	Pasture	41.67	41.67
	Squash-unstored	61.11	66.67
	Diabetes	73.56	75.09
	Vehicle	72.92	73.26
	Eucalyptus	21.20	26.40
	Credit	75.88	76.18
	Audit	96.32	96.76
	Soybean	75.86	76.29
7	Pasture	41.67	41.67
	Squash-unstored	61.11	66.67
	Breast-cancer	72.16	74.23
	Diabetes	73.56	73.95
	Vehicle	72.92	73.61
	Eucalyptus	21.20	29.20
	Credit	75.88	76.47
	Audit	96.32	97.21
3	Pasture	41.67	41.67
	Squash-unstored	61.11	66.67
	Diabetes	73.56	74.71
	Eucalyptus	21.20	24.00
	Sonar	00.00	01.41
	Credit	75.88	76.76
	Audit	96.32	97.06
	Soybean	75.86	77.15

(continued)

Table 5 (continued)

Scenario	Dataset	Accuracy classic RF (%)	Accuracy DRF (%)
9	Pasture	41.67	41.67
	Squash-unstored	61.11	77.78
	breast-cancer	72.16	74.23
	Vehicle	72.92	73.96
	Eucalyptus	21.20	28.80
	Sonar	00.00	01.41
	Audit	96.32	97.06
	Soybean	75.86	85.78
4	Pasture	41.67	41.67
	Squash-unstored	61.11	72.22
	Breast-cancer	72.16	75.26
	Eucalyptus	21.20	26.40
	Credit	75.88	78.82
	Audit	96.32	96.91
	Soybean	75.86	78.02
2	Pasture	41.67	41.67
	Breast-cancer	72.16	74.23
	Eucalyptus	21.20	24.00
	Credit	75.88	76.18
	Audit	96.32	97.06
	Soybean	75.86	78.02
8	Pasture	41.67	41.67
	Squash-unstored	61.11	77.78
	Vehicle	72.92	74.65
	Eucalyptus	21.20	29.60
	Audit	96.32	96.91
	Soybean	75.86	78.88
5	Pasture	41.67	41.67
	Breast-cancer	72.16	74.23
	Vehicle	72.92	73.26
	Eucalyptus	21.20	24.00
	Audit	96.32	97.21

learning was another domain of RD [12]. Finally, in social networks, [22] used RD to understand the origin of social norms and dominant behavioral and cultural trends in social networks.

Table 6 Third experiments set results

Scenario	Wins	Losses	Ties
4	6	3	6
2	4	2	9
7	4	2	9
10	4	2	9
5	4	3	8
1	4	5	6
6	3	3	9
9	3	3	9
3	3	3	9
8	2	3	10

Table 7 Third experiments set winning datasets accuracy

Scenario	Dataset	Accuracy before (%)	Accuracy after (%)
4	Squash-unstored	66.67	77.78
	Squash-stored	50.00	55.56
	Vote	96.62	97.30
	Vehicle	70.83	72.92
	Eucalyptus	20.00	34.00
	Soybean	72.41	73.28
2	Squash-unstored	61.11	77.78
	Vehicle	72.22	73.26
	Eucalyptus	24.80	28.40
	Soybean	72.41	75.00
7	Diabetes	73.18	74.33
	Vehicle	71.87	72.92
	Eucalyptus	22.80	25.20
	Audit	96.76	96.91
10	Squash-unstored	55.56	66.67
	Eucalyptus	24.40	26.80
	Audit	96.62	96.76
	Soybean	77.59	86.21
5	Squash-unstored	61.11	66.67
	White clover	61.90	66.67
	Eucalyptus	17.60	27.20
	Soybean	77.59	79.74

(continued)

Table 7 (continued)

Scenario	Dataset	Accuracy before (%)	Accuracy after (%)
1	Squash-unstored	55.56	61.11
	Squash-stored	50.00	55.56
	Eucalyptus	19.20	19.60
	Audit	96.03	97.06
6	Squash-unstored	61.11	66.67
	Eucalyptus	22.80	24.00
	Audit	96.32	96.76
9	Squash-unstored	61.11	66.67
	Eucalyptus	19.60	24.00
	Audit	96.62	96.76
3	Squash-unstored	50.00	55.56
	Eucalyptus	16.40	16.80
	Soybean	78.02	80.17
8	Eucalyptus	22.00	27.20
	Audit	96.32	96.91

Table 8 Fourth experiments set results

Scenario	Wins	Losses	Ties
5	7	3	5
10	7	5	3
9	6	4	5
1	5	6	4
2	5	6	4
3	5	7	3
4	5	7	3
6	4	3	8
7	4	5	6
8	4	5	6

7 Conclusion and Future Work

We have demonstrated in this paper how to apply *Replicator Dynamics* (RD) to boost the performance Diversified Random Forest (DRF). DRF is built using random subspacing with each subspace used to construct a subforest. DRF was used to impose diversity in Random Forests. Noting that subspaces may vary in their discriminative power, we applied RD to perform collective feature engineering, allowing stronger subspaces to grow and weaker ones to shrink, and potentially disappear.

As was demonstrated in Tables 2, 4, 6, and 8, favourable results were obtained demonstrating the potential of the proposed method. In our experiments, we have

Table 9 Fourth experiments set winning datasets accuracy

Scenario	Dataset	Accuracy classic RF (%)	Accuracy DRF (%)
5	Pasture	41.67	41.67
	Squash-unstored	61.11	66.67
	Vehicle	72.92	73.26
	Eucalyptus	21.20	27.20
	Credit	75.88	76.47
	Audit	96.32	96.91
	Soybean	75.86	79.74
10	Pasture	41.67	41.67
	Squash-unstored	61.11	66.67
	Breast-cancer	61.11	75.26
	Diabetes	73.56	73.95
	Eucalyptus	21.20	26.80
	Audit	96.32	96.76
	Soybean	75.86	86.21
9	Squash-unstored	61.11	66.67
	Breast-cancer	72.16	74.23
	Car	61.90	62.58
	Eucalyptus	21.20	24.00
	Audit	96.32	96.76
	Soybean	75.86	76.72
1	Breast-cancer	72.16	76.29
	Diabetes	73.56	74.71
	Car	61.90	64.28
	Vehicle	72.92	73.61
	Audit	96.32	97.06
2	Squash-unstored	61.11	77.78
	Breast-cancer	72.16	76.29
	Vehicle	72.92	73.26
	Eucalyptus	21.20	28.40
	Audit	96.32	96.91
3	Breast-cancer	72.16	73.19
	Diabetes	73.56	73.95
	Sonar	00.00	01.41
	Audit	96.32	97.06
	Soybean	75.86	80.17
4	Squash-unstored	61.11	77.78
	Eucalyptus	20.00	34.00
	Sonar	00.00	01.41
	Credit	75.88	76.47
	Audit	96.32	97.20

(continued)

Table 9 (continued)

Scenario	Dataset	Accuracy classic RF (%)	Accuracy DRF (%)
6	Squash-unstored	61.11	66.67
	Breast-cancer	72.16	73.19
	Eucalyptus	21.20	24.00
	Audit	96.32	96.76
7	Diabetes	73.56	74.33
	Eucalyptus	21.20	25.20
	Credit	75.88	76.76
	Audit	96.32	96.91
8	Squash-unstored	61.11	66.67
	Eucalyptus	21.20	27.20
	Audit	96.32	96.91
	Soybean	75.86	76.72

used a subspace factor of 2 and 4 %, a size of 500 trees for the DRF to be created, and 70 % of the features in each subspace.

In the future, we will attempt different values for these parameters. We envisage to apply the proposed method on higher dimensional datasets. RD has also the potential to address the concept drift problem in data stream ensembles, as growing and shrinking can take place according to the drifted concept.

References

1. Adeva, J.J.G., Beresi, U., Calvo, R.: Accuracy and diversity in ensembles of text categorisers. CLEI Electron. J. **9**(1) (2005)
2. Bache, K., Lichman, M.: Uci machine learning repository (2013)
3. Bomze, I.M.: Lotka-volterra equation and replicator dynamics: new issues in classification. Biol. Cybern. **72**(5), 447–453 (1995)
4. Breiman, L.: Bagging predictors. Mach. Learn. **24**(2), 123–140 (1996)
5. Breiman, L.: Random forests. Mach. Learn. **45**(1), 5–32 (2001)
6. Brown, G., Wyatt, J., Harris, R., Yao, X.: Diversity creation methods: a survey and categorisation. Inf. Fusion **6**(1), 5–20 (2005)
7. Cuzzocrea, A., Francis, S.L., Gaber, M.M.: An information-theoretic approach for setting the optimal number of decision trees in random forests. In: Systems, Man, and Cybernetics (SMC), 2013 IEEE International Conference on, pp. 1013–1019. IEEE (2013)
8. Fawagreh, K., Gaber, M.M., Elyan, E.: Diversified random forests using random subspaces. In: Intelligent Data Engineering and Automated Learning–IDEAL 2014, pp. 85–92. Springer (2014)
9. Fawagreh, K., Gaber, M.M., Elyan, E.: Random forests: from early developments to recent advancements. Syst. Sci. Control Eng: Open Access J. lbf 2(1), 602–609 (2014)
10. Fernández-Delgado, M., Cernadas, E., Barro, S., Amorim, D.: Do we need hundreds of classifiers to solve real world classification problems? J. Mach. Learn. Res. **15**, 3133–3181 (2014)

11. Freund, Y., Schapire, R.E.: A decision-theoretic generalization of on-line learning and an application to boosting. J. Comput. Syst. Sci. **55**(1), 119–139 (1997)
12. Galstyan, A.: Continuous strategy replicator dynamics for multi-agent Q-learning. Auton. Agents Multi-agent Syst. **26**(1), 37–53 (2013)
13. Hauert, C.: Replicator dynamics of reward & reputation in public goods games. J. Theor. Biol. **267**(1), 22–28 (2010)
14. Hauert, C., De Monte, S., Hofbauer, J., Sigmund, K.: Replicator dynamics for optional public good games. J. Theor. Biol. **218**(2), 187–194 (2002)
15. Hilbe, C.: Local replicator dynamics: a simple link between deterministic and stochastic models of evolutionary game theory. Bull. Math. Biol. **73**(9), 2068–2087 (2011)
16. Hofbauer, J., Sigmund, K.: Evolutionary game dynamics. Bull. Am. Math. Soc. **40**(4), 479–519 (2003)
17. Hutson, V., Schmitt, K.: Permanence and the dynamics of biological systems. Math. Biosci. **111**(1), 1–71 (1992)
18. Kuncheva, L.I., Whitaker, C.J.: Measures of diversity in classifier ensembles and their relationship with the ensemble accuracy. Mach. Learn. **51**(2), 181–207 (2003)
19. Lohmann, G., Bohn, S.: Using replicator dynamics for analyzing fMRI data of the human brain. IEEE Trans. Med. Imag. **21**(5), 485–492 (2002)
20. Maclin, R., Opitz, D.: Popular ensemble methods: an empirical study. arXiv:1106.0257 (2011) (preprint)
21. Nowak, M.A., Sigmund, K.: Evolutionary dynamics of biological games. Science **303**(5659), 793–799 (2004)
22. Olfati-Saber, R.: Evolutionary dynamics of behavior in social networks. In: Decision and Control, 2007 46th IEEE Conference on, pp. 4051–4056. IEEE (2007)
23. Polikar, R.: Ensemble based systems in decision making. Circuits Syst. Mag. IEEE **6**(3), 21–45 (2006)
24. Roca, C.P., Cuesta, J.A., Sánchez, A.: Evolutionary game theory: temporal and spatial effects beyond replicator dynamics. Phys. Life Rev. **6**(4), 208–249 (2009)
25. Rokach, L.: Ensemble-based classifiers. Artif. Intell. Rev. **33**(1–2), 1–39 (2010)
26. Tang, E.K., Suganthan, P.N., Yao, X.: An analysis of diversity measures. Mach. Learn. **65**(1), 247–271 (2006)
27. Taylor, P.D., Jonker, L.B.: Evolutionary stable strategies and game dynamics. Math. Biosci. **40**(1), 145–156 (1978)
28. Wolpert, D.H.: Stacked generalization. Neural Netw. **5**(2), 241–259 (1992)
29. Yan, W., Goebel, K.F.: Designing classifier ensembles with constrained performance requirements. In: Defense and Security, pp. 59–68. International Society for Optics and Photonics (2004)

A Directed Acyclic Graph Based Approach to Multi-Class Ensemble Classification

Esra'a Alshdaifat, Frans Coenen and Keith Dures

Abstract In this paper a novel, ensemble style, classification architecture is proposed as a solution to the multi-class classification problem. The idea is to use a non-rooted Directed Acyclic Graph (DAG) structure which holds a classifier at each node. The potential advantage offered is that a more accurate and reliable classification can be obtained when the classification process is conducted progressively, starting with groups of class labels that are repeatedly refined into smaller groups until individual labels are arrived at. Reported experimental results indicated that the proposed DAG classification model can improve classification performance, in terms of average accuracy and average AUC (Area Under the receiver operating Curve), in the context of some data sets.

1 Introduction

Classification is concerned with the generation of a model, using pre-labelled "training" data, that can be used to allocate labels (classes) to previously unseen data. The nature of the classification problem is characterised by two factors: (i) the number of class labels that can be assigned to an instance (single-label versus multi-label classification), and (ii) the number of classes from which the class labels may be drawn (binary versus multi-class classification). In single-label classification a classifier model is generated using a set of training examples where each example is associated with a single class label c taken from a set of disjoint class labels C ($|C| > 1$). If $|C| = 2$ we have a binary classification problem; if $|C| > 2$, we have a multi-class classification problem. The distinction between single-label and

E. Alshdaifat (✉) · F. Coenen · K. Dures
Department of Computer Science, University of Liverpool, Ashton Building, Ashton Street, Liverpool L69 3BX, UK
e-mail: esraa@liv.ac.uk

F. Coenen
e-mail: coenen@liv.ac.uk

K. Dures
e-mail: dures@liv.ac.uk

M. Bramer and M. Petridis (eds.), *Research and Development in Intelligent Systems XXXII*, DOI 10.1007/978-3-319-25032-8_3

multi-label classification is that in multi-label classification the examples are each associated with a set of class labels Z, $Z \subseteq C$. In the work presented in this paper we focus on the multi-class single-label classification problem where examples are associated with exactly one element of the set of class labels C. For simplicity, throughout this work, we will refer to this as "multi-class" classification.

The main issue with multi-class classification is that as the number of classes increases the classification performance tends to degrade [9]. There are three main mechanisms for addressing multi-class classification: (i) using a single classifier, (ii) utilising a collection of binary classifiers such and adopting a One-Versus-All (OVA) [15] or One-Verses-One (OVO) strategy [16], and (iii) using an "ensemble"of classifiers. The ensemble strategy has been shown to work well in the context of the multi-class problem [9, 14, 17]. The ensemble strategy involves using a collection of classifiers typically arranged in either: (i) a "concurrent" form, such as "Baggin" [6]; or (ii) a "sequential" form, such as "Boosting" [8]. In more recent work on ensemble classification, hierarchical arrangements of classifiers have been used [4, 10, 11]. A commonly adopted structure is a binary tree constructed in either a bottom-up or top-down manner [5, 10].

The proposed work is directed at hierarchical ensemble classification where classifiers are arranged in a more sophisticated structure, than a binary tree structure, namely a non-rooted Directed Acyclic Graph (DAG) structure. Nodes at leaves hold classifiers designed to distinguish between individual class labels while the remaining nodes hold classifiers designed to discriminate between groups of class labels. The intuition here is that if we start off with a "coarse grained" classification, moving down to a more "fine grained" classification, a more effective classification model can be produced. This is based on the observation that as the number of classes increases the classification performance tends to degrade as noted above [9]. In this context the advantage offered by the DAG structure is that it enables the inclusion of a greater number of possible class label combinations at each level than in the case of a binary structure. The main challenges associated with this kind of classification are: (i) how to distribute (divide) classes between nodes within the DAG model, (ii) how to determine the starting node among the set of nodes available at the first level in the DAG, and (iii) how to handle the general drawback associated with hierarchical forms of ensemble classification, the "successive miss-classification" problem, whereby if a record is miss-classified early on in the process it will continue to be miss-classified at deeper levels, regardless of the classifications proposed at lower level nodes. To address the first issue a "combination technique" is proposed to assign classes to nodes within the DAG. To address the second challenge the use of Bayesian classifiers is proposed, one per DAG node, which offers the advantage that the probability values produced can be used to determine the best starting first level node. To address the "successive miss-classification" issue two strategies are proposed: (i) following multiple paths within the DAG by utilising the probability values generated by the Naive Bayesian classification model to determine where single or multiple paths should be followed; and (ii) a pruning scheme, applied during the generation process, to eliminate the *weak* classifiers that might affect eventual classification accuracy.

The rest of this paper is organised as follows. Section 2 gives a review of related work on multi-class classification. Section 3 describes the process for generating and operating the proposed DAG ensemble classification model. Section 4 presents an evaluation of the proposed DAG classification model as applied to a range of different data sets. Section 5 concludes the work presented in this paper.

2 Literature Review

This section provides a review of "Ensemble" methods for solving the multi-class classification problem. An ensemble model is a composite model comprised of a number of learners (classifiers), called *base learners* or *weak learners*, that are used together to obtain a better classification performance than can be obtained when using a single "stand alone" model. If the base learners in an ensemble model are all comprised of the same classification algorithm the ensemble model is referred to as an *homogeneous learner*, while when different classification algorithms are used the ensemble model is referred to as *heterogeneous learner* [17].

Depending on the relationships between the classifiers forming the ensemble, two main structures can be identified: concurrent (parallel) [6], and sequential (serial) [8]. The hierarchical ensemble methodology is a much more recent approach to solving the multi-class classification problem which involves the generation of a hierarchical "meta-algorithm" [4, 10, 11]. Work to date has been mostly directed at binary tree based hierarchical ensemble classification. Using a binary tree hierarchical classification model the classifiers are arranged in a binary tree formation such that the classifiers at the leaves conduct fine-grained (binary) classifications while the classifiers at non-leaf nodes further up the hierarchy conduct coarse-grained classification directed at categorising records using groups of labels. An example binary tree hierarchy is presented in Fig. 1. At the root we distinguish between two groups of class

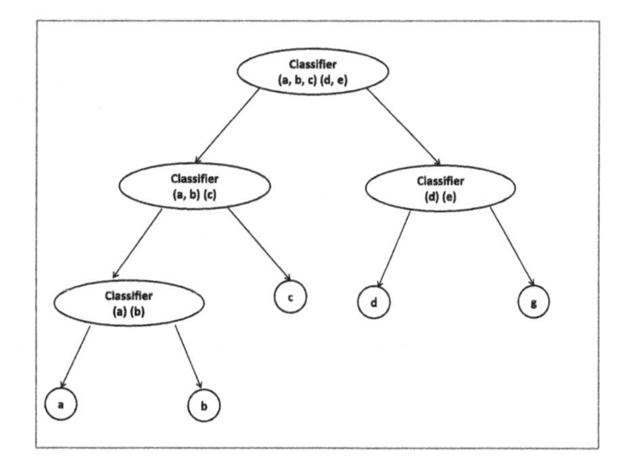

Fig. 1 Binary tree hierarchical classification example

labels $\{a, b, c\}$ and $\{d, e\}$. At the next level we discriminate between smaller groups, and so on, until we reach classifiers that can assign a single class label to the record.

Although the desired binary tree structure can be constructed in either a bottom-up or a top-down manner, top down construction is the most widely used because it tends to produce a more balanced structure and because it is easier to implement [4]. Using the top down approach the process is as follows: Starting at the root of the tree, a grouping technique is used to segment the records into two clusters, each cluster is labelled with a group-class label. Then, a classifier is trained to classify records using the two group-classes. The process continues recursively until classifiers are arrived at that can assign single class labels to individual records. For classifying a new record a "path" is followed from the root, according to the classification at each hierarchy node, until a leaf node associated with a single class label is reached. As noted in the introduction to this paper successive miss-classification is a general drawback associated with hierarchical classification, where a miss-classification near the root of the tree is passed on down the hierarchy. To address this problem a multiple path strategy, as previously suggested by the authors in [1, 3], can be adopted whereby multiple paths can be followed within the binary tree hierarchy. However, experiments also reported in [1, 3], indicated that this did not provide an entirely satisfactory solution to the problem.

3 Directed Acyclic Graph (DAG) Classification Model Framework

The nature of the proposed DAG hierarchical classification model is presented in this section. As noted in the introduction to this paper the model is founded on the idea of arranging the classifiers into a hierarchical form by utilising a DAG structure where each node in the DAG holds a classifier. Classifiers at leaves act as binary classifiers while the remaining classifiers (at the beginning and intermediate nodes) are directed at groupings of class labels. An example non-rooted DAG classifier for four class labels $C = \{a, b, c, d\}$, is presented in Fig. 2. In this case, the nodes at the first level are assigned with classes of size three, all possible combination of size $|C| - 1$, while nodes at the second level are assigned with classes of size two ($|C| - 2$). Classifiers at the first level distinguish between three groups of class labels, while classifiers at the leaves discriminate between two individual class labels. The rest of this section is organised as follow. Section 3.1 below explains the generation of the DAG ensemble model in detail. While Sect. 3.2 presents the operation of the proposed model.

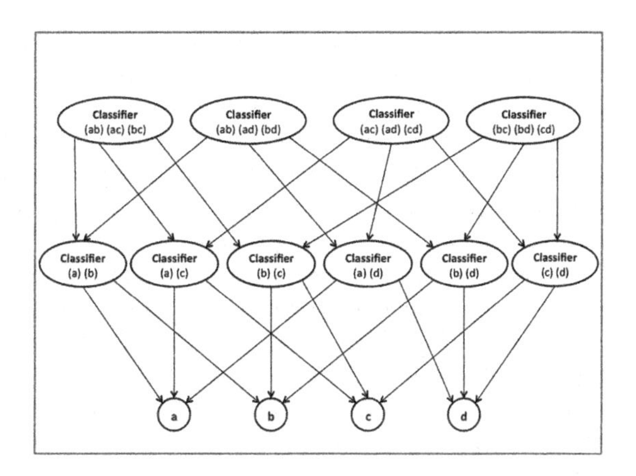

Fig. 2 DAG classification model example

3.1 DAG Generation

To create the desired DAG model, a classifier needs to be generated for each node in the DAG using an appropriate training set. At "start up" the training set D comprises a set of n records $D = \{r_1, r_2, \dots .r_n\}$ such that each record has a class label associated with it taken from the set C. The process requires a class label grouping mechanism. One way of doing this is to apply a clustering mechanism such a k-means (k-means is particularly well suited because the value of k can be pre-specified). In previous work conducted by the authors [1–3], in the context of binary hierarchies, it was found that this clustering approach did not work well because similar classes were grouped together early on in the process so that entire branches ended up dealing with very similar classes, ideally we would like individual branches to deal with very different classes so that highly discriminative classifiers can be built at each leaf node. However, identifying such groups is also not straight forward. Instead, the use of a combination mechanism is proposed that covers all potential groupings. The class groupings (sub sets) at each level are determined by finding all possible classes combinations of size $|C| - i$ (where i is the level number, initially $i = 1$). As the process proceeds i is increased by one and consequently the "combination size" is decreased by one. The process continues until the combination size reaches two. The process was used to generate the DAG given in Fig. 2. The number of classifiers to be trained in order to generate the DAG classification model can be calculated using Eq. 1.

$$Number\ Of\ Classifiers = 2^N - N - 2 \tag{1}$$

where N is the number of class labels in a given data set.

However, using the above mechanism, the number of classifiers to be generated is large. Breadth and depth pruning mechanisms are thus proposed to reduce the number of classifiers, as well as, to improve the performance of the suggested model.

Starting with breadth pruning, in which *weak* classifiers, at each DAG level, are eliminated. The *weak* classifiers are identified by evaluating the classifiers at the first level and pruning the classifiers associated with an *AUC* value of less than a predefined threshold α. The pruning is transmitted to the remaining levels by only including nodes that are, directly or indirectly, referenced by the first level nodes. Note here that the pruning process at the first level is post pruning, while the pruning at the remanning levels is pre-pruning.

Algorithm 1 presents the proposed DAG generation process with breadth pruning. The input to the algorithm is the training data set D, the set of class labels C, and the breadth pruning threshold α. The DAG is created in a top down manner starting with $k = |C| - 1$ (where k is the combination size) to $k = 2$. Because of the nature of the breadth pruning mechanism the algorithm comprises two procedures: *dagFirstLevelGen*, and *dagNlevelGen*. Starting with *dagFirstLevelGen* procedure, which is responsible for generating and pruning the first level in the DAG model. The process commences (with reference to Algorithm 1 (a)) by finding the set of size k class combinations, the set C_k (line 10). We then loop through this set (line 12) and on each iteration: (i) find the set of records D_i that feature the combination $C_i \in C_k$ (line 13); (ii) identify the training and evaluating records, T_i and E_i (lines 14, and 15); (iii) generate a classifier G_i using T_i (line 16); (iv) evaluate the classifier G_i using E_i to produce an *AUC* value (line 17); (v) create a new DAG node, *node*, and add the new node to the set of accumulated level k nodes so far, *NodeSet* (line 18). The next stage is the pruning stage, where the nodes in the *NodeSet* are arranged according to their associated *AUC* values (Line 20). We the loop through this set of ordered nodes (line 21): if the node associated with a particular *AUC* value is less than α (line 22), and its classes are included in the remaining nodes in *NodeSet* (line 23); then the node will be pruned (line 24). (We want to ensure that each class is the subject of at least one classifier.)

After generating and pruning the first level in the DAG model, the next step is to generate the remaining levels. The remaining levels in the DAG are created in a recursive manner using the *dagNlevelGen* procedure (Algorithm 1 (b)). The procedure is invoked with two parameters: (i) the combination size, k, and a reference to the nodes in the current level of the DAG, *CurrentNodes*. For each call to *dagNlevelGen* the set of size k class combinations, the set C_k, is calculated (line 30), then pruning is applied to this set (lines 31-40), where the class combination is only considered if it is a subset of one or more of the previous levels nodes' class sets. We then loop through the pruned combination set (line 42) and on each iteration: (i) find the set of training set records T_i that feature the combination $C_i \in C_k$ (line 43), (ii) generate a classifier G_i using T_i (line 44); (iii) create a new DAG node, *node* (line 45); and (iv) add the new node to the set of accumulated level k nodes so far, *NodeSet* (line 46). We then loop through the set of current nodes (from the previous iteration) and add a link from each current node $CurrentNode_j$ to the new node *node* whenever the set of class labels associated with the new node (C_i) is included in the set of class labels associated with a current node ($C_i \subset CurrentNodes_j.C$). Finally, if k has not yet reached 2, we repeat (line 53).

Algorithm 1 (a) DAG Generation

```
1: INPUT: D = The input training data set, C = The set of Classes featured in D,
2:        α = Breadth pruning threshold value
3: OUTPUT: The generated DAG
4: Start
5: k = |C| − 1
6: dagFirstLevelGeneration(k)
7: dagNlevelGen(k − 1, NodeSet)
8: End
9: procedure dagFirstLevelGen(k)
10:    C_k = Set of size k combinations in C
11:    NodeSet = { }
12:    for i = 1 to i = |C_k| do
13:        D_i = Set of records in D that feature C_i (D_i ⊂ D)
14:        T_i= Set of records in D_i for training G_i
15:        E_i = Set of records in D_i for evaluating G_i
16:        G_i = Classifier for C_i built using training set T_i
17:        AUC_i = Evaluation of G_i using E_i
18:        NodeSet = NodeSet ∪ new Node(G_i, C_i, AUC_i)
19:    end for
20:    Arrange nodes in NodeSet in ascending order of associated AUC value
21:    for i = 1 to i = |NodeSet| do
22:        if (node_i.auc < α) then
23:            if (classesCovered (node.C_i, NodeSet)) then
24:                Delete node_i
25:            end if
26:        end if
27:    end for
28: end procedure
```

Because of the flexibility of: (i) the DAG structure and (ii) the combination technique used; it is fairly straightforward (in addition to breadth pruning) to apply depth pruning. Depth pruning in this context is achieved by limiting the number of levels in the DAG (pre-pruning). Note that the minimum number of levels is always 2. For simplicity, in this paper we considered the generation of the all levels of the DAG model (no depth pruning) and the generation of only 2 levels of the DAG (maximum depth pruning). The idea here is that by applying depth pruning the performance of the suggested model might be improved because the number of classifiers that are required to be learned and evaluated will be decreased; consequently scalability, efficiency, and effectiveness might be improved. For simplicity, and throughout the rest of this paper, we will refer to these approaches as the standard DAG and the two-levels DAG approaches respectively.

Algorithm 1 (b) DAG Generation

29: **procedure** *dagNlevelGen*(k, *CurrentNodes*)
30: C_k = Set of size k combinations in C
31: **for** $i = 1$ to $i = |C_k|$ **do**
32: **for** $j = 1$ to $j = |CurrentNodes|$ **do**
33: **if** $C_i \subset CurrentNodes_j.C$ **then**
34: *flag*=true, break
35: **end if**
36: **end for**
37: **if** flag == false **then**
38: Delete C_i
39: **end if**
40: **end for**
41: $NodeSet = \{\}$
42: **for** $i = 1$ to $i = |C_k|$ **do**
43: T_i = Set of training records in D that feature C_i ($T_i \subset D$)
44: G_i = Classifier for C_i built using training set T_i
45: $node = new\ Node(G_i, C_i)$
46: $NodeSet = NodeSet \cup node$
47: **for** $j = 1$ to $j = |CurrentNodes|$ **do**
48: **if** $C_i \subset CurrentNodes_j.C$ **then**
49: $CurrentNodes_j.childNodes = CurrentNodes_j.childNodes \cup node$
50: **end if**
51: **end for**
52: **end for**
53: **if** $k > 2$ **then**
54: *dagNlevelGen*($k - 1$, *NodeSet*)
55: **end if**
56: **end procedure**

3.2 DAG Operation

Section 3.1 above described the process for generating the proposed DAG classification model. After the model has been generated it is ready for use. In this section the operation of the suggested model is explained. As noted earlier, two main challenges are associated with the operation of the proposed DAG model: (i) how to determine the "start node" among the set of nodes available at the first level, and (ii) how to address the general drawback associated with hierarchical forms of ensemble classification, the "successive miss-classification" problem. To address these issues Naive Bayesian probabilities were utilised to determine the best starting node for the DAG model, and also to decide whether a single or a multiple path should be followed at each node; consequently, two strategies are considered for classifying individual records: single path and multiple path.

The single path strategy is the most straight-forward, and involves following a single path through the DAG to classify the record. The classification process commences with the evaluation of all classifiers at the first level and selecting the node who's classifier generates the highest probability value to be the start node, the process continues as directed by the individual node classifications, until a leaf node is arrived at. Leaf nodes, as already noted, hold binary classifiers; thus when a leaf node is reached a binary classification can be conducted and a single class label can be assigned to the record. However, as also already noted, the issue with the single path strategy is that if a misclassification occurs early on in the process there is no opportunity for addressing this situation later on in the process. The multiple path strategy is designed to handle this problem by allowing more than one path to be followed within the DAG according to a predefined probability threshold σ ($0 \leq \sigma < 1$). In this paper we suggest following up to two branches from each DAG node as a maximum. An alternative strategy might be to follow more than two branches per node, however, this will require more computational resource. Following only two branches also, of course, allows us to make comparisons with the operation of binary tree based hierarchical ensemble classifiers. In cases where more than one path is followed we may end up with a number of alternative class labels at the leaf nodes of the DAG, thus we have a set of "candidate class labels". In order to determine a final classification we take into consideration the Bayesian probability values identified along the paths from each relevant leaf node back to the root node, and produce a set of *accumulated* values. The class with the highest accumulated Bayesian probability value is then selected.

When using the multiple-path strategy, the process commences with the evaluation of the classifiers at the first level and selecting one or two nodes to be the start of the DAG depending on the associated probability (p) for each. This is achieved by considering the two nodes with highest probability values, if both probabilities are greater than a predefined threshold σ then both nodes will be considered to be start nodes, otherwise the node with the highest associated probability value will be considered to be the start node. The process is facilitated as follows. At each DAG node the two class groups associated with the highest probabilities are identified, then if their probabilities are greater than σ both branches will be explored, otherwise the branch with the highest associated p value will be selected. In order to decide the final class label to be assigned to the record among the set of candidate classes, the class associated with the highest accumulated probability value will be selected. The accumulated probability for each candidate class is calculated by summing the probability values identified along the path and then dividing the total by the number of classifiers that were invoked along the path.

4 Experiments and Evaluation

The effectiveness of the suggested DAG classification model was evaluated using twelve different data sets taken from the UCI machine learning repository [13], and pre-processed using the LUCS-KDD-DN software [7]. Ten-fold Cross Validation (TCV) was used throughout. The evaluation measures used were average accuracy and average AUC (Area Under the receiver operating Curve). We will discuss the results in terms of average AUC for simplicity and because of the inclusion of unbalanced data sets within the considered evaluation data sets. For comparison purposes the data sets were also classified using:

1. A "stand alone" **Naive Bayes** classifier [12], the objective being to compare the proposed DAG model with a single conventional model. Naive Bayes was chosen because this was also used in the context of the DAG.
2. A **Bagging** ensemble [6] using Naive Bayes classifiers as the base classifiers, the objective being to compare the proposed DAG model with alternative forms of ensemble.
3. A **Binary Tree Hierarchical Ensemble** classifier of the form described in Sect. 2, with a Naive Bayes classifier generated for each tree node, and data segmentation to distribute class labels between nodes within the tree, both single-path and multi-path strategies were used, the objective being to compare the proposed DAG model with a simple binary tree model.
4. A **One-Versus-One** (OVO) classification mechanism using support vector machines as the base classifiers [16]. The objective being to compare the proposed DAG model with a classification mechanism founded on the use of a set of binary classifiers for solving the multi-class classification problems.

4.1 Comparison Between DAG Approaches

This section presents the results obtained using the DAG classification approaches, standard DAG and two-level DAG, coupled with the Single-Path and Multi-Path strategies with respect to the ten different data sets considered in the evaluation. The objective of the comparison is to determine the best DAG approach, and to determine whether following more than one path within the DAG classification model could address the successive miss-classification issue noted earlier. With respect to the Multi-Path strategy a very low threshold of $\sigma = 0.1 \times 10^{-4}$ was used. In previous experiments, not reported in this paper, a range of alternative σ values were considered and it was found that $\sigma = 0.1 \times 10^{-4}$ produced the best performance. The results, in terms of average accuracy and average AUC, are presented in Table 1 (best values highlighted in bold). Considering the two different approaches for the DAG model, standard DAG and two-level DAG, from the table it can be noted that the two-level DAG produced better results than the standard DAG; the suggested reason for this is that two-level DAG is simpler and may therefore feature less overfitting. In addition, it can be observed that using the multi-path strategy can significantly improve the classification accuracy with respect to the single path strategy, especially for the standard DAG approach where the number of levels are higher and consequently the probability of miss-classification is higher.

Table 1 Average accuracy and AUC values obtained using proposed DAG models coupled with the single and multiple path strategies

Data set	Classes	Single-path strategy				Multi-path strategy			
		Standard		Two-level		Standard		Two-level	
		Acc.	AUC	Acc.	AUC	Acc.	AUC	Acc.	AUC
Nursery	5	79.83	0.40	**91.44**	0.54	82.32	0.41	90.02	**0.58**
Heart	5	57.01	0.39	**59.91**	**0.40**	56.25	0.37	59.64	**0.40**
PageBlocks	5	91.83	0.53	92.02	0.49	91.87	**0.54**	**92.05**	0.47
Dermatology	6	**87.23**	**0.85**	86.09	0.84	**87.23**	**0.85**	85.51	0.84
Glass	7	69.81	0.46	57.58	0.48	**71.16**	**0.50**	57.18	0.49
Zoo	7	92.18	0.58	**93.18**	**0.59**	92.18	**0.59**	**93.18**	**0.59**
Ecoli	8	**84.43**	**0.41**	82.40	0.40	82.26	0.38	80.89	0.39
Led	10	75.66	**0.76**	**75.75**	**0.76**	75.53	**0.76**	75.66	**0.76**
PenDigits	10	83.59	**0.84**	**83.84**	**0.84**	83.59	**0.84**	**83.84**	**0.84**
Soybean	15	**90.57**	**0.92**	90.04	**0.92**	**90.57**	**0.92**	90.04	**0.92**
Mean		81.21	0.61	81.23	**0.63**	**81.30**	0.62	80.80	**0.63**

4.2 Comparison Between Stand-Alone Classification, Bagging, Binary Tree, OVO SVM and DAG Ensemble Classification

In this section a comparison between the proposed DAG classification model and conventional classification models is presented. In order to conduct a "consistent" comparison between the DAG and existing conventional models, comparison was conducted using the same classifier generator. More specifically, comparison between the operation of a stand alone Naive Bayes classification, Bagging of Naive Bayes classifiers, binary tree hierarchical classification with a Naive Bayes classifier generated for each tree node, and Naive Bayes DAG classification was considered. In addition, a "non-consistent" comparison between the Naive Bayes DAG classification model and OVO SVM was conducted. Recall that the objective of this last comparison was to compare the suggested model with one of the state of the art methods for multi-class classification.

Commencing with the comparison between stand alone classification, Bagging classification, binary tree hierarchical classification, and DAG classification. Table 2 presents the results obtained. Note here that the presented DAG results are the results obtained using the two-level DAG approach coupled with the multiple-path strategy, because the foregoing section established that the two-level DAG multiple path strategy produces a better performance. Because of the lower memory requirements for the two-level DAG compared to the standard DAG, two additional data sets (Chess KRvK, and Letter recognition) were included in addition to those considered in Table 1. From the average AUC given in the table, for each method with respect

Table 2 Average Accuracy and AUC values obtained using "stand-alone" Naive Bayes classification, Bagging, Binary Tree Based hierarchical ensemble classification, and the proposed DAG classification model

Data set	Classes	Naive Bayes Single model		Bagging Ensemble		Binary Tree Model		DAG Model	
		Acc.	AUC	Acc.	AUC	Acc.	AUC	Acc.	AUC
Nursery	5	**90.22**	0.45	89.96	0.46	89.09	0.58	90.02	**0.58**
Heart	5	54.60	0.34	51.28	0.30	53.77	0.36	**59.64**	**0.40**
PageBlocks	5	**92.69**	**0.52**	92.62	**0.52**	91.27	0.48	92.05	0.47
Dermatology	6	**86.66**	**0.85**	81.00	0.81	84.60	0.84	85.51	0.84
Glass	7	**67.83**	0.49	55.28	0.46	55.28	**0.51**	57.18	0.49
Zoo	7	92.27	0.59	**94.27**	**0.62**	92.18	0.58	93.18	0.59
Ecoli	8	81.70	0.38	**82.56**	**0.39**	64.15	0.27	80.89	**0.39**
Led	10	75.59	**0.76**	75.50	**0.76**	61.13	0.61	**75.66**	**0.76**
PenDigits	10	**84.94**	**0.85**	84.57	**0.85**	81.18	0.81	83.84	0.84
Soybean	15	**91.11**	**0.93**	86.83	0.89	83.71	0.83	90.04	0.92
ChessKRvK	18	**36.32**	0.33	35.66	0.34	33.88	**0.37**	35.36	0.36
LetterRecog	26	**57.37**	**0.57**	56.93	**0.57**	53.44	0.53	55.84	0.56
Mean		**75.94**	0.59	73.87	0.58	70.31	0.56	74.93	**0.60**

to the twelve data sets considered, it can be noted that the operation of the proposed DAG classification model is superior to that for stand alone Naive Bayse classification, Bagging ensemble classification, and simple binary tree hierarchical classification where the average AUCs obtained from these methods were 0.59, 0.58, 0.56 respectively, while for the DAG model it was 0.60. The proposed DAG classification model improved the classification accuracy for four data sets (Nursery, Heart, Ecoli, and ChessKRvK), although for one data set (Led) the same result obtained from using stand-alone Naive Bayes classifier.

Comparing with the use of the binary tree based structure a significant improvement was recorded when using the DAG model. The suggested reason for this is that the DAG model provides for greater flexibility than in the case of the binary tree model because of the overlap between class groups represented by nodes at the same level in the hierarchy. The consequence of this is that the overlap partly mitigates against the early miss-classification issue. In addition, pruning the weak classifiers from the DAG model results in a better classification accuracy than in the case of the binary tree structure where all the classifiers are used.

With respect to the comparison with stand alone Naive Bayes classification and Bagging ensemble classification, it is interesting to note that the proposed DAG classification model tends to improve the classification effectiveness with respect to unbalanced datasets such as: Nursery, Heart, Ecoli, and ChessKRvK. It is conjectured that the combination techniques, used to distribute class labels between nodes within the DAG, helps in the handling of unbalanced datasets. With respect to the

Table 3 Average Accuracy and AUC values obtained using the proposed Naive Bayes DAG coupled with the multi-path strategy, and One-versus-One using SVM as the base classifier

Data set	Naive Bayes		OVO	
	DAG		SVM	
	Acc.	AUC	Acc.	AUC
Nursery	90.02	0.58	99.69	**0.64**
Heart	59.91	**0.40**	53.01	0.22
PageBlocks	92.02	0.49	92.58	**0.50**
Dermatology	86.09	0.84	88.73	**0.86**
Glass	57.18	**0.49**	72.04	0.47
Zoo	93.18	**0.59**	94.00	0.58
Ecoli	82.40	**0.40**	82.95	0.36
Led	75.75	**0.76**	75.62	**0.76**
PenDigits	83.84	0.84	98.60	**0.99**
Soybean	90.04	**0.92**	92.54	0.91
ChessKRvK	35.36	0.36	86.40	**0.81**
LetterRecog	55.85	0.56	82.92	**0.83**
Mean	75.14	0.60	84.92	**0.66**

"non-consistent" comparison between Naive Bayes DAG and OVO SVM, Table 3 presents the results obtained in terms of average accuracy and average AUC (best results highlighted in bold font). From the table it can be observed that the Naive Bayes DAG produced the best classification accuracy with respect to six of the twelve data sets considered (Heart, Glass, Ecoli, Zoo, Led, and Soybean), although for one data set (Led) the same result was produced using OVO SVM. In the remaining six cases, the OVO SVM produced the best result.

4.3 Note on Efficiency

The number of classifiers required to be generated or evaluated with respect to the suggested DAG approaches can not be determined in advance because of the application of the breadth pruning scheme, but the efficiency can be evaluated according to the generation and classification time. Unfortunately space limitations preclude the presentation of a detailed run time analysis here, however, the analysis of run times indicates that following multiple paths within the hierarchical classification model consumes more run time than in the case of following only a single path, regardless of the adopted structure (binary tree or DAG structure). With respect to the proposed DAG approaches, the two level DAG approach, in which depth and breadth pruning were adopted, requires less run time. Regarding comparison between the DAG structure and binary tree structure, as excepted, the binary tree approach requires less run time because the proposed DAG structure is more complex. Of course, with respect to comparison with the conventional methods, single Naive classifier and Bagging classification, the proposed DAG approach requires more run time. However, the generation of the model needs to be done only once. Regarding the comparison with OVO SVM, the stat-of-the-art approach for multi-class classification, the DAG classification model is more efficient than OVO SVM according to the recorded generation and classification run times.

5 Conclusion

A hierarchical ensemble classification model for multi-class classification using a Directed Acyclic Graph (DAG) structure has been presented. The DAG structure facilitated the use of three mechanisms to address the successive miss-classification problem associated with hierarchical classification. The first mechanism was the combination technique used to group classes across nodes at individual levels in the DAG so that an overlap exists between the class groups; unlike in the case of binary tree based ensemble classifiers where this option is not available. The second mechanism was the option to follow multiple paths down the hierarchy by utilising the probability values generated by the Naive Bayes classifiers generated for each node in the DAG. The third mechanism was the pruning scheme applied to eliminate the weak classifiers that can affect classification effectiveness. From the reported evaluation it was demonstrated that the proposed DAG classification model could be successfully used to classify data in a more effective manner than when alternative conventional methods were used, such as Naive Bayes, Bagging, and OVO SVM with respect to some of the data sets considered in the evaluation. In addition, it was demonstrated that following more than one path in the DAG tended to produce a better classification effectiveness with respect to the majority of the data sets considered. The evaluation also indicated that the overall performance of the DAG structure was clearly superior to a simple binary tree structure.

References

1. Alshdaifat, E., Coenen, F., Dures, K.: Hierarchical classification for solving multi-class problems: a new approach using naive bayesian classification. In: Proceedings of the Ninth International Conference on Advanced Data Mining and Applications (ADMA'13), pp. 493–504. Springer (2013)
2. Alshdaifat, E., Coenen, F., Dures, K.: Hierarchical single label classification: an alternative approach. In: Proceedings of the Thirty-third BCS SGI International Conference on Innovative Techniques and Applications of Artificial Intelligence (BCS SGAI'13), pp. 39–52. Springer (2013)
3. Alshdaifat, E., Coenen, F., Dures, K.: A multi-path strategy for hierarchical ensemble classification. In: Proceedings of the Tenth International Conference on Machine Learning and Data Mining in Pattern Recognition (MLDM'14), pp. 198–212. Springer (2014)

4. Athimethphat, M., Lerteerawong, B.: Binary classification tree for multiclass classification with observation-based clustering. In: Proceedings of the Ninth International Conference on Electrical Engineering/Electronics, Computer, Telecommunications and Information Technology (ECTI-CON'12), pp. 1–4. IEEE (2012)
5. Beygelzimer, A., Langford, J., Ravikumar, P.: Multiclass classification with filter trees (2007). Preprint, available at http://hunch.net/jl/projects/reductions/mc_to_b/invertedTree.pdf
6. Breiman, L.: Bagging predictors. Mach. Learn. **24**(2), 123–140 (1996)
7. Coenen, F.: The LUCS-KDD discretised/normalised arm and carm data library. http://www.csc.liv.ac.uk/frans/KDD/Software/LUCS_KDD_DN
8. Freund, Y., Schapire, R.: A short introduction to boosting. J. Japn. Soc. Artif. Intell. **14**(5), 80–771 (1999)
9. Han, J., Kamber, M., Pei, J.: Data Mining: Concepts and Techniques. Morgan Kaufmann, Burlington (2011)
10. Kumar, S., Ghosh, J., Crawford, M.: Hierarchical fusion of multiple classifiers for hyperspectral data analysis. Pattern Anal. Appl. **5**(2), 210–220 (2002)
11. Lei, H., Govindaraju, V.: Half-against-half multi-class support vector machines. In: Proceedings of the Sixth International Workshop on Multiple Classifier Systems (MCS'05), pp. 156–164. Springer (2005)
12. Leonard, T., Hsu, J.: Bayesian Methods: An Analysis for Statisticians and Interdisciplinary Researchers. Cambridge Series in Statistical and Probabilistic Mathematics. Cambridge University Press, Cambridge (2001)
13. Lichman, M.: UCI machine learning repository. http://archive.ics.uci.edu/ml
14. Oza, N., Tumer, K.: Classifier ensembles: select real-world applications. Inf. Fusion **9**(1), 4–20 (2008)
15. Rifkin, R., Klautau, A.: In defense of one-vs-all classification. J. Mach. Learn. Res. **5**, 101–141 (2004)
16. Tax, D., Duin, R.: Using two-class classifiers for multiclass classification. In: Proceedings of the Sixteenth International Conference on Pattern Recognition, pp. 124–127. IEEE (2002)
17. Zhou, Z.H.: Ensemble learning. In: Encyclopedia of Biometrics, pp. 270–273. Springer (2009)

CLUB-DRF: A Clustering Approach to Extreme Pruning of Random Forests

Khaled Fawagreh, Mohamed Medhat Gaber and Eyad Elyan

Abstract Random Forest (RF) is an ensemble supervised machine learning technique that was developed by Breiman over a decade ago. Compared with other ensemble techniques, it has proved its superiority. Many researchers, however, believe that there is still room for enhancing and improving its performance accuracy. This explains why, over the past decade, there have been many extensions of RF where each extension employed a variety of techniques and strategies to improve certain aspect(s) of RF. Since it has been proven empirically that ensembles tend to yield better results when there is a significant diversity among the constituent models, the objective of this paper is twofold. First, it investigates how data clustering (a well known diversity technique) can be applied to identify groups of similar decision trees in an RF in order to eliminate redundant trees by selecting a representative from each group (cluster). Second, these likely diverse representatives are then used to produce an extension of RF termed *CLUB-DRF* that is much smaller in size than RF, and yet performs at least as good as RF, and mostly exhibits higher performance in terms of accuracy. The latter refers to a known technique called ensemble pruning. Experimental results on 15 real datasets from the UCI repository prove the superiority of our proposed extension over the traditional RF. Most of our experiments achieved at least 92 % or above pruning level while retaining or outperforming the RF accuracy.

K. Fawagreh · M.M. Gaber (✉) · E. Elyan
School of Computing Science and Digital Medial, Robert Gordon University,
Garthdee Road, Aberdeen AB10 7GJ, UK
e-mail: m.gaber1@rgu.ac.uk

K. Fawagreh
e-mail: k.fawagreh@rgu.ac.uk

E. Elyan
e-mail: e.elyan@rgu.ac.uk

M. Bramer and M. Petridis (eds.), *Research and Development in Intelligent Systems XXXII*, DOI 10.1007/978-3-319-25032-8_4

1 Introduction

Ensemble classification is an application of ensemble learning to boost the accuracy of classification. Ensemble learning is a supervised machine learning paradigm where multiple models are used to solve the same problem [28, 39, 41]. Since single classifier systems have limited predictive performance [33, 39, 41, 52], ensemble classification was developed to yield better predictive performance [33, 39, 41]. In such an ensemble, multiple classifiers are used. In its basic mechanism, majority voting is then used to determine the class label for unlabeled instances where each classifier in the ensemble is asked to predict the class label of the instance being considered. Once all the classifiers have been queried, the class that receives the greatest number of votes is returned as the final decision of the ensemble.

Three widely used ensemble approaches could be identified, namely, boosting, bagging, and stacking. Boosting is an incremental process of building a sequence of classifiers, where each classifier works on the incorrectly classified instances of the previous one in the sequence. AdaBoost [14] is the representative of this class of techniques. However, AdaBoost is prone to overfitting. The other class of ensemble approaches is the Bootstrap Aggregating (Bagging) [6]. Bagging involves building each classifier in the ensemble using a randomly drawn sample of the data, having each classifier giving an equal vote when labeling unlabeled instances. Bagging is known to be more robust than boosting against model overfitting. Random Forest (RF) is the main representative of bagging [8]. Stacking (sometimes called stacked generalization) extends the cross-validation technique that partitions the dataset into a held-in dataset and a held-out dataset; training the models on the held-in data; and then choosing whichever of those trained models performs best on the held-out dataset. Instead of choosing among the models, stacking combines them, thereby typically getting performance better than any single one of the trained models [51]. Stacking has been successfully used on both supervised learning tasks (regression) [7], and unsupervised learning (density estimation) [46].

The ensemble method that is relevant to our work in this paper is RF. RF has been proved to be the state-of-the-art ensemble classification technique. Since RF algorithms typically build between 100 and 500 trees [50], in real-time applications, it is of paramount importance to reduce the number of trees participating in majority voting and yet achieve performance that is at least as good as the original ensemble. In this paper, we propose a data clustering approach to prune RF ensembles where only a small subset of the ensemble is selected. We cluster trees according to their classification behavior on a subset of the dataset. Then we choose only one tree from each cluster, motivated by the fact that the tree is a representative of its cluster. At voting time, the number of voting trees is reduced significantly yielding classification accuracy at least as good as all voting trees. A thorough experimental study is conducted, with multiple of 5 clusters ranging from 5 through to 40, over an RF of 500 trees. Results show the potential of our technique for deployment in real-time systems, yielding higher predictive accuracy than traditional RF, with 17 to 100 times faster classification per instance resulting from pruning levels in the range of 94–99 %.

This paper is organized as follows. First we discuss related work in Sect. 2. Section 3 formalizes our approach and corresponding algorithm. Experiments, results and analysis demonstrating the superiority of the proposed pruned RF over the traditional RF are detailed in Sect. 4. The paper is finally concluded with a summary and pointers to future directions in Sect. 5.

2 Related Work

Several enhancements have been made in recent years in order to produce a subset of an ensemble that performs as well as, or better than, the original ensemble. The purpose of ensemble pruning is to search for such a good subset. This is particularly useful for large ensembles that require extra memory usage, computational costs, and occasional decreases in effectiveness. Grigorios et al. [49] recently amalgamated a survey of ensemble pruning techniques where they classified such techniques into four categories: ranking based, clustering based, optimization based, and others. Clustering based methods, that are relevant to us in this paper, consist of two stages. In the first stage, a clustering algorithm is employed in order to discover groups of models that make similar predictions. Pruning each cluster then follows in the final stage. In this stage, several approaches have been used. One approach by [4] was to train a new model for each cluster, using the cluster centroids as values of the target variable. Another interesting approach was proposed by [16] that involved selecting from each cluster the classifier that is most distant to the rest of the clusters. A yet different approach by [29] that does not guarantee the selection of a single model from each cluster was by iteratively removing models from the least to the most accurate, until the accuracy of the entire ensemble starts to decrease. Selecting the most accurate model from each cluster was proposed by [40].

More recent work in ensemble pruning was also reported. Without a significant loss of prediction accuracy, a combination of static and dynamic pruning techniques were applied on Adaboost ensembles in order to yield less memory consumption and improved classification speed [47]. A pruning scheme for high dimensional and large sized benchmark datasets was developed by [11], In such a scheme, an extended feature selection technique was used to transform ensemble predictions into training samples, where classifiers were treated as features. Then a global heuristic harmony search was used to select a smaller subset of such artificial features.

It is worth pointing out that there are several techniques available in the literature that use clustering-based approaches to reduce the number of trees in the ensemble (see [26] for a good review). Unlike such techniques, our clustering approach in this paper is different in many aspects. First of all, to the best of our knowledge, none of the clustering-based approaches was developed for RF ensembles. For example, all the approaches proposed by the respective authors in [4, 16, 29, 40] were developed for neural network ensembles. Furthermore, our approach differs from theirs in two other ways. For the selection of a representative from each cluster, we have selected the best performing representative on the out-of-bag (OOB) instances and this selection method was not used by anyone of them. As will be discussed in Sect. 3, using OOB samples to evaluate a tree gives an unbiased estimate of its predictive accuracy since, unlike about 64 % of the training data that was seen by the tree when it was built, OOB data was not seen and therefore, it is a more accurate measure of the tree's predictive accuracy. Furthermore, at the experimental level, we have used 15 datasets from the UCI repository, however, very few datasets were used by them: 2 in [4], 1 in [16], 4 in [29], and 4 in [40].

2.1 Diversity Creation Methods

Because of the vital role diversity plays on the performance of ensembles, it had received a lot of attention from the research community. The work done to date in this domain was summarized in [9] from two main perspectives. The first is a review of the various attempts that were made to provide a formal foundation of diversity. The second, which is more relevant to this paper, is a survey of the various techniques to produce diverse ensembles. For the latter, two types of diversity methods were identified: implicit and explicit. While implicit methods tend to use randomness to generate diverse trajectories in the hypothesis space, explicit methods, on the other hand, choose different paths in the space deterministically. In light of these definitions, bagging and boosting in the previous section are classified as implicit and explicit respectively.

Ensemble diversity techniques were also categorized by [9] into three categories: starting point in hypothesis space, set of accessible hypotheses, and manipulation of training data. Methods in the first category use different starting points in the hypothesis space, therefore, influencing the convergence place within the space. Because of their poor performance of achieving diversity, such methods are used by many authors as a default benchmark for their own methods [33]. Methods in the second category vary the set of hypotheses that are available and accessible by the ensemble. For different ensembles, these methods vary either the training data used or the architecture employed. In the third category, the methods alter the way space is traversed. Occupying any point in the search space, gives a particular hypothesis. The type of the ensemble obtained will be determined by how the space of the possible hypotheses is traversed.

2.2 Diversity Measures

Regardless of the diversity creation technique used, diversity measures were developed to measure the diversity of a certain technique or perhaps to compare the diversity of two techniques. Tang et al. [48] presented a theoretical analysis on six existing diversity measures: disagreement measure [45], double fault measure [15], KW variance [25], inter-rater agreement [13], generalized diversity [38], and measure of difficulty [13]. The goal was not only to show the underlying relationships between them, but also to relate them to the concept of margin, which is one of the contributing factors to the success of ensemble learning algorithms.

We suffice to describe the first two measures as the others are outside the scope of this paper. The disagreement measure is used to measure the diversity between two base classifiers (in RF case, these are decision trees) h_j and h_k, and is calculated as follows:

$$dis_{j,k} = \frac{N^{10} + N^{01}}{N^{11} + N^{10} + N^{01} + N^{00}}$$

where

- N^{10}: means number of training instances that were correctly classified by h_j, but are incorrectly classified by h_k
- N^{01}: means number of training instances that were incorrectly classified by h_j, but are correctly classified by h_k
- N^{11}: means number of training instances that were correctly classified by h_j and h_k
- N^{00}: means number of training instances that were incorrectly classified by h_j and h_k

The higher the disagreement measure, the more diverse the classifiers are. The double fault measure uses a slightly different approach where the diversity between two classifiers is calculated as:

$$DF_{j,k} = \frac{N^{00}}{N^{11} + N^{10} + N^{01} + N^{00}}$$

The above two diversity measures work only for binary classification (AKA binomial) where there are only two possible values (like Yes/No) for the class label, hence, the objects are classified into exactly two groups. They do not work for multiclass (multinomial) classification where the objects are classified into more than two groups. In Sect. 3.2, we propose a simple diversity measure that works with both binary and multiclass classification.

3 Proposed Extension: CLUB-DRF

In this section, we propose an extension of RF called CLUB-DRF that spawns a child RF that is (1) much smaller in size than the parent RF and (2) has an accuracy that is at least as good as that of the parent RF. In the remainder of this paper, we will refer to the parent/original traditional Random Forest as *RF*, and refer to the resulting child RF based on our method as *CLUB-DRF*.

3.1 Clustering-Based Diverse Random Forest (CLUB-DRF)

The CLUB-DRF extension applies a clustering-based technique to produce diverse groups of trees in the RF. Assume that the trees in the RF are denoted by the vector RF $=< t_1, t_2, \ldots, t_n >$ (where n is number of trees in the RF), and the training set is denoted by T $= \{r_1, r_2, \ldots, r_m\}$. Each tree in the RF casts a vote to classify each record in the training set to determine the class label c. We use C(t_i, T) (where t_i $\in$ RF) to denote a vector of class labels obtained after having t_i classify the training set T. That is C(t_i, T) $=< c_{i1}, c_{i2}, \ldots, c_{im} >$. The result obtained of having each tree classify the training records will therefore be a super vector $\mathring{A}$ containing class labels vectors as classified by each tree (that is, $\mathring{A}$ is a vector of vectors):

$\mathring{A} = \mathrm{C}(t_1, \mathrm{T}) \cup \mathrm{C}(t_2, \mathrm{T}) \cup \ldots \mathrm{C}(t_n, \mathrm{T})$

This set will be fed as input to a clustering algorithm as shown in Fig. 1. When clustering is completed, we will have a set of clusters where each cluster contains vectors that are similar and likely to have the least number of discrepancies. For example, using a training set of 5 records with the class label being a boolean (Y/N), the vectors $< Y, Y, Y, N, N >$ and $< Y, Y, Y, Y, N >$ are likely to fall in the same cluster. However, the vectors $< Y, Y, Y, N, N >$ and $< Y, N, N, Y, Y >$ are likely to appear in different clusters because there are many discrepancies in the class labels. As defined above, since a cluster contains objects which are similar between them but are dissimilar to other objects belonging to other clusters, a vector of class labels as classified by a tree in one cluster will be dissimilar (different) from another vector belonging to another cluster, and this means that the two vectors are diverse.

When using a clustering algorithm that requires the number of clusters to be specified in advance (as in K-means), one interesting and challenging question that immediately comes to mind is how to determine this number so that it is not high and it is not low. Mardi [35] proposed a simple rule of thumb: *number_of_clusters* $\approx \sqrt{\frac{n}{2}}$, where n refers to the size of data points to be clustered.

Based on Fig. 1, we formalize the CLUB-DRF algorithm as shown in Algorithm 1 where T stands for the training set. The constant k refers to the number of clusters to be created which we define as a multiple of 5 in the range 5 to 40 (for experimental purposes). This way and as we shall see in the experimental study section, we can compare the performance of a CLUB-DRF ensemble of different sizes with that of RF. As shown in the algorithm, the process starts by creating a random forest in the traditional way. Then, using the labels produced from the trees over the training data, clustering of trees using k-median takes place. Finally, a representative from each cluster (a tree) is chosen to form the pruned and diversified random forest, which the output of our CLUB-DRF method.

It is important to remember that the size of the resulting CLUB-DRF is determined by the number of clusters used. For example, if the number of the clusters is 5, then the resulting CLUB-DRF will have size 5, and so on.

When selecting a representative from each cluster (refer to Algorithm 1), we will consider the best performing representative on the out-of-bag (OOB) instances.

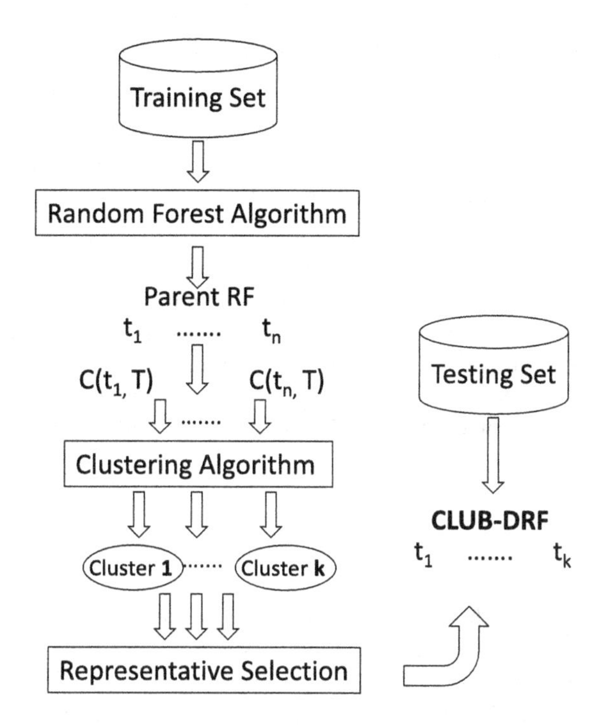

Fig. 1 CLUB-DRF approach

Algorithm 1: CLUB-DRF Algorithm

{User Settings}
input T, k
{Process}
Create an empty vector $\overrightarrow{classLabels}$
Create an empty vector $\overrightarrow{CLUB-DRF}$
Using T, call Algorithm to create RF
for $i = 1 \rightarrow RF.getNumTrees()$ **do**
  $\overrightarrow{classLabels} \Leftarrow \overrightarrow{classLabels} \cup$ C(RF.tree(i), T)
end for
Cluster $\overrightarrow{classLabels}$ into a set of k clusters:
$cluster_1 \ldots cluster_k$
From each cluster, select a representative tree RF.tree(j) that corresponds to the instance C(RF.tree(j), T) in the cluster
Add $RF.tree(j)$ to $\overrightarrow{CLUB-DRF}$
{Output}
A vector of trees $\overrightarrow{CLUB-DRF}$

These are the instances that were not included in the sample with replacement that was used to build the tree, and they account for about 36 % of the total number of instances. Using the OOB samples to evaluate a tree gives an unbiased estimate of its predictive accuracy since, unlike training data that was seen by the tree when it was built, OOB data was not seen and therefore, it is a more accurate measure of the tree's predictive accuracy.

3.2 Diversity Measure

Here we propose a simple diversity measure to measure the diversity of two classifiers that works with binary and multiclass classification. In other words, we used this technique to determine how much dissimilar two classifiers (trees) are in the clustering process. Given two classifiers h_j and h_k and a training set T of size n. Let $C(t_l, s_i)$ denote the class label obtained after having t_l classify the sample s_i in the training set T. Utilizing the following dissimilarity measure

$$diversity_{j,k} = \frac{\sum_{i=1}^{n} \delta(C(t_j, v_i), C(t_k, v_i))}{n} \tag{1}$$

The higher the number of discrepancies between the two classifiers, the higher the diversity is. For example, assume that we have a training set consisting of 10 training samples $T = \{s_1, s_2, s_3, s_4, s_5, s_6, s_7, s_8, s_9, s_{10}\}$, and two classifiers t_1 and t_2. Assume also that there are 3 possible values for the class label {a,b,c}. Let $C(t_1, T) = < a, a, b, c, c, a, b, c, b, b >$ and $C(t_2, T) = < a, a, b, b, a, a, b, c, c, c >$. According to Eq. 1 above, the diversity between the two classifiers is therefore 4/10 or 40 %.

4 Experimental Study

For our experiments, we have used 15 real datasets with varying characteristics from the UCI repository [3]. To use the holdout testing method, each dataset was divided into 2 sets: training and testing. Two thirds (66 %) were reserved for training and the rest (34 %) for testing. Each dataset consists of input variables (features) and an output variable. The latter refers to the class label whose value will be predicted in each experiment. For the RF in Fig. 1, the initial RF to produce the CLUB-DRF had a size of 500 trees, a typical upper limit setting for RF [50]. We chose the upper limit for two main reasons. First, the more trees we have, the more diverse ones we can get. Secondly, when we have many clusters, the more trees we have, the more unlikely that we wind up with empty clusters [37].

The CLUB-DRF algorithm described above was implemented using the Java programming language utilizing the API of Waikato Environment for Knowledge Analysis (WEKA) [18]. We ran the CLUB-DRF algorithm 10 times on each dataset where a new RF was created in each run. We then calculated the average of the 10 runs for each resulting CLUB-DRF to produce the average for a variety of metrics including ensemble size after pruning, accuracy rate, minimum accuracy rate, maximum accuracy rate, standard deviation, F-measure, and AUC as shown in Table 1. For the RF, we just calculated the average accuracy rate, F-Measure, and AUC as shown in the last 3 columns of the table.

4.1 Results

Table 1 compares the performance of CLUB-DRF and RF on the 15 datasets. To show the superiority of CLUB-DRF, we have only listed CLUB-DRFs that have achieved highest accuracy after applying parameters tuning. In other words, these results are based on the best k value (number of clusters). However, it is worth pointing out that most results across different settings outperformed RF, as summarized in Fig. 2. Taking a closer look at this table, we find that CLUB-DRF performed at least as good as RF on 13 datasets. Interestingly enough, of the 13 datasets, CLUB-DRF, regardless of its size, completely outperformed RF on 6 of the datasets, namely, *breast-cancer*, *pasture*, *eucalyptus*, *glass*, *sonar*, and *vehicle*. For the datasets that lost to RF (namely, *audit* and *vote*), we listed the best performing CLUB-DRFs. For *audit*, the best performing CLUB-DRF achieved an accuracy of 96.47 %; with a difference of only 0.06 % from RF. For *vote*, 2 best performing CLUB-DRFs achieved an accuracy of 95.81 %; with a difference of only 0.14 % from RF. The results are statistically significant, with the p-value for the paired t-test being 0.001087 with 95 % confidence.

4.2 Analysis

By counting the number of datasets each was superior on, Fig. 2 compares the performance of RF and CLUB-DRF using different sizes of CLUB-DRF. With one exception of size 35, the figure clearly shows that CLUB-DRF indeed performed at least as good as RF. Notably, the lesser the size of the ensemble, the better performance our method exhibits. This can be attributed to the higher diversity the ensemble has. Having 5 or 10 representatives guarantees a further away trees chosen to form the ensemble. When we move to a higher number of clusters, and consequently larger ensemble sizes, CLUB-DRF moves towards the original RF. This can be especially true when tree behaviors are no longer distinguishable, and creating more clusters does not add any diversity to the ensemble.

4.3 Pruning Level

By applying the above proposed clustering technique, we managed to achieve two objectives. First, CLUB-DRF ensembles with diverse trees were produced. Second

Table 1 Performance data table

Dataset	Size	AVG	MIN	MAX	SD	F-measure	AUC	AVG_{RF}	$F\text{-measure}_{RF}$	AUC_{RF}
Breast-cancer	20	**72.58**	67.01	75.26	2.45	0.64	0.59	69.18	0.63	0.58
Audit	20	96.47	95.58	97.05	0.41	0.91	0.87	**96.53**	0.92	0.88
Credit	5	**79.68**	72.06	87.94	5.06	0.68	0.62	77.47	0.67	0.61
Pasture	10	**55.83**	33.33	75.00	13.46	0.45	0.58	41.67	0.43	0.57
Squash-unsorted	10	**80.00**	75.00	83.33	4.08	0.64	0.73	72.50	0.58	0.70
Squash-sorted	10	**65.00**	50.00	83.33	9.72	0.55	0.62	55.83	0.50	0.58
White-clover	30	**79.29**	57.14	92.86	92.82	0.62	0.57	77.14	0.60	0.55
Eucalyptus	10	**56.20**	46.63	63.80	5.74	0.52	0.71	48.40	0.51	0.70
Soybean	15	**84.11**	80.79	88.08	2.05	0.76	0.89	82.32	0.75	0.89
Diabetes	20	**81.42**	79.31	83.14	1.24	0.71	0.67	81.26	0.71	0.67
Glass	5	**76.16**	64.38	84.93	5.62	0.65	0.76	67.53	0.63	0.75
Car	5	**64.17**	62.41	67.52	1.33	0.56	0.78	62.26	0.56	0.78
Sonar	5	**12.25**	7.04	18.31	3.34	0.26	0.00	0.14	0.29	0.00
Vehicle	10	**72.43**	64.58	81.25	5.22	0.66	0.77	69.90	0.65	0.77
Vote	30	95.81	95.27	96.62	0.41	0.91	0.94	**95.95**	0.91	0.94

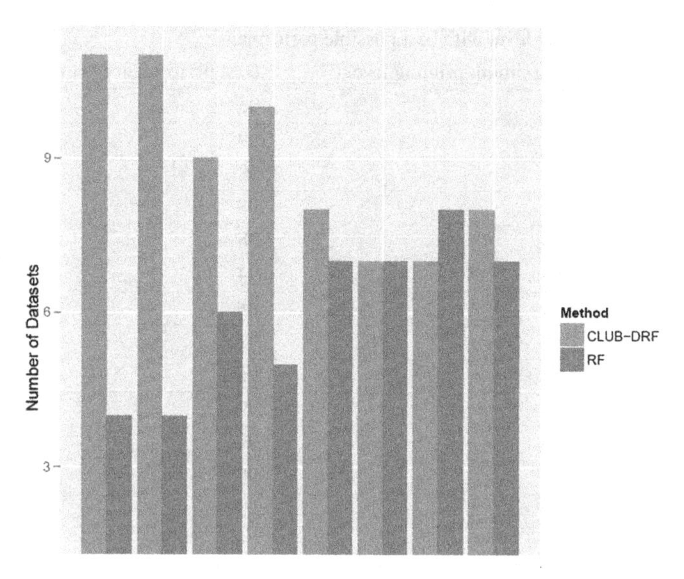

Fig. 2 Performance comparison of RF & CLUB-DRF by number of winning datasets

and more importantly, we manged to significantly reduce the size of RF. The resulting pruned CLUB-DRF ensembles performed at least as good as the original RF, but mostly outperformed the original RF, as previously discussed. In ensemble pruning, a pruning level refers to the reduction ratio between the original ensemble and the pruned one. The first column in this table shows the maximum possible pruning level for an CLUB-DRF that has outperformed RF, and the second column shows the pruning level of the best performer CLUB-DRF. We can see that with extremely healthy pruning levels ranging from 94 to 99 %, our technique outperformed RF. This makes CLUB-DRF a natural choice for real-time applications, where fast classification is an important desideratum. In most cases, 100 times faster classification can be achieved with the 99 % pruning level, as shown in the table. In the worst case scenario, only 16.67 times faster classification with 94 % pruning level in the *white clover* dataset. Such estimates are based on the fact that the number of trees traversed in the RF is the dominant factor in the classification response time. This is especially true, given that RF trees are unpruned bushy trees (Table 2).

4.4 Performance Comparison with Pruned Neural Network Ensemble

In a research by Lazarevic and Obradovic [29] where clustering was also used to prune neural network ensembles, the researchers used *diabetes* and *glass* datasets, which we also used in our experiments. Table 3 depicts the accuracy of their entire

Table 2 Maximum pruning level with best possible performance

Dataset	Maximum pruning level (%)	Best performance pruning level (%)
Breast-cancer	99	96
Credit	99	99
Pasture	99	98
Squash-unstored	98	98
Squash-stored	99	98
White-clover	94	94
Eucalyptus	99	98
Soybean	99	97
Diabetes	96	96
Glass	99	99
Car	99	99
Sonar	99	99
Vehicle	99	98

Table 3 Performance comparison between entire and pruned ensemble [29] with our RF and CLUB-DRF

Dataset	Entire ensemble (%)	Pruned ensemble (%)	RF	CLUB-DRF (%)
Diabetes	77.10	77.90	81.26	**81.42**
Glass	68.60	69.10	67.53	**76.16**

and pruned ensembles with RF and our CLUB-DRF. For both datasets, our CLUB-DRF was superior to their pruned ensemble. Notably with the *glass* dataset, CLUB-DRF has made a clear progression in predictive accuracy with a healthy 7.06 % increase over the pruned neural network.

5 Conclusion and Future Directions

Research conducted in this paper was based on how diversity in ensembles tends to yield better results [1, 9, 28, 48]. We have used clustering to produce groups of similar trees and selected a representative tree from each group. These likely diverse trees are then used to produce a pruned ensemble of the original one which we called CLUB-DRF. As was demonstrated in the experiments section, CLUB-DRF, in most cases, performed at least as good as the original ensemble.

As a future research direction, we can consider using another method when selecting representative trees from the clusters. Instead of picking the best performing representative, a randomly selected representative from each cluster can be picked instead. It would be interesting to compare the results of both methods to determine the one that yields better results. We can also try different sizes for the initial RF and also different cluster increments.

Another interesting research direction would be to use other clustering algorithms like DBSCAN [12], CLARANS [36], BIRCH [53], and/or CURE [17]. The way clusters are formed by each algorithm may well have an impact on the performance of CLUB-DRF. This can happen when representatives selected from the clusters of one algorithm are more/less diverse than others selected from clusters produced by another algorithm.

References

1. Adeva, J.J.G., Beresi, U., Calvo, R.: Accuracy and diversity in ensembles of text categorisers. CLEI Electron. J. **9**(1) (2005)
2. Amit, Y., Geman, D.: Shape quantization and recognition with randomized trees. Neural Comput. **9**(7), 1545–1588 (1997)
3. Bache, K., Lichman, M.: Uci Machine Learning Repository. University of California, Irvine (2013)
4. Bakker, B., Heskes, T.: Clustering ensembles of neural network models. Neural Netw. **16**(2), 261–269 (2003)
5. Bernard, S., Heutte, L., Adam, S.: On the selection of decision trees in random forests. In: International Joint Conference on Neural Networks. IJCNN 2009, pp. 302–307. June 2009
6. Breiman, L.: Bagging predictors. Mach. Learn. **24**(2), 123–140 (1996)
7. Breiman, L.: Stacked regressions. Mach. Learn. **24**(1), 49–64 (1996)
8. Breiman, L.: Random forests. Mach. Learn. **45**(1), 5–32 (2001)
9. Brown, G., Wyatt, J., Harris, R., Yao, X.: Diversity creation methods: a survey and categorisation. Inf. Fusion **6**(1), 5–20 (2005)
10. Brown, R.D., Martin, Y.C.: An evaluation of structural descriptors and clustering methods for use in diversity selection. SAR QSAR Environ. Res. **8**(1–2), 23–39 (1998)
11. Diao, R., Chao, F., Peng, T., Snooke, N., Shen, Q.: Feature selection inspired classifier ensemble reduction. Cybern. IEEE Trans. **44**(8), 1259–1268 (2014)
12. Ester, M., Kriegel, H.-P., Sander, J., Xiaowei, X.: A density-based algorithm for discovering clusters in large spatial databases with noise. KDD **96**, 226–231 (1996)
13. Fleiss, J.L., Levin, B., Cho Paik, M.: Statistical Methods for Rates and Proportions. Wiley, New York (2013)
14. Freund, Y., Robert, E.: A decision-theoretic generalization of on-line learning and an application to boosting. J. Comput. Syst. Sci. **55**(1), 119–139 (1997)
15. Giacinto, G., Roli, F.: Design of effective neural network ensembles for image classification purposes. Image Vis. Comput. **19**(9), 699–707 (2001)
16. Giacinto, G., Roli, F., Fumera, G.: Design of effective multiple classifier systems by clustering of classifiers. In: Proceedings of 15th International Conference on Pattern Recognition, vol. 2, pp. 160–163. IEEE (2000)
17. Guha, S., Rastogi, R., Shim, K.: Cure: an efficient clustering algorithm for large databases. In: ACM SIGMOD Record, vol. 27, pp. 73–84. ACM (1998)
18. Hall, M., Frank, E., Holmes, G., Pfahringer, B., Reutemann, P., Witten, I.H.: The WEKA Data Mining Software: An Update, vol. 11. ACM, New York (2009)
19. Ho, T.H: Random decision forests. In: Proceedings of the Third International Conference on Document Analysis and Recognition, vol. 1, pp. 278–282. IEEE (1995)
20. Ho, T.K.: The random subspace method for constructing decision forests. Pattern Anal. Mach. Intell. IEEE Trans. **20**(8), 832–844 (1998)

21. Huang, Z.: Extensions to the k-means algorithm for clustering large data sets with categorical values. Data Min. Knowl. Discov. **2**(3), 283–304 (1998)
22. Huang, Z., Ng, M.K.: A fuzzy k-modes algorithm for clustering categorical data. Fuzzy Syst. IEEE Trans. **7**(4), 446–452 (1999)
23. Jain, A.K.: Data clustering: 50 years beyond k-means. Pattern Recognit. Lett. **31**(8), 651–666 (2010)
24. Kohavi, R., et al.: A study of cross-validation and bootstrap for accuracy estimation and model selection. IJCAI **14**, 1137–1145 (1995)
25. Kohavi, R., Wolpert, D.H., et al.: Bias plus variance decomposition for zero-one loss functions. In: ICML, pp. 275–283 (1996)
26. Kulkarni, V.Y., Sinha, P.K.: Pruning of random forest classifiers: a survey and future directions. In: International Conference on Data Science Engineering (ICDSE), pp. 64–68, July 2012
27. Kuncheva, L.I., Hadjitodorov, S.T.: Using diversity in cluster ensembles. In: IEEE International Conference on Systems, Man and Cybernetics, vol. 2, pp. 1214–1219. IEEE (2004)
28. Kuncheva, L.I., Whitaker, C.J.: Measures of diversity in classifier ensembles and their relationship with the ensemble accuracy. Mach. Learn. **51**(2), 181–207 (2003)
29. Lazarevic, A., Obradovic, Z.: Effective pruning of neural network classifier ensembles. In: Proceedings of International Joint Conference on Neural Networks. IJCNN'01, vol. 2, pp. 796–801. IEEE (2001)
30. Lee, J., Sun, Y., Nabar, R., Lou, H.-L.: Cluster-based transmit diversity scheme for mimo ofdm systems. In: IEEE 68th Vehicular Technology Conference, VTC 2008-Fall, pp. 1–5. IEEE (2008)
31. Leo, B., Friedman, J.H., Olshen, R.A., Stone, C.J.: Classification and regression trees. Wadsworth Int. Group (1984)
32. Li, J., Yi, Ke., Zhang, Q.: Clustering with diversity. In: Automata, Languages and Programming, pp. 188–200. Springer (2010)
33. Maclin, R., Opitz, D.: Popular ensemble methods: an empirical study. J. Artif. Intell. Res. **11**(1–2), 169–198 (1999)
34. MacQueen, J., et al.: Some methods for classification and analysis of multivariate observations. In: Proceedings of the Fifth Berkeley Symposium on Mathematical Statistics and Probability, vol. 1, p. 14. California (1967)
35. Mardia, K.V., Kent, J.T., Bibby, J.M.: Multivariate analysis (1980)
36. Ng, R.T., Han, J.: Clarans: a method for clustering objects for spatial data mining. Knowl. Data Eng. IEEE Trans. **14**(5), 1003–1016 (2002)
37. Pakhira, M.K.: A modified k-means algorithm to avoid empty clusters. Int. J. Recent Trends Eng. **1**(1), 1 (2009)
38. Partridge, D., Krzanowski, W.: Software diversity: practical statistics for its measurement and exploitation. Inf. Softw. Technol. **39**(10), 707–717 (1997)
39. Polikar, R.: Ensemble based systems in decision making. Circuits Syst. Mag. IEEE **6**(3), 21–45 (2006)
40. Qiang, F., Shang-Xu, H., Sheng-Ying, Z.: Clustering-based selective neural network ensemble. J. Zhejiang Univ. Sci. A **6**(5), 387–392 (2005)
41. Rokach, L.: Ensemble-based classifiers. Artif. Intell. Rev. **33**(1–2), 1–39 (2010)
42. San, O.M., Huynh, V.-N., Nakamori, Y.: An alternative extension of the k-means algorithm for clustering categorical data. Int. J. Appl. Math. Comput. Sci. **14**(2), 241–248 (2004)
43. Sharpton, T., Jospin, G., Wu, D., Langille, M., Pollard, K., Eisen, J.: Sifting through genomes with iterative-sequence clustering produces a large, phylogenetically diverse protein-family resource. BMC Bioinform. **13**(1), 264 (2012)
44. Shemetulskis, N.E., Dunbar Jr, J.B., Dunbar, B.W., Moreland, D.W., Humblet, C.: Enhancing the diversity of a corporate database using chemical database clustering and analysis. J. Comput.-Aided Mol. Des. **9**(5), 407–416 (1995)
45. Skalak, D.B.: The sources of increased accuracy for two proposed boosting algorithms. In: Proceedings of American Association for Artificial Intelligence, AAAI-96, Integrating Multiple Learned Models Workshop, vol. 1129, p. 1133. Citeseer (1996)

46. Smyth, P., Wolpert, D.: Linearly combining density estimators via stacking. Mach. Learn. **36**(1–2), 59–83 (1999)
47. Soto, V., Garcia-Moratilla, S., Martinez-Munoz, G., Hernández-Lobato, D., Suarez, A.: A double pruning scheme for boosting ensembles. Cybern. IEEE Trans. **44**(12), 2682–2695 (2014). Dec
48. Tang, EKe, Suganthan, P.N., Yao, X.: An analysis of diversity measures. Mach. Learn. **65**(1), 247–271 (2006)
49. Tsoumakas, G., Partalas, I., Vlahavas, I.: An ensemble pruning primer. In: Applications of supervised and unsupervised ensemble methods, pp. 1–13. Springer (2009)
50. Williams, G.: Use R: Data Mining with Rattle and R: the Art of Excavating Data for Knowledge Discovery. Springer, New York (2011)
51. Wolpert, D.H.: Stacked generalization. Neural Netw. **5**(2), 241–259 (1992)
52. Yan, W., Goebel, K.F.: Designing classifier ensembles with constrained performance requirements. In: Defense and Security, International Society for Optics and Photonics, pp. 59–68 (2004)
53. Zhang, T., Ramakrishnan, R., Livny, M.: Birch: an efficient data clustering method for very large databases. In: ACM SIGMOD Record, vol. 25, pp. 103–114. ACM (1996)

Machine Learning and Knowledge Acquisition

Fast Handwritten Digit Recognition with Multilayer Ensemble Extreme Learning Machine

Hossein Ghodrati Noushahr, Samad Ahmadi and Arlene Casey

Abstract Conventional artificial neural networks and convolutional neural networks perform well on the task of automatic handwriting recognition. But, they suffer from long training times and their complex nature. An alternative learning algorithm called Extreme Learning Machine overcomes these shortcomings by determining the weights of a neural network analytically. In this paper, a novel classifier based on Extreme Learning Machine is proposed that achieves competitive accuracy results while keeping training times low. This classifier is called multilayer ensemble Extreme Learning Machine. The novel classifier is evaluated against traditional backpropagation and Extreme Learning Machine on the well-known MNIST dataset. Possible future work on parallel Extreme Learning Machine is shown up.

1 Introduction

Artificial Neural Networks (ANN) have been successfully applied for the difficult task of handwritten digit recognition. However, ANN that train their weights with the traditional backpropagation (BP) algorithm suffer from slow learning speed. This has been a major bottleneck for ANN applications in the past.

Recently, Extreme Learning Machine (ELM) has been proposed as an alternative to BP for the task of training ANN [8]. ELM follows an approach that aims to reduce human invention, increase learning accuracy, and to reduce the time to train an ANN. This is done by randomly initiating the weights, then fixing the weights of the hidden layer nodes and subsequently determining the weights of the output layer analytically.

H. Ghodrati Noushahr (✉) · S. Ahmadi · A. Casey
Centre for Computational Intelligence, School of Computer Science and Informatics, De Montfort University, The Gateway, Leicester LE1 9BH, UK
e-mail: hossein@ghodrati.net

S. Ahmadi
e-mail: sahmadi@dmu.ac.uk

A. Casey
e-mail: arlene.casey@myemail.dmu.ac.uk

M. Bramer and M. Petridis (eds.), *Research and Development in Intelligent Systems XXXII*, DOI 10.1007/978-3-319-25032-8_5

Human invention is reduced as hyper-parameters, such like the learning rate and the momentum of traditional BP do not have to be determined manually.

ELM was applied successfully on a variety of classification and function approximation tasks [7]. In this paper, a novel classifier based on ELM will be presented that achieves competitive accuracy results while keeping the training time low and limiting human invention.

The remainder of this paper is structured as follows. In Sect. 2 the handwriting recognition problem and the corresponding dataset will be explained in detail. In Sect. 3 recent classifier for this task will be reviewed. This includes conventional ANN, but also very successful variations of ANN called convolutional neural networks (CNN) that fall into the research area of *Deep Learning*. Section 4 introduces ELM and ELM ensemble models. Furthermore, the novel classifier will be presented in the same section. The results of the experimental work conducted will be shown in Sect. 5, and Sect. 6 concludes the paper.

2 Problem Definition

Automatic handwriting recognition is a challenging problem that caught academic and commercial interest. Some commercial applications are: letter sorting in post offices, personal check reading in banks, or large-scale digitization of manuscripts [5]. The Mixed National Institute of Standards and Technology Database (MNIST) is the most widely used benchmark for isolated handwritten digit recognition [10]. It consists of 70,000 images from approx. 250 writers. 60,000 images represent the training sample, and the remaining 10,000 images the test sample for evaluation. The images have $28 \times 28 = 784$ gray-scale pixels (0: background—255: maximum foreground intensity). Figure 1 shows examples of the ten digits in the MNIST database.

3 Related Work

3.1 Single Hidden Layer Feedforward Neural Networks

ANN are massively parallel distributed processors made up of simple processing units. ANN are inspired by the human brain and the way it processes information.

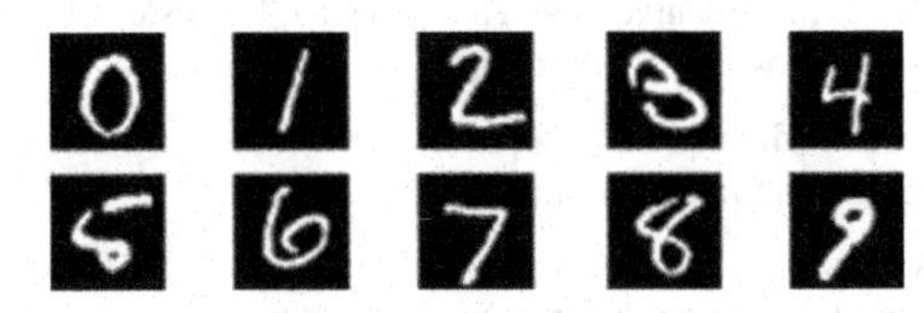

Fig. 1 Examples from the MNIST database

One of the main benefits of ANN is their ability to detect nonlinear relations and patterns. Single Hidden Layer Feedforward Neural Networks (SLFN) are ANN with only one hidden layer. Conventional SLFN train the weights of the ANN with the BP algorithm. BP is a gradient-based learning algorithm that tunes iteratively all parameters of the SLFN.

LeCun et al. [9] evaluated SLFN against the MNIST database. A SLFN with 300 hidden layer nodes had an error rate of 4.7 % on the test set. A SLFN with 1,000 hidden layer nodes achieved a slightly better error rate of 4.5 % on the test set.

3.2 Multiple Hidden Layer Feedforward Neural Networks

Multiple Hidden Layer Feedforward Neural Networks (MLFN) are identical to SLFN, but with the main difference that they have more than one hidden layer. Although it is proven that SLFN are universal approximators [6], MLFN were used in the past for the handwritten digit recognition problem. In [9] error rates as low as 3.05 and 2.95 % were achieved with a 300-100-10 and a 500-150-10 MLFN.

3.3 Convolutional Neural Networks

LeCun et al. [9] proposed CNN with a focus on automatic learning and higher order feature selection. CNN combine three architectural ideas to ensure shift, scale, and distortion invariance: local receptive fields, shared weights, and spatial subsampling.

A node in the hidden layer is not connected to all inputs from the previous layer, but only to a subregion. The advantage of local receptive fields is that they reduce dramatically the number of weights compared to fully connected hidden layer. Furthermore, this approach is computationally less expensive.

Hidden layer nodes are organized in so called feature maps that share the same weights. As each hidden layer node within a feature map has a different local receptive field, the same pattern can be detected across the whole receptive field. Each feature map is specialized to recognize a different pattern by having different weights. The architectural idea of weight sharing reduces even more the number of weights.

The idea of spatial subsampling refers to the reduction of the receptive field resolution. In the case of LeNet-5 [9] a non-overlapping 2x2 neighborhood in the previous layer is aggregated to a single output. The aggregation could be either the maximum, or the average of the 2x2 neighborhood. The subsampling layer reduces the number of inputs by the factor 4. Spatial subsampling provides invariance to local translations.

Convolutional layer implement the local receptive field concept and also the weight sharing. Subsampling layer realize the idea of spatial subsampling.

Figure 2 illustrates the architecture of LeNet-5 [9]. It consists of 6 layers and convolutional layers are labeled Cx, subsampling layers Sx, and fully connected layers

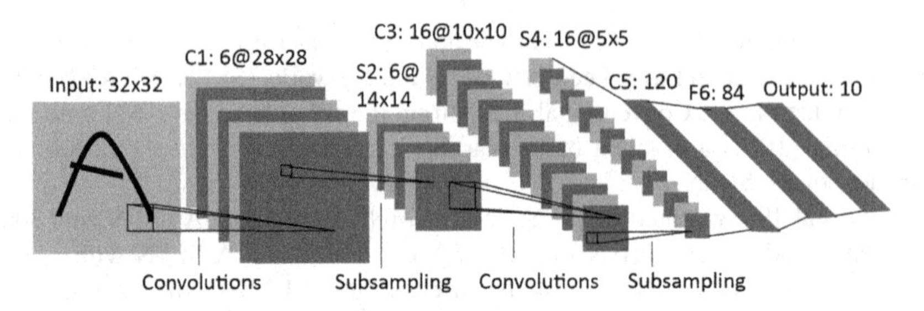

Fig. 2 Architecture of the CNN LeNet-5 [9]

Fx, where x is the layer index. The first convolutional layer *C1* has 6 28x28 feature maps followed by the subsampling layer *S2* with also 6 feature maps that reduce the size to 14x14. Layer *C3* is a 10x10 convolutional layer having 16 feature maps. Layer *S4* is again a subsampling layer that reduces the size further down to 5x5. Layer *C5* is a convolutional layer with 120 1x1 feature maps. Layer *F6* is a fully connected layer that computes a dot product between the input vector and the weight vector and a bias like in traditional ANN.

It can be summarized that CNN scan automatically the input image for higher order features. The exact position of these higher order features is less relevant, only the relative position to other higher order feature is relevant. In the case of the number 7 a CNN would look for the endpoint of a horizontal element in the upper left area, a corner in the upper right area, and an endpoint of a roughly vertical segment in the lower portion of the image.

LeNet-5 could reach a test error rate of 0.95 % [9] on the MNIST dataset, more recent classifier based on CNN could reach test error rates as low as 0.23 % [4]. These error rates are comparable to error rates of humans performing the same task [11].

3.4 Other Approaches

Image deformation is a technique that was applied in some studies for the problem discussed in this paper. Ciresan et al. [5] deformed the 60,000 training images to get more training samples. They combined rotation, scaling, horizontal shearing, and elastic deformations in order to emulate uncontrolled oscillations of hand muscles and trained with the original and additional training samples several MLFN. A test error of 0.32 % was reached. Although, 12.1 M weights had to be trained with a total training time of 114.5 h (Table 1).

Alonso-Weber et al. [1] followed a similar approach. In addition to the above mentioned deformations, noise was fed into the MLFN by wiping out a proportion of pixels and also adding pixels randomly. The MLFN had a topology of 300-200-10. A test error rate of 0.43 % was achieved. No statistics about the training times were provided (Fig. 3).

Table 1 Error rates on MNIST database in [5]

Model	Topology	Best test error (%)	Training time	Weights (M)
1	1000-500-10	0.44	23.4h	1.34
2	1500-1000-500-10	0.40	44.2h	3.26
3	2000-1500-1000-500-10	0.39	66.7h	6.69
4	2500-2000-1500-1000-500-10	0.32	114.5h	12.11

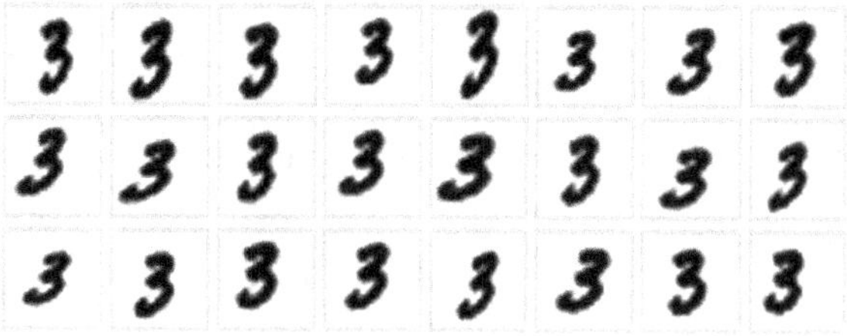

Fig. 3 Image deformation examples applied in [1]

4 Extreme Learning Machines

4.1 Review of Extreme Learning Machine

Huang et al. [8] proposed ELM as a new learning algorithm to train SLFN. The original design objectives and key advantages of ELM compared to conventional BP are: least human invention, high learning accuracy, and fast learning speed [7].

Due to the slow learning speed, BP has been a major bottleneck for SLFN applications in the past decades [8]. ELM follows a very different approach: hidden layer weights are chosen randomly and the output layer weights determined analytically by solving a linear system using the least square method. Hence, no hyper-parameters such like the learning rate or the momentum need to be determined manually compared to traditional BP.

For N arbitrary distinct training samples $\{(x_i, t_i)\}_{i=1}^N$, where $x_i \in R^d$ and $t_i \in R^m$, the output of a SLFN with L hidden nodes is:

$$f_L(x) = \sum_{i=1}^{L} \beta_i h_i(x) = h(x)\beta \tag{1}$$

where $\beta = [\beta_1, \ldots, \beta_L]^T$ is the output weight vector between L hidden layer nodes and $m \geq 1$ output nodes. $h_i(x)$ is the output in form of a nonlinear activation function of the ith hidden node for the input x. Table 2 lists the most common activation functions.

Table 2 Activation functions in ELM

Sigmoid function	$h(x) = \frac{1}{1+\exp(-(\omega x+b))}$
Hyperbolic tangent function	$h(x) = \frac{1-\exp(-(\omega x+b))}{1+\exp(-(\omega x+b))}$
Gaussian function	$h(x) = \exp(-b \lVert x - \omega \rVert)$

For all N training samples, Eq. 1 can be written in an equivalent compact form as:

$$T = H\beta \tag{2}$$

where H is the hidden layer output matrix:

$$H = \begin{bmatrix} h(x_1) \\ \vdots \\ h(x_N) \end{bmatrix} = \begin{bmatrix} h_1(x_1) & \dots & h_L(x_1) \\ \vdots & \ddots & \vdots \\ h_1(x_N) & \dots & h_L(x_N) \end{bmatrix} \tag{3}$$

and T is the training sample target matrix:

$$T = \begin{bmatrix} t_1 \\ \vdots \\ t_N \end{bmatrix} = \begin{bmatrix} t_{11} & \dots & t_{1m} \\ \vdots & \ddots & \vdots \\ t_{N1} & \dots & t_{Nm} \end{bmatrix} \tag{4}$$

The output layer weights β are determined by minimizing the squared approximation error:

$$\min_{\beta \in R^{L\times m}} \lVert H\beta - T \rVert^2 , \tag{5}$$

The optimal solution to 5 is given by

$$\hat{\beta} = H^{+}T, \tag{6}$$

where H^{+} denotes the Moore-Penrose generalized inverse of matrix H. Algorithm 1 summarizes the ELM learning algorithm.

Algorithm 1 ELM

Input: N arbitrary training samples $\{(x_i, t_i)\}_{i=1}^{N}$

1: Assign randomly hidden layer weights
2: Calculate the hidden layer output matrix H
3: Calculate the output layer weights β : $\hat{\beta} = H^{+}T$

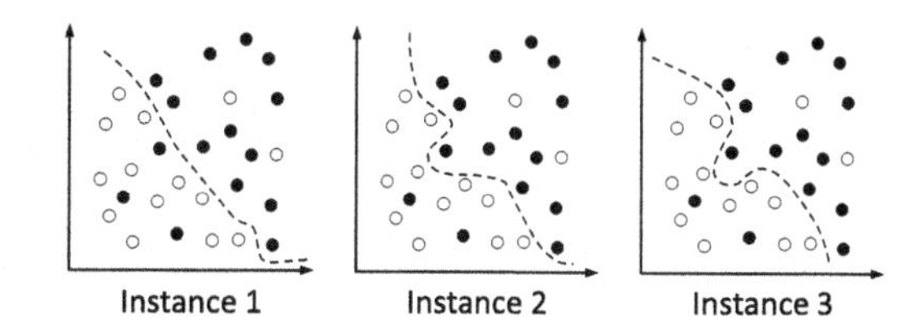

Fig. 4 ELM instances with different random parameters

4.2 Ensemble ELM

The combination of multiple classifiers reduces the risk of overfitting and leads to better accuracy compared to single model classifiers. Such combined classifiers are referred to as ensemble classification models. Based on promising results of ensemble models for ELM presented in [3], an ensemble model called Ensemble ELM (EELM) will be built and evaluated against the MNIST database.

ELM constructs a nonlinear separation boundary in classification problems. Samples that are located near the classification boundary might be misclassified by one single ELM model. Figure 4 illustrates on the left hand side such a misclassification near the boundary. The classification boundary depends on the randomly initiated weights of the hidden layer neurons. As these weights are not changed during the training phase, the classification boundary remains as initialized. The majority vote of several ELM that are initialized with independent random weights reduces the misclassification of samples near the classification boundary. Algorithm 2 summarizes the EELM algorithm.

Algorithm 2 EELM

Input:

N arbitrary training samples $\{(x_i, t_i)\}_{i=1}^{N}$;
N^{test} test samples $\{(x_i^{test}, t_i^{test})\}_{i=1}^{N^{test}}$;
Odd number of indepentent ELM: K

Training phase:
1: Create K ELM classifier with independent weights
2: Train all K independent ELM classifier with the training samples
Test phase:
3: For each test sample predict the class as the majority vote of all K independent ELM classifier

---

Figure 4 illustrates for a $K = 3$ EELM model the correct classification of a sample near the classification boundary due to a majority vote of two ELM models. Additional to [3], further ELM ensembles are mentioned in [7].

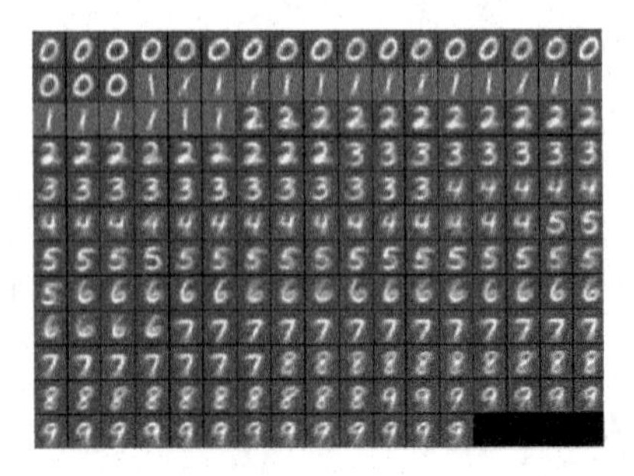

Fig. 5 ELM autoencoder output weight visualization

4.3 Multilayer ELM

Kasun et al. [2] proposed a multilayer ELM (ML-ELM) based on the idea of autoencoding. That is, extract higher order features by reducing the high dimensional input data to a lower dimensional feature space similar to CNN. This is done as follows: unsupervised learning is performed by setting the input data as the output data $t = x$. The random weights are chosen to be orthogonal as it tends to improve the generalization performance. Figure 5 visualizes the output weights of an 784-20-784 ELM autoencoder. It can be seen that the autoencoder is able to extract digit specific patterns. ML-ELM consists of several stacked ELM autoencoder. In [2] the presented model has a topology of 784-700-700-15000-10 achieving a test error rate of 0.97 % while it took only 7.4 m to train the model on a system with an Intel i7-3740QM CPU at 2.7 GHz and 32 GB RAM.

4.4 Multilayer Ensemble ELM

The novel classifier called Multilayer Ensemble ELM (ML-EELM) presented in this paper combines concepts of CNN, ensemble models, and the ELM training algorithm. The architectural idea of shift invariance is realized by spatial subsampling. In order to classify correctly a digit, it is not necessary to know the grey-scale intensity of each pixel. Instead, it is sufficient to know the approximate position of high intensity activations on the receptive field. Hence, a convolution layer reduces the image size from 28x28 down to 26x26 by applying an aggregation function to overlapping 3x3 regions. A subsampling layer then halves the image resolution further down to 13x13 by applying an aggregation function to non-overlapping 2x2 regions. The image is then fed into an EELM model with K instances. The topology of a single instance is illustrated on Fig. 6.

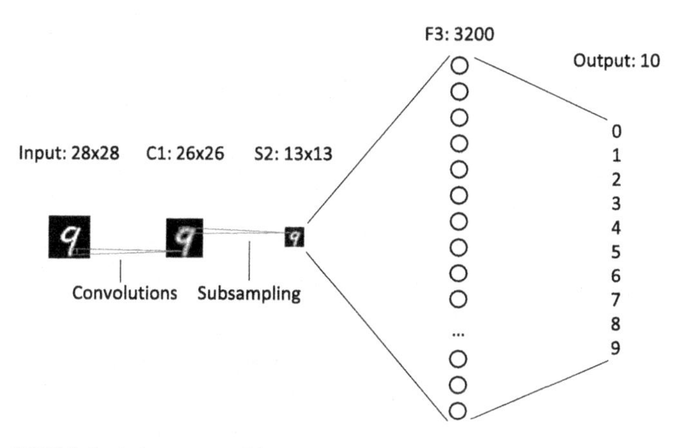

Fig. 6 ML-EELM single instance architecture

5 Experimental Setup and Evaluation

Experimental work is focused on ELM, EELM, and the novel classifier: Multilayer EELM.

Four ELM models with different number of nodes in the hidden layer were created and tested. The number of hidden layer nodes ranges from 800 to 3,200. Each model was trained and tested ten times in order to validate the robustness of the model. The training and testing time, the error rate on the training and test sample, as well as the standard deviations (SD) for all ten iterations were measured. Table 3 lists all statistics for the ELM model evaluation. A model with 3,200 hidden nodes could

Table 3 ELM evaluation results for MNIST database

Model	Hidden Layer Nodes	Iterations	Training Time (min)	SD Training Time (min)	Training Error Rate (%)	SD Training Error Rate (%)
1	800	10	0.5743	0.0289	8.9837	0.1396
2	1,600	10	1.6786	0.0262	6.2192	0.0832
3	2,400	10	3.4801	0.0326	4.7735	0.0647
4	3,200	10	6.3509	0.1024	3.8955	0.0543
Model	Hidden Layer Nodes	Iterations	Test Time (min)	SD Test Time (min)	Test Error Rate (%)	SD Test Error Rate (%)
1	800	10	0.0218	0.0019	8.9410	0.1449
2	1,600	10	0.0400	0.0033	6.6860	0.1786
3	2,400	10	0.0576	0.0043	5.6630	0.1189
4	3,200	10	0.1035	0.0079	5.0350	0.0920

Table 4 EELM evaluation results for MNIST database

Model	Hidden Layer Nodes	K	Training Time (min)	Training Error Rate (%)	Test Time (min)	Test Error Rate (%)
1	800	3	1.6222	7.8967	0.0694	7.8900
2	800	9	4.8298	6.9233	0.2174	6.7400
3	800	15	8.0541	6.6283	0.3502	6.5700
4	1600	3	4.9866	5.2933	0.1292	5.9500
5	1600	9	14.9735	4.7617	0.3843	5.2900
6	1600	15	25.2254	4.6367	0.6596	5.2400
7	2400	3	10.4370	4.1533	0.1747	5.0800
8	2400	9	31.2249	3.7583	0.5169	4.6300
9	2400	15	52.2731	3.6500	0.9365	4.5300
10	3200	3	19.5807	3.3900	0.2915	4.5900
11	3200	9	57.7446	3.0017	1.2527	4.1600
12	3200	15	97.1288	2.9550	1.5457	4.0700

achieve a test error rate of 5.04 % (SD: 0.09 %). It took on average 6.4 min (SD: 0.1 min) to train the ELM.

Twelve EELM models were trained and evaluated. The hidden layers of the models have 800 to 3,200 nodes. The parameter K for the number of independent models per EELM was set to 3, 9, and 15. Table 4 lists the training and testing times, as well as the training and test error rates for all twelve EELM models. The EELM model #12 (3,200 nodes, $K = 15$) was trained in 97 min achieving a test error of 4.07 %. The model #11 (3,200 nodes, $K = 9$) however, could be trained after only 58 min having only a slightly higher test error of 4.16 %.

The novel classifier ML-EELM was built and evaluated after having determined the aggregation functions applied in the convolution and subsampling layers. This was done by training a simple ELM model with one additional 2x2 subsampling layer. The average, standard deviation, maximum, and minimum value of all 4 pixels were evaluated as aggregation functions. Table 5 lists the results. The average function achieved the lowest test error rate and was subsequently used as the aggregation function in the convolution and subsampling layers of the ML-EELM model. A ML-EELM model with $K = 15$ and 3,200 hidden layer nodes achieved a test error rate as low as 2.73 % and was trained in only 96 min.

It was observed that with increasing number of hidden layer nodes, and in the case of EELM with increasing K, the test error rate decreases. The training time grows linearly with K, and exponentially with the number of hidden layer neurons.

Although the accuracy on the training data set becomes very high with more hidden layer nodes, the test error rate does not increase. The effect of improving accuracy on the training data and decreasing accuracy on the test data, known as overfitting could not be observed which speaks for the good generalization performance of ELM.

Table 5 ELM evaluation results for MNIST database with 2x2 subsampling

Aggregation	Hidden Layer Nodes	Iterations	Training Time (min)	SD Training Time (min)	Training Error Rate (%)	SD Training Error Rate (%)
Avg.	3,200	10	7.1067	0.4953	3.0330	0.0567
Std.dev.	3,200	10	6.9656	0.4880	9.5905	0.1104
Max.	3,200	10	6.7650	0.6209	3.5038	0.0284
Min.	3,200	10	7.1850	0.3130	8.7338	0.1090
Aggregation	Hidden Layer Nodes	Iterations	Training Time (min)	SD Test Time (min)	Training Error Rate (%)	SD Test Error Rate (%)
Avg.	3,200	10	0.0677	0.0052	3.9120	0.1125
Std.dev.	3,200	10	0.0766	0.0061	12.1140	0.1393
Max.	3,200	10	0.0693	0.0072	4.5010	0.1718
Min.	3,200	10	0.0704	0.0056	10.8430	0.1059

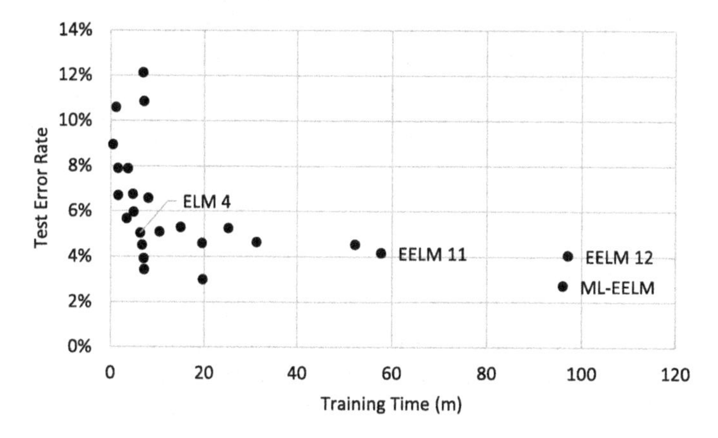

Fig. 7 Graphical evaluation of the results

Other models presented in the literature outperform with regards to the test error rate the SLFN ELM models presented in this paper. However, the experimental results confirm the initial design objectives of ELM: least human invention, high learning accuracy and fast learning speed. No training times were provided in most of the papers mentioned previously. The training time of 114 h in [5] acts as a guiding value for the training time of the other models identified in the literature.

ML-EELM, first introduced in this paper, achieves competitive test error rates on the MNIST database while requiring only fractions of training time on commodity hardware. The results confirm the conceptual ideas of CNN. Due to the convolution and subsampling layers in ML-EELM, the feature space could be reduced from 784 down to 169 leading to further improved accuracy rates (Fig. 7).

Table 6 Model comparison (NA = Not available)

Source	Model	Accuracy (%)	Training Time
[4]	CNN	0.23	NA
[5]	MLFN	0.32	114 h
[1]	MLFN	0.43	NA
[9]	CNN	0.95	NA
[2]	ML-ELM	0.97	7.4 m
This paper	ML-EELM	2.73	96 m
This paper	EELM	4.07	97 m
[9]	SLFN	4.70	NA
This paper	ELM	5.04	6.4 m

ML-ELM presented in [2] achieves slightly higher test error rates than CNN, but training the model in less time. In general, models trained with ELM outperform all other models with regards to training time. This is inline with experimental results from Huang et al. [8]. Table 6 lists a comparison of the models.

Matlab R2014a (8.3.0.532) was used for the computation of the ELM, EELM, and ML-EELM models on a Windows 7 64 bit system with 8 GB RAM and an Intel Core i5-2310 CPU at 2.90 GHz.

6 Conclusion

ELM could successfully be applied for the task of handwritten digit recognition based on the MNIST dataset. Competitive results were achieved with a test error rate of only 2.73 % with a novel multilayer ensemble ELM model presented first in this paper. While these results cannot beat the results of CNN, classifier based on ELM are relatively easy to create, have good generalization performance and most important, have fast learning speed. An ELM model with 3,200 hidden layer nodes can be trained in just about 6 min on a standard commodity desktop PC.

Moreover, Huang et al. [8] applied ELM to a variety of classification and function approximation problems and found that ELM learns up to hundreds times faster than BP. Furthermore, they observed that ELM does not face BP specific issues like local minima, improper learning rates and overfitting. The results presented in this paper confirm the initial design objectives of ELM.

ELM has great potential for applications where the training data changes frequently and hence the models need to be re-trained often. This could be the case when writing styles of different individuals have to be learned in high frequencies.

Moreover, the parallel computation possibility of EELM and ML-EELM models with a large number of K individual instances could further improve the accuracy of ELM based classifier while keeping the training time low. Wang et al. [12] have made first efforts to implement parallel ELM using MapReduce on some classification problems. A parallel implementation of EELM or ML-EELM is recommended as possible future research.

References

1. Alonso-Weber, J., Sesmero, M., Sanchis, A.: Combining additive input noise annealing and pattern transformations for improved handwritten character recognition. Expert Syst. Appl. **41**(18), 8180–8188 (2014)
2. Cambria, E., Huang, G.B., Kasun, L.L.C., Zhou, H., Vong, C.M., Lin, J., Yin, J., Cai, Z., Liu, Q., Li, K., Leung, V.C., Feng, L., Ong, Y.S., Lim, M.H., Akusok, A., Lendasse, A., Corona, F., Nian, R., Miche, Y., Gastaldo, P., Zunino, R., Decherchi, S., Yang, X., Mao, K., Oh, B.S., Jeon, J., Toh, K.A., Teoh, A.B.J., Kim, J., Yu, H., Chen, Y., Liu, J.: Extreme learning machines [Trends & Controversies]. IEEE Intell. Syst. **28**(6), 30–59 (2013)
3. Cao, J., Lin, Z., Huang, G.B., Liu, N.: Voting based extreme learning machine. Inf. Sci. **185**(1), 66–77 (2012)
4. Ciresan, D., Meier, U., Schmidhuber, J.: Multi-column deep neural networks for image classification. In: 2012 IEEE Conference on Computer Vision and Pattern Recognition, pp. 3642–3649 (2012)
5. Ciresan, D.C., Meier, U., Gambardella, L.M., Schmidhuber, J.: Deep, big, simple neural nets for handwritten digit recognition. Neural Comput. **22**(12), 3207–3220 (2010)
6. Hornik, K., Stinchcombe, M., White, H.: Multilayer feedforward networks are universal approximators. Neural Netw. **2**(5), 359–366 (1989)
7. Huang, G., Huang, G.B., Song, S., You, K.: Trends in extreme learning machines: a review. Neural Netw. **61**, 32–48 (2015)
8. Huang, G.B., Zhu, Q.Y., Siew, C.K.: Extreme learning machine: theory and applications. Neurocomputing **70**(1–3), 489–501 (2006)
9. LeCun, Y., Bottou, L., Bengio, Y., Haffner, P.: Gradient-based learning applied to document recognition. Proc. IEEE **86**(11), 2278–2323 (1998)
10. LeCun, Y., Jackel, L., Bottou, L., Brunot, A., Cortes, C., Denker, J., Drucker, H., Guyon, I., Müller, U., Säckinger, E., Simard, P., Vapnik, V.: Comparison of learning algorithms for handwritten digit recognition. In: International Conference on Artificial Neural Networks, pp. 53–60 (1995)
11. Schmidhuber, J.: Deep learning in neural networks: an overview. Neural Netw. **61**, 85–117 (2015)
12. Wang, B., Huang, S., Qiu, J., Liu, Y., Wang, G.: Parallel online sequential extreme learning machine based on MapReduce. Neurocomputing **149**, 224–232 (2015)

Stylochronometry: Timeline Prediction in Stylometric Analysis

Carmen Klaussner and Carl Vogel

Abstract We examine stylochronometry, the question of measuring change in linguistic style over time within an authorial canon and in relation to change in language in general use over a contemporaneous period. We take the works of two prolific authors from the 19th/20th century, Henry James and Mark Twain, and identify variables that change for them over time. We present a method of analysis applying regression on linguistic variables in predicting a temporal variable. In order to identify individual authors' effects on the model, we compare the model based on the novelists' works to a model based on a 19th/20th century American English reference set. We evaluate using R^2 and *Root mean square error (RMSE)*, that indicates the average error on predicting the year. On the two-author data, we achieve an RMSE of $\pm$7.2 years on unseen data (baseline: $\pm$13.2); for the larger reference set, our model obtains an RMSE of $\pm$4 on unseen data (baseline: $\pm$17).

1 Introduction

In authorship analysis, it is a natural idealization to treat different works of an author as synchronous events even though this is tantamount to the impossibility that they were all written at the same instant. The assumption is made despite the fact that the works of prolific authors are partially ordered over their lifetimes: some works will have been composed in a non-overlapping sequential manner, while others, largely in parallel over more or less the same duration. Therefore, this takes into account neither the individual changes that an author's style might undergo over time, nor the general underlying language change influencing all contemporaneous writers.

C. Klaussner (✉)
ADAPT Centre, School of Computer Science and Statistics, Trinity College Dublin,
Dublin, Ireland
e-mail: klaussnc@tcd.ie

C. Vogel
Centre for Computing and Language Studies, Trinity College Dublin,
Dublin, Ireland
e-mail: vogel@tcd.ie

M. Bramer and M. Petridis (eds.), *Research and Development in Intelligent Systems XXXII*, DOI 10.1007/978-3-319-25032-8_6

Ignoring the distinctiveness of an author with respect to other authors, it is relevant to consider the variables that separate each period of composition for an author from other periods for the same author. For example, if we consider an author such as Henry James, who is widely perceived to have changed his style considerably from his early to late works [2, 11], the variables for which he remained consistent might be as interesting to examine as those which may be quantified as having undergone great change. The external factors that may have influenced whether variables are in one category or the other may be of great interest.

Enormous amounts of human ingenuity have been applied over the centuries to the task of temporal classification of text authorship.[1] Methods such as are explored here contribute to semi-automatic methods that draw on text-internal analysis to support stylochronometry. These are generalizations of authorship attribution problems. In the present work, rather than trying to learn features that discriminate two or more authors in synchronic terms, analyzing each one's collection of works against the others' works, we mean to identify elements that are not only prevalent over time, but also provide good indicators of the year a text originated in. In this, one needs to differentiate between individual style change of particular authors as opposed to general language change over time independent of any individual writer. For this purpose, we build regression models based on the works of two prolific authors of the late 19th to early 20th century, Henry James and Mark Twain, as well as models based on a reference corpus corresponding to language use at that time.

In Sect. 2, we situate our work with respect to other contributions in the literature. The details of the corpus collection and treatment are outlined in Sect. 3. In this section, we also present a methodology for conducting this sort of analysis in general. The data treatment and methods of each individual experiment are outlined in Sect. 4, and the results are presented. The outcomes are discussed in Sect. 5. Finally, in Sect. 6 we conclude.

2 Previous Work

Language change is ever present and complicates analysis and comparison of works of different temporal origin. Apart from being of interest in terms of style change over time, this also presents an issue for synchronic analyses of style, as discussed in Daelemans [7]: unless style is found to be invariant for an author and does not change with age and experience, temporality can be a confounding factor in stylometry and authorship attribution.[2] Stamou [19] reports on various studies in the domain and suggests applying more common methodologies to make comparisons between studies in stylochronometry more feasible.

[1] See Coleman [6], Frontini et al. [9] for discussion of attempts in the 15th century to date a text purported to be from the 3rd but shown to be most likely from (circa) the 8th. The former depends on manual methods and the latter, semi-automatic methods.

[2] Early-Wittgenstein may be stylistically as well as conceptually distinct from Late-Wittgenstein.

There have been longitudinal studies on linguistic change with respect to grammatical complexity and idea density, contrasting participants who were to develop dementia against those who were not [13], showing that both variables declined over time for both groups although at different rates.

Recent research concentrated on detecting changes in writing styles of two Turkish authors, Cetin Altan and Yasar Kemal, in old and new works [4]. The study looked at three different style markers: type and token length and the frequency of the most frequent word unigrams. Employing different methods, such as linear regression, PCA and ANOVA, they found that word types are slightly better discriminators than type and token length.[3] That study is similar to the current one in that it also used regression analysis to evaluate the relationship between the age of a work and particular variables, although token length was used rather than words' relative frequencies as we do here. The authors report a strong relationship between average token length and age of text in Altan's works, although an R^2 value of 0.24 indicates that there are likely to be other factors involved.[4]

Regarding temporal style analysis with respect to an author considered here as well, Hoover [11] investigates changes in James' style using word unigrams (100–4000 most frequent) with different methods, such as Cluster Analysis, Burrows' *Delta*, Principal Component Analysis and Distinctiveness Ratio.[5] Three different divisions in early (1877–1881), intermediate (1886–1890) and late style (1897–) (that have also been identified by literary scholars [2]) are identified, although there are transitions inbetween with, for instance, the first novels of the late period being somewhat different from the rest of them. The results on the 100 words with the largest Distinctiveness Ratio either increasing or decreasing over time show that James appears to have increased in his use of *-ly* adverbs and also in his use of more abstract diction, preferring more abstract terms over concrete ones. This work on James' style brought the writer to our attention as an interesting candidate for a temporal analysis of style. In contrast to the previous study, the work we present here focuses on a seamless interpretation of style over time rather than classification into different periods along the timeline of an author's works.

---

[3]ANalysis Of VAriance (ANOVA) is a collection of methods developed by R.A. Fisher to analyze differences within and between different groups. Principal Component Analysis (PCA) is an unsupervised statistical technique to convert a set of possibly related variables to a new uncorrelated representation or principal components.

[4]The coefficient of determination R^2 indicates how well a model fits the observed data ranging from 0 to 1–0 indicating a poor fit and 1 a perfect one; in the case of evaluating predictions against the outcome (test set) values can also range from -1 to 1;–in the case of negative values, the mean of the data provides a better fit.

[5]Distinctiveness Ratio: Measure of variability defined by the rate of occurrence of a word in a text divided by its rate of occurrence in another.

3 Data and Methods

In Sect. 3.1, we describe the data sets used and the feature preprocessing applied. We outline a general method for preparing this kind of text data for temporal analysis and introduce time-oriented analysis using explanatory regression models in Sect. 3.2.

3.1 Corpora

For this study, we consider works of individual authors, Mark Twain and Henry James—both who wrote during the late 19th century to the early 20th century—as well as a reference corpus comprising language of that time. Even though James' and Twain's timelines are not completely synchronous, they largely overlap, which renders them suitable candidates for a combined temporal analysis. In addition, they seem to have been, although both considered to be highly articulate and creative writers, contrasting in temperament and in their art [5, p. xii], yet each conscious of the other [1, 3]. It is interesting to see to what extent perceived differences are apparent in predictive models based on their data.

Tables 1 and 2 show James' and Twain's main works, 31 and 20 works respectively[6] collected from the *Project Gutenberg*[7] and the *Internet Archive*.[8] Project Gutenberg is the better source in terms of text formatting, but works are not always labelled with publication date, and especially for Henry James, who is known to have revised many works, one has to be sure of the exact version used. Ideally, collected pieces should be close to the original publication date to avoid confounding factors; otherwise, the collected piece might not be the same as the one originally published, and this may introduce irregularities into time-oriented analysis.

The reference corpus is an extract from the *The Corpus of Historical American English (COHA)* [8] which comprises samples of American English from 1810–2009 from different sources, such as fiction and news articles. For the purpose of the current experiments, we consider texts starting from the 1860s to the 1910s in order to cover both authors' creative life span. There are 1000–2500 files for each decade, spread over the individual years and genre. Models built on the basis of this data are likely to be more complete than the authorial data sets, as this collection is more balanced without gaps in the timeline.

In order to extract the features of interest from the texts, we build *R* scripts to lowercase all text before extracting context sensitive word unigrams by using Part-Of-Speech (POS) tagging from the R *koRpus* package (that uses TreeTagger POS tagger)

[6]Here, we only include the main works/novels for reasons of text length and genre homogeneity.

[7]http://www.gutenberg.org/-lastverifiedAugust2015.

[8]https://archive.org/-lastverifiedAugust2015.

Table 1 Henry James' main works

Title	1st Pub.	Version	Size	Genre
The American	1877	1877	721	Novel
Watch and Ward	1871	1878	345	Novel
Daisy Miller	1879	1879	119	Novella
The Europeans	1878	1879	346	Novel
Hawthorne	1879	1879	314	Biography
Confidence	1879	1880	429	Novel
Washington Square	1880	1881	360	Novel
Portrait of a Lady	1881	1882	1200	Novel
Roderick Hudson	1875	1883	750	Novel
The Bostonians	1886	1886	906	Novel
Princess Casamassima	1886	1886	1100	Novel
The Reverberator	1888	1888	297	Novel
The Aspern Papers	1888	1888	202	Novella
The Tragic Muse	1890	1890	1100	Novel
Picture and Text	1893	1893	182	Essays
The Other House	1896	1896	406	Novel
What Maisie Knew	1897	1897	540	Novel
The Spoils of Poynton	1897	1897	376	Novel
Turn of the Screw	1898	1898	223	Novella
The Awkward Age	1899	1899	749	Novel
Little Tour in France	1884	1900	418	Travel writings
The Sacred Fount	1901	1901	407	Novel
The Wings of the Dove	1902	1902	1,003.7	Novel
The Golden Bowl	1904	1904	1100	Novel
Views and Reviews	1908	1908	279	Literary criticism
Italian Hours	1909	1909	711	Travel essays
The Ambassadors	1903	1909	890	Novel
The Outcry	1911	1911	304	Novel
The Ivory Tower (unfinished)	1917	1917	488	Novel
The Sense of the Past (unfinished)	1917	1917	491	Novel
In the Cage	1893	1919	191	Novella

Showing *Title*, the original publication date (1st *Pub.*), version collected (*Version*), *Size* in kilobytes and *Genre*

[15, 17, 18]. Thus, we distinguish between different function/syntactic contexts of one lexical representation: e.g. without taking the context into account, the item LIKE could refer to the verb LIKE or the preposition LIKE. Since we would consider these to be separate entities despite them sharing the same lexical representation, we create

Table 2 Collected Mark Twain's works

Title	1st Pub.	Version	Size	Genre
Innocents Abroad	1869	1869	1100	Travel novel
The Gilded Age: A Tale of Today	1873	1873	866	Novel
The Adventures of Tom Sawyer	1876	1876	378	Novel
A Tramp Abroad	1880	1880	849	Travel literature
Roughing It	1880	1880	923	Semi-autobiography
The Prince and the Pauper	1881	1881	394	Novel
Life on the Mississippi	1883	1883	777	Memoir
The Adventures of Huckleberry Finn	1884	1885	586	Novel
A Connecticut Yankee in King Arthur's Court	1889	1889	628	Novel
The American Claimant	1892	1892	354	Novel
The Tragedy of Pudd'nhead Wilson	1894	1894	286	Novel
Tom Sawyer Abroad	1894	1894	182	Novel
Tom Sawyer Detective	1896	1896	116	Novel
Personal Recollections of Joan Arc	1896	1896	796	Historical novel
Following the Equator: A Journey Around the World	1897	1897	1000	Travel novel
Those Extraordinary Twins	1894	1899	1200	Short story
A Double Barrelled Detective Story	1902	1902	103	Short story
Christian Science	1907	1907	338	Essays
Chapters from My Autobiography	1907	1907	593	Autobiography
The Mysterious Stranger (finished by ed.)	1908	1897–1908	192	Novel
Is Shakespeare Dead?	1909	1909	121	Semi-autobiography

Showing *Title*, the original publication date (1st *Pub.*), version collected (*Version*), *Size* in kilobytes and *Genre* type separate entries for these, i.e. ⟨LIKE.VB⟩ and ⟨LIKE.IN⟩.[9] Punctuation and sentence endings are also included as features and in relativization (discussed in Sect. 3.2).

3.2 Timeline Compression and Analysis

As can be observed from the data in Tables 1 and 2, both authors composed works over the span of around forty years each, with overlap for about twenty years. However, in each case works are unevenly distributed with some years giving rise to more than one work. In the present context, where we aim to predict the year on the basis of word features, we combine different works in a year into one.[10] In the following experiments, we sometimes combine all available data for a year or if we investigate different sources (authors) we process these separately and differentiate between them by adding a CLASS attribute indicating the author that is a categorical variable rather than the ordinal YEAR or a continuous lexical variable.[11] Thus, in the context of style analysis, we examine a particular variable v over time by considering its relative frequency distribution, e.g. we count the occurrence of that particular word and relativize by the total number of occurrences of all words in that document (or document bin for multiple works in the same temporal span).[12] Building models on the basis of individual authors might lead to less stable models for prediction, since not all years will have given rise to a publication, and the resulting models will need more interpolation than aggregating yearly bins from both authors' works.

This study is motivated by quantitative forecasting analysis that monitors how a particular variable (or variables) changes over time and uses that information to predict how that variable is likely to behave in future [14]. Thus, the (future) value of a particular variable v is predicted by considering a function over a set of other variable values. One differentiates between the use of a *time-series* and *explanatory models*. Time-series analysis considers the prediction of the value the variable v_i takes at a future time point $t + 1$ based on a function f over its values (or errors) at previous distinct points of time $(v_i^t, v_i^{t-1} \ldots v_i^{t-n})$, as shown in example (1).

$$v_i^{t+1} = \mathrm{f}(v_i^t, v_i^{t-1}, v_i^{t-2} \ldots, error) \tag{1}$$

In contrast, *explanatory models* assume that the variable to be predicted has an explanatory relationship with one or more independent variables. Therefore, prediction of a variable v_i is on the basis of a function f over a set of distinct variables:

[9]The separate entries are created using the POS tags assigned by the tagger to the individual word entity in its context.

[10]This is without loss of generality to the bag-of-words analysis of texts in which sentence structures are not used subsequent to POS tagging.

[11]Lexical features are continuous here because we use relative frequencies.

[12]This applies if it is meaningful to count instances of the variable, as it is for token n-grams: such relativization does not apply, for example, to average word lengths.

$v_1, v_2 \ldots v_n = V$, with $v_i \notin V$ at the same time point t, as shown in example (2).

$$v_i^t = \mathrm{f}(v_1^t \ldots v_n^t, error) \tag{2}$$

Thus, a time-series involves considering prediction on the basis of a variable at distinct time points, explanatory models, which we employ here, consider distinct variables at the same time point; here the latter are taking the shape of multiple regression models predicting the year of publication of a particular text.

4 Experiments

In Sect. 4.1, we first present details of the data preparation and the way we constructed the regression models (Sect. 4.2). We present our analyses for the data sets in Sect. 4.3.

4.1 Data Preparation

In order to build a model to predict the year of a work's publication from the relative frequencies of lexical variables, all data is compressed to an interval level of one year, meaning that counts for features in works of the same year are joined and relativized over the entire token count for that author for that year. In addition, all instances receive a label indicating the YEAR of publication; the two-author data is also marked by a CLASS label. In the case of the two-author model (Sect. 4.3.1), empty years, i.e. those where neither author has published anything, are omitted. This results in thirty-nine cases for all main experiments here. These rows are unique with respect to author and year; there might be cases where both authors have published during the same year, which would result in there being two entries for a particular year; however these are distinct for the CLASS variable. Generally, we only consider features that occur in all YEAR instances in the training corpus to ensure the selection of consistent and regular predictors later on.[13] However, for the two-author experiments, we consider those types that appear in the majority of all instances. Since that data set is much smaller than the reference set, the constant feature selection is more prone to overfit on the training set and would be worse at test set generalization.

[13]This is not to argue that complementary categories (e.g. relativized counts of features that are not shared between both authors over the entire duration or features that are never shared by the authors over the duration, etc.) are uninteresting. However, for this work we are addressing change in language shared by the two authors and relative to change in background language of their time, thinking that this provides an interesting perspective on their distinctiveness from each other and everyone else.

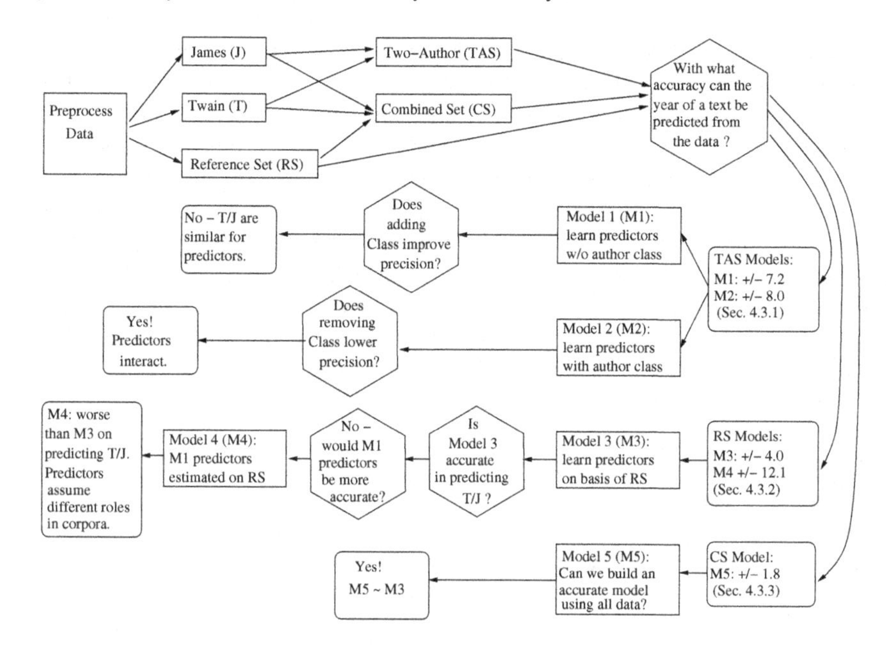

Fig. 1 Preparation and sequence of experiments

The reference corpus was preprocessed the same way as the other corpora; however all files belonging to a particular year were joined together, ordering the files arbitrarily, leaving 60 individual year entries spanning from 1860 to 1919 as a basis for calculating feature values.

There are two possible outcomes for selecting features to predict the year of a text in the two-author case. Either the CLASS attribute is among those considered helpful, meaning that there is a difference for authors for the other/some of the variables in the model or it is excluded, indicating that it did not help prediction in combination with the other features selected. Those features are arguably more representative of the language use shared by the authors rather than temporal change in any of the authors considered individually.

For all of the following experiments the data was randomly separated into training and test set to evaluate model generality; the split is 75 and 25 % for the training and test set respectively (using the *caret* package in R [12]).

4.2 Variable Selection and Model Evaluation

In order to predict the year of a particular work, we consider multiple linear regression models—these however require some pre-selection of features. Even after discarding less constant features, a fair number of possible predictors of about 200–13,000 are left. In order to rank variables according to predictive power with respect to the response variable, we use the *filterVarImp* function in *caret*; this evaluates the relationship between each predictor and the response by fitting a linear model and calculating the absolute value of the t-value for the slope of the predictor.[14] This is evaluating whether there is a systematic relationship between predictor and response rather than only chance variation. A higher absolute t-value would signal a higher probability of there being a non-random relationship between the two variables.

For the final selection of model predictors, we use *backward* variable selection, whereby the first step tests the full model and then iteratively removes the variable that decreases the error most until further removal results in an error increase.[15] Backward selection might have an advantage over forward selection, which although arguably computationally more efficient, cannot assess the importance of variables in the context of other variables not included yet [10]. Moreover, some of our exploratory experiments showed that forward selection was more prone to overfitting on the training data.

Model fit is assessed using the adjusted version of the coefficient of determination R^2 (henceforth denoted as: $\bar{R}^2$), which takes into account the number of explanatory variables and thus does not automatically increase when an additional predictor is added; it only increases if the model is improved more than would be expected by chance. R^2 should be evaluated in connection to an F-test assessing the reliability of the result. The F-test evaluates the null-hypothesis that all coefficients in the model are equal to zero versus the alternative that at least one is not—if significant it signals that R^2 is reliable.[16] We also consider the root mean squared error (RMSE), which is the square root of the variance of the residuals between outcome and predicted value.[17] The baseline model for all training/test set divisions is reported on as well; this equates fitting a model where all regression coefficients are equal to zero: this reduces the model to an intercept through the data tested (i.e. the arithmetic mean).[18]

In the following, we only report on models that fulfil the model assumptions measured by the *gvlma* package in R [16]: kurtosis, skewness, nonlinear link function (for testing linearity), heteroscedasticity and global statistics. Thus, any models reported on here will have been found acceptable by this test, and we dispense with reporting acceptability for each individual case.

---

[14]The t-value measures the size of the difference between an observed sample statistic and its hypothesized population parameter relative to the variation in the sample data. The further the t-value falls on either side of the t-distribution, the greater the evidence against the null hypothesis that there is no significant difference between hypothesized and observed value.

[15]In this case, the Akaike information criterion (AIC) is used to evaluate the model: $AIC = -2 * logL + 2k$, where L is the likelihood and k the number of estimated parameters in the model. Thus, AIC rewards goodness-of-fit, but penalizes the number of parameters in the model.

[16]All models reported on here had reliable $\bar{R}^2$ values at a level of a p-value <0.0001 associated to them, so we dispense with reporting on this in each individual case.

[17]$RMSE = \sqrt{\frac{\sum_{t=1}^{n}(\hat{y}_t - y)^2}{n}}$.

[18]This might not be an entirely realistic scenario in that most predictors, even randomly selected ones, will bear some kind of relation with the response. However, in the case of the test set, the wrong predictors can also have a worse effect than the null-model, so this might be an acceptable approximation.

4.3 Results

Here we present our predictive models (Fig. 1); the ones based on only James and Twain are discussed in Sect. 4.3.1. Further in Sect. 4.3.2, we evaluate two models on the basis of the reference set and in terms of how well they classify works of the individual authors. Finally, we combine both data sets to investigate the effects on the model (Sect. 4.3.3).

4.3.1 Two-Author Models

For the first experiment, we consider the lexical features of the two-author training set corpus, which contains 273 terms after only retaining features present in most year instances (28 of 31 instances of both Twain (13) and James (18)); the features are then ranked according to predictive power. The baseline model for the training and test data are an estimate of 1892 (RMSE: ±11.3) and 1893 (RMSE: ±13.2) respectively. Thus, the average error in prediction is 11 and 13 years respectively.[19]

One of the best models (a trade-off between training set and test set accuracy) is shown in (3)—this is the result of using the ten highest rated features. $\bar{R}^2$ is 0.71 (RMSE: ±5.5) on the training set and R^2 on the test set is 0.70 (RMSE: ±7.2). All except one predictor are significant with respect to the response variable. In addition, one can check for multicollinearity, i.e. whether the predictors are likely to be correlated: all of them seem only slightly correlated (all values<2).[20]

$$year = intercept + required.vbn + lay.vbd + received.vbd + put.vbp + fail.vb \quad (3)$$

In this model, both authors' data was used in unison without taking the individual author of a year instance into account. This implies that the rate at which each of them was using the predictors is unlikely to have been different—these predictors are thus likely to be good indicators of when a piece of text was published, but not necessarily distinctive with respect to either James or Twain. If we manually add the CLASS attribute to the existing model and re-train it, the results change almost imperceptibly on both training and test set by 0.003–0.015 points around 0.70/0.71 for $R^2/\bar{R}^2$ and a 0.2 rise for the RMSE. Thus adding authorship information seems to neither support nor to add conflicting information to the current model. One might interpret this to mean that there is very little difference between the two authors for these predictors. Inspecting the corresponding VIF confirms this in so far as that CLASS does not seem to be particularly related to any of the other predictors.

[19]The system reports estimates and predictions as decimals; we dispense with reporting these here, as texts were only ordered according to year rather than exact month, which renders those numbers meaningless. R^2 and RMSE are on the basis of rounded versions of predictions.

[20]This can be tested by using the *variable inflation factor (VIF)* that measures how much the variance of the estimated coefficients in regression is inflated compared to when the predictors are not linearly related; a value of 1–4 indicating low correlation and 5–10 high correlation.

In order to inspect a model where the CLASS was important, we retain all those features present in 29 of the instances in the corpus (333) and rank these as done previously. The resulting model based on subjecting the best ten features to backward selection is shown in example (4). This model is distinct from the previous one with respect to all predictors. $\bar{R}^2$ on the training set is 0.72 (RMSE: $\pm$5.2) and R^2 on the test set is 0.63 (RMSE: $\pm$8). If we exclude the CLASS attribute from this model, all evaluation parameters deteriorate on both sets. $\bar{R}^2$ drops to 0.62 (RMSE: $\pm$6.3) while the test set's R^2 reduces to 0.49 (RMSE: $\pm$9.5). Thus, the CLASS attribute seemed to somewhat interact with the other predictors in the model.

$$\begin{aligned} &year = intercept + class + floor.nn + dressed.vbn + blue.jj + waited.vbd+ \\ &space.nn + sufficiently.rb \end{aligned} \tag{4}$$

4.3.2 Reference Set Model

Here we investigate how the YEAR is predicted using the reference set rather than the two authors' data. The model is built as before by first creating a random split into training and test data and then discarding features not present in all year instances. The remaining 10,504 features are ranked with respect to the response YEAR and the best five are used in backward selection. The baseline model for training and test set are estimates of 1890 (RMSE: $\pm$17.4) and 1889 (RMSE: $\pm$17) respectively.

The resulting model is shown in example (5). The use of the comma seems to be very telling as it is highly significant as predictor. $\bar{R}^2$ on the training set is 0.96 (RMSE: $\pm$3.2), while R^2 on the test set is comparable with 0.94 (RMSE: c. $\pm$4). There does not seem to be an overlap with the previous models in terms of predictors. Although the model assumptions are met, predictors seem to be somewhat related: ⟨OUTSIDE.IN⟩ seems to be slightly related to ⟨,.COMMA⟩; when it is dropped from the model, the VIF of ⟨,.COMMA⟩ decreases by at least 2 points. This could indicate that these form common collocations, however, this would have to be quantified as part of a concordance analysis.

$$year = intercept+, .comma + later.rbr + outside.in + planned.vbn \tag{5}$$

One question that emerges from this is to what extent the reference model is able to classify James' and Twain's works. Taking each author's year averages separately as test sets (16 for Twain and 23 for James), the reference set model performs quite poorly for both; R^2 is -0.79 (RMSE: $\pm$15.4) / R^2 is -2.1 (RMSE: $\pm$20.3) for Twain and James respectively. The baseline model for James' and Twain's sets are 1889 (RMSE: $\pm$11.5) and 1894 (RMSE: $\pm$11.5) respectively. In this case, the mean through the data provides a better prediction than the reference model.

The predictors that are most reliable for estimating the year for the reference set are not effective for Twain or James. There might be common reliable predictors, but these are not among the ones chosen for the reference set alone, it seems. In

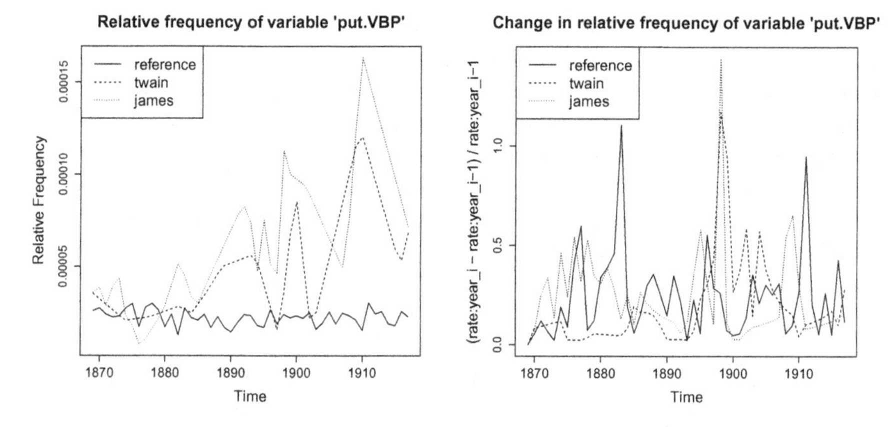

Fig. 2 Depicting predictor frequency and change of this predictor across all three corpora

this, James' data seems to be harder to classify than Twain's; both his scores are considerably worse than Twain's–this might be an indication that James' works differ more from the general style of that time. In order to see whether the reverse is true; reliable predictors for YEAR on the basis of James' and Twain's corpus performing worse on the reference corpus, we use the very first model's predictors to build a model based on the reference corpus data. Thus, the predictors are the same, but the instantiations might be different because of possible deviations in terms of word frequencies. Considering the results of $\bar{R}^2$ of 0.47 (RMSE: ± 11.9) on the training set and R^2 on the test set of 0.49 (RMSE: ± 12.1) indicate that Twain's and James' predictors are less successful for the reference set data.

Again, taking the two-author's data as test sets returns even worse results than previously: R^2 decreases to $-12/-14$ (RMSE = c. $\pm$ 42.2/44.7). Inspecting the model parameters shows that the estimates for the predictors are quite different for the two-author model and the reference corpus; thus these seem to be taking on genuinely different roles in each corpus; this is further depicted in Fig. 2, where we show one Twain/James predictor over time for each subcorpus separately, showing considerably more variation for Twain and James (partly interpolated).

4.3.3 Combining Models

Here, we present a final model built on all the data available, i.e. the reference corpus and the two author data. Thus, all data is aggregated together without reference to the source—James' and Twain's individual year data is added to that of the reference corpus before relativization. After discarding features not in all year instances, 13,245 features remain. As would be expected, adding more data should yield more constant instances than before; thus James and Twain might have constant features that are not present in all of the reference corpus data. The addition of their data contributed to a rise of c. 2,700 more constant features that would not have been constant over the reference corpus on its own. The baseline model for the training set here is the same as for the reference model, as we are only considering the average year over the data sets rather than any features within, the estimates also do not change from the previous ones. Considering the same number of highest ranked features as in the previous reference set model yields the model shown in example (6). This is rather similar to the previous one, except for the features ⟨WANT.NN⟩ and ⟨ATTITUDE.NN⟩ rather than the predictor ⟨PLANNED.VBN⟩. The model's $\bar{R}^2$ on the training set is slightly higher than previously: 0.97 (RMSE: $\pm$2.8). The test set's R^2 is also higher with 0.988 (correspondingly RMSE: $\pm$1.8).

Thus, James' and Twain's data might be adding different information in terms of constant features that complement the reference set. The model estimates are somewhat different from previously indicating that the two author data might be creating a shift there as well. Increasing the number of input features causes an improvement on the training set, but slightly less accurate results on the test set.

$$year = intercept +, .comma + later.rbr + outside.in + attitude.nn + want.nn \quad (6)$$

5 Discussion

The results of these experiments show it possible to accurately predict the year of a publication in the two-author case and in particular in the case where we have a larger (reference) corpus at our disposal. The exact predictor selection is subject to the underlying data set, although the more data is available, the more stable this process seems to become. The results obtained seem to indicate that the model built on the basis of the two-authors (Sect. 4.3.1) has to approximate two potentially rather different styles. Using a more balanced corpus in terms of authors and genre seems to create a better approximation to a general style of that time. In order to truly account for the differences between models only built using James and Twain and those built on the larger reference set, one would need to examine the development of those features within in detail in order to see in what way the individual authors deviate from the *general* style. Future work should address those features not attested in all yearly bins in order to investigate differences to constant features examined here as well as individual and general language change, i.e. are some features abandoned over time and does this happen gradually or abruptly. Apart from the word features examined here, one might also consider syntactic shift and in what way prolific authors, such as James and Twain differ from the general style.

6 Conclusion

The stylochronometric analysis reported here supports qualitative assessments of the texts analyzed: despite differences noted between James and Twain, when using their novels to predict year of authorship, their mutual discriminability dissipates. A contribution of this work is to introduce methods of preparation and analysis for the temporal analysis of stylometry. We have shown that it is possible to predict the year of a publication relatively accurately from lexical features whether one is analyzing individual authors or a general reference set of the time. Future work includes the analysis of structural patterns, general and individual ones.

Acknowledgments This research is supported by Science Foundation Ireland through the CNGL Programme (Grant 12/CE/I2267) in the ADAPT Centre (www.adaptcentre.ie) at Trinity College Dublin.

References

1. Ayres, A.: The Wit and Wisdom of Mark Twain. Harper Collins, New York (2010)
2. Beach, J.W.: The Method of Henry James. Yale University Press, New Haven (1918)
3. Brooks, V.W.: The ordeal of mark twain. William Heineman, London (1922)
4. Can, F., Patton, J.M.: Change of writing style with time. Comput. Humanit. **38**(1), 61–82 (2004)
5. Canby, H.S.: Turn West, Turn East: Mark Twain and Henry James. Biblo & Tannen Publishers, New York (1951)
6. Coleman, C.: The Treatise Lorenzo Valla on the Donation of Constantine: Text and Translation. First published 1922. Russell & Russell, New York (1971)
7. Daelemans, W.: Explanation in computational stylometry. In: Computational Linguistics and Intelligent Text Processing, pp. 451–462. Springer, New York (2013)
8. Davies, M.: The Corpus of Historical American English: 400 million words,1810–2009. http://corpus.byu.edu/coha/, vol. 24, p. 2011 (2010). Accessed 24 Aug 2015
9. Frontini, F., Lynch, G., Vogel, C.: Revisiting the donation of constantine. In: Kibble, R., Rauchas. S. (eds.) Artificial Intelligence and Simulation of Behavior—Symposium: Style in Text, vol. 2008, pp. 1–9 (2008)
10. Guyon, I., Elisseeff, A.: An introduction to variable and feature selection. J. Mach. Learn. Res. **3**, 1157–1182 (2003)
11. Hoover, D.L.: Corpus stylistics, stylometry, and the styles of Henry James. In: Style 41.2 (2007)
12. Jed Wing, M.K.C., et al.: caret: Classification and Regression Training. R package version 6.0–30, http://CRAN.R-project.org/package=caret (2014). Accessed 24 Aug 2015
13. Kemper, S., et al.: Language decline across the life span: findings from the Nun Study. In: Psychology and aging 16.2, p. 227 (2001)
14. Makridakis, S., Wheelwright, S.C., Hyndman, R.J.: Forecasting Methods and Applications. Wiley, New York (2008)
15. Michalke, M.: koRpus: An R Package for Text Analysis. Version 0.04-40. http://reaktanz.de/?c=hacking&s=koRpus (2013). Accessed 24 Aug 2015
16. Pena, E.A., Slate, E.H.: gvlma: Global Validation of Linear Models Assumptions. R package version 1.0.0.2. http://CRAN.R-project.org/package=gvlma (2004). Accessed 24 Aug 2015
17. R Core Team. R: A Language and Environment for Statistical Computing. R Foundation for Statistical Computing. Vienna, Austria. http://www.r-project (2014). Accessed 24 Aug 2015

18. Schmid, H.: Probabilistic part-of-speech tagging using decision trees. In: Proceedings of international conference on new methods in language processing, vol. 12, pp. 44–49. Manchester (1994)
19. Stamou, C.: Stylochronometry: stylistic development, sequence of composition, and relative dating. Lit. Linguist. Comput. **23**(2), 181–199 (2008)

Semantic Analysis for Document Similarity and Search Queries

Charlene Cassar and Samad Ahmadi

Abstract Document similarity is a popular topic in natural language processing fields, especially in areas of Information Retrieval. Current systems are often limited to keyword search due to the complexities of handling free text search. Accurate results are rarely achieved. In this research we propose a combination of different semantic analysis methods for using with free text searches. Our aim is to enable the user to describe their search more freely. We will be using semantic analysis tools to understand the context of search and find more relevant results. Amongst several different domains which could be chosen for proof of concept, for our case study we have chosen extracting information from dream interpretation websites. In this case study our research aims to return results based on specific dreams input from users. This is part of an ongoing research and results so far are satisfactory.

1 Introduction

Most current search engines base their results on indexing keywords or expressions and are sensitive to the specific keywords used. The aim of this research is to test the hypothesis of whether or not the use of NLP techniques, such as keyword extraction and semantic analysis can be used to correct and enrich a search query in the form of text descriptions into relevant search results acceptable to humans. Our aim is to provide relevant results based on large amounts of text rather than keyword submissions. For proof of concept we have chosen dream interpretations web

---

C. Cassar (✉) · S. Ahmadi
Faculty of Technology, De Montfort University, Leicester, UK
e-mail: charlenecas@gmail.com

S. Ahmadi
e-mail: sahmadi@dmu.ac.uk

M. Bramer and M. Petridis (eds.), *Research and Development in Intelligent Systems XXXII*, DOI 10.1007/978-3-319-25032-8_7

sites and dictionaries where user queries naturally come as descriptions rather than keywords. In order to satisfactorily find interpretations for users descriptions of their dreams, steps are: to accept a piece of text depicting a dream, process this text and return a suitable list of interpretations.

2 Previous Research

Due to the nature of semantic analysis on search queries, we focus on query processing for a specific domain, dream interpretation. In the case of Question Answering (QA) a query must be submitted by a user, the query is then processed, and a large amount of documents are retrieved and ranked according to likeliness of containing the correct answer. A close similarity between QA field and dream interpretations can be seen where the user submit their dream (in the case of dream interpretation); this text is processed using NLP techniques; and from a list of interpretations (documents retrieved from a database), the similar documents to the query are returned as the interpretation.

For the scope of this work, the evaluation method will focus on the information retrieval metrics. The users would have the chance to enter a piece of text about their dream and rate the result. If they find that the documents returned are relevant to the query, the user will then give the system a positive rating; otherwise a negative rating is expected.

[2] mention two evaluation techniques which are useful to evaluate information retrieval which are known as the retrieval performance evaluation. Two of these popular techniques are known as the recall and precision evaluation methods.

Recall
The recall method measures the amount of documents that were retrieved in comparison to the amount of documents that were relevant and should have been retrieved. This can be defined as:

$$\text{Recall} = \frac{|\text{Ra}|}{|\text{R}|}$$

where Ra is the number of documents retrieved (which were relevant) and R is the total number of documents which should have been retrieved.

Precision
Precision is a measure of the amount of relevant documents retrieved in comparison to the total number of documents (regardless of whether they are relevant or not) retrieved. Precision is defined as:

$$\text{Precision} = \frac{|\text{Ra}|}{|\text{A}|}$$

Similar to the recall method mentioned above, Ra is the total number of relevant documents retrieved. A is the total number of documents retrieved.

For the scope of this research, precision and the recall metrics are used. The text input will be short mainly due to time constraints, since it would take a very long time to sift through all of the documents available to find the relevant documents which should have been retrieved. In this research ten different dream queries will be used for evaluation purposes [4].

In [5] the researchers divide the question answering task into two main stages; the syntactic analysis and the semantic analysis stage. The first stage takes the query and transforms this into an array of keywords. The semantic analysis obtains a semantic representation with the use of domain-specific ontologies to derive their relationships.

[7] concentrates on processing a user query (in the case of the current research, the dream submitted will be the query). Several Natural Language Processing techniques are used in the four layers which Pasca defines to be necessary.

In the first layer Pasca focuses on obtaining the lexical terms. Each term within the text is considered separate, bearing no relation to the other words in the textual query. The stemming algorithm is then applied to each individual term (Porter's Stemming Algorithm) to obtain the root word from which the term was derived [7].

Each term in the query is given a part-of-speech (POS) tag, which basically identifies the type of word (noun, verb, etc.). The position of the term within the query is also obtained. After this information has been extracted and the terms have been tagged, the terms are divided into two groups; content and non-content lexicons [7].

Content words are the terms which contribute to identifying a relevant document. Content words are usually tagged as nouns, verbs, adverbs or adjectives whilst non-content words are usually modal verbs, auxiliary verbs, prepositions, conjunctions, pronouns and interjections. For clarification purposes we will take an example; "The President of Malta was wearing a coat". The content words would be President, Malta, wearing and coat [7].

In the second layer the relationships between the content words are obtained. In the example given above we can easily see that there is a clear relationship between the terms President and Malta. This enabled Pasca to obtain a clearer idea of what the user required [7].

In the third layer, this will not be necessary in the case of dream interpretation, deals with the question stem and the expected answer type. Question stems are terms such as Who, What, How, Which, Why, Where, When Name, Whom, etc. Pasca found that How questions were amongst the most difficult questions to answer due to the ambiguity these questions pose. The expected answer type would be what answer type we expect to obtain from the question posed, for instance, person. However this layer does not related at all to the dream interpretation system described in this work [7].

In the fourth layer Pasca gives importance to semantic constraints. Semantic constraints occur when the meanings of words pose certain constraints due to the relationship between the terms for example, United Kingdom [7].

Once a significant number of keywords and their relationships have been obtained, the next step is to use information retrieval techniques to retrieve a set of relevant documents [7]. In the scope of this project these documents were the dream interpretations obtained.

For information retrieval, an inverted index may be used. The retrieved set of documents is taken, split into tokens (where a token is a word or phrase). The tokens are then stemmed to their normalized format. The inverted index is applied to a dictionary with a postings list and the dictionary contains the frequency of the term. The postings list is a list of all the documents containing the term [6].

In [2], the inverted index is also used to index relevant documents. The three most popular methods of Information Retrieval mentioned in [2] are the Boolean, Vector and Probabilistic methods.

The Boolean retrieval method tags documents as being relevant or not; relevant documents are tagged with a "1" and non-relevant documents are tagged with a "0" [1]. The Boolean method accepts user queries and matches these against a dictionary. For each query term, a postings list is obtained and an intersection is performed on these lists [2].

The Boolean Retrieval method is the most popular method of Information Retrieval since it can be simply represented formally and due to the simplicity of its use. This method is known to be the least likely to yield relevant results from the set than the other popular methods since the document can either be relevant or non-relevant without a rating for documents in between [2].

Due to the limitations of the Boolean Retrieval method, there is also the extended Boolean model, which caters for some of the original method's shortcomings. The extended Boolean retrieval method is found to retrieve more related documents than that of its ancestor [3]. This method is an extension of the Boolean method with the inclusion of the vector based method (described below) [2].

Another retrieval method is the Vector Space Model, which is also known as the cosine similarity measure [4]. The purpose of this method was to give a non-boolean (not limited to only 0 or 1) weight value to documents to give a clear indication of the degree to which the document is relevant.

This method also makes use of term frequency to acknowledge whether a document is relevant or not relevant to the query posed. The term frequency, which also goes by the name of tf factor, defines the document using terms as a feature. In other words, the tf factor defines the number of occurrences of that term within a particular document. The inverse document frequency uses a differentiating feature to define the document. This is the frequency rate of the term within the set of documents as a whole. If all of the documents contain a particular term, then the term can hardly be considered as a distinguishing feature [4].

After the query has been processed and the relevant documents have been retrieved, the user may then be presented with the results.

3 Dream Interpretations

Various psychologists such as Sigmund Freud and Carl Jung have based their careers on dream analysis [8]. Many people are interested in getting an insight to why they are experiencing certain dreams and the hidden meaning that may be behind these dreams. The current stance of the advancement of dream analysis interpreters is still quite premature.

Using natural language processing (NLP) we believe that a proper semantic analysis can be performed on the dream text input by the user and thus extract the meaning of the text (to differentiate from the dream interpretation, here we simply mean the semantic value of the text). From this, information retrieval algorithms are applied to the text sources retrieved from dream dictionaries and the highest matching documents are chosen to be the possible interpretation of the dream.

4 Semantic Analysis of Search Queries

This section describes details of our algorithms in our proposed system. The system accepts a user's text input and the language of the text is detected. If it is English it may proceed to the next stage of the process. The language of the text is detected using Google Translation API.[1]

The next step of the system is to ensure that the spelling of the text is correct. The NHunspell API[2] is used to check the spelling of the words and to suggest relevant recommendations when matched against its database. Due to time constraints, only a limited amount of slang word conversions were provided. The slang words are corrected when matching against the NHunspell database. Words such as "r" which is often used in place of the word "are" in slang are automatically converted.

To enrich the text with additional information, the OpenCalais API[3] is used. This library enables the semantic context of the text to be derived. These keywords would be added to the content word list derived from OpenCalais to increase the relevant documents found (Fig. 1).

In the next step, the lexical terms are derived, as stated in [7]. We used Porter's stemming algorithm to obtain the stem of the tokens in the text but realised that this gave us results such as "comput" instead of "computer" which might affect the retrieval of documents since they are indexed by their proper term "computer"

[1]https://cloud.google.com/translate/docs.

[2]http://www.crawler-lib.net/nhunspell.

[3]https://www.drupal.org/project/opencalais_api.

```
var openCalais = new Calais.CalaisDotNet(apiKey, content);
String doc = openCalais.Call<CalaisJsonDocument>().RawOutput.ToString();
List<String> tags = webScraper.GetSemantics(doc);
```

Fig. 1 Snippet of code calling OpenCalais API[4]

rather than their stem "comput". From this list of lexical terms, the content terms and non-content terms were identified. Part of this process involved automatically removing the stop words from the content words list. Stop words are terms which carry no semantic weight and thus do not change the meaning of the text, such as the words "a", "the" or "that", which definitely do not need to be used in the information retrieval process.

Once the content word list has been obtained, along with the additional information, the next step is to retrieve the documents. The interpretations for this system were obtained from the dreammoods website.[5] A process of webscraping was used in order to obtain the documents which relate to the content terms obtained. The page was parsed and the relevant interpretations were retrieved according to the queries set. Had there been more time available, a database could have been set up and the information could be loaded there. This would ensure less dependency on the website.

In more cases than not, there would be more documents retrieved than is necessary. A lot of these documents might simply contain one or two of these terms but in reality, have nothing to do with the dream inputted. Thus, the vector space model was chosen to handle this problem and only return the relevant documents to the user.

The vector space model was chosen because it considers the frequency of the occurrence of the terms in the documents. The term frequency and the inverse document frequency are calculated for each of the documents. Finally when this process is done the TF IDFs (multiplication of the TF value and the IDF value) is calculated. The last step of the vector space model is to apply the cosine similarity rule. A predefined threshold value is set and any document exceeding this value is added to the list and returned to the user [4].

If the documents were obtained from an unreliable source, they would need to be spell checked and have its slang corrected. However, since the documents are coming from a trusted website, we did not feel that this was necessary.

The following figure gives a representation of the algorithm flow for the semantic analysis of search queries (Fig. 2).

[4]http://calaisdotnet.codeplex.com/

[5]http://dreammoods.com/.

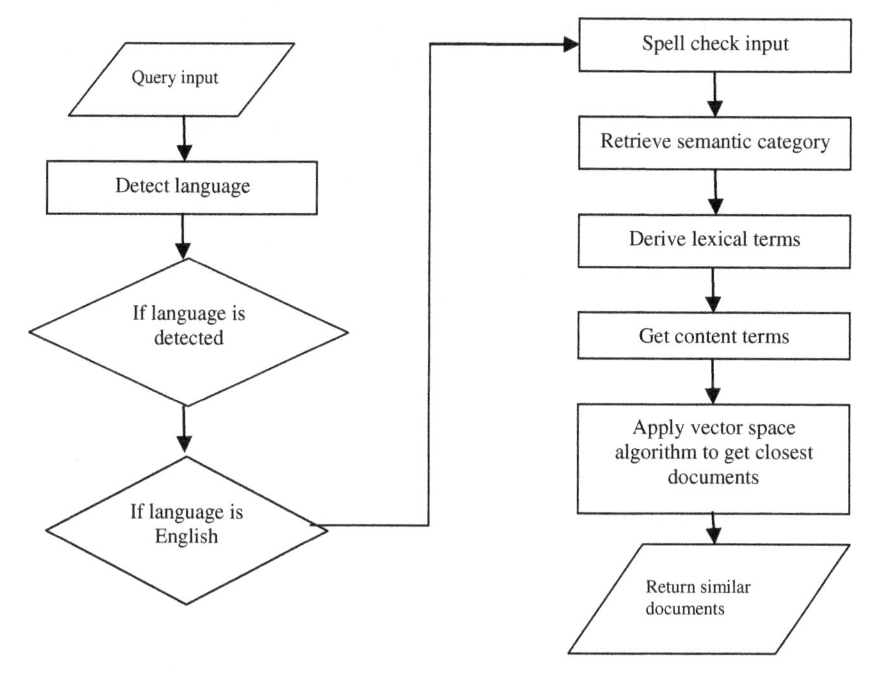

Fig. 2 Semantic analysis for search queries algorithm

5 Results

In most cases the program proved to yield relevant results but also had the tendency of returning more results than necessary.

For the purpose of the evaluation study, short but consistent dream queries are submitted to the system. This enables us to calculate the recall and precision rates in a reasonable time frame.

Initially the threshold was set to 0.15.

Precision
The results for the ten test cases can be seen below (Table 1):

The average precision rate for the ten documents was 0.3775; meaning 37.75 % of the documents retrieved by the system and returned to the user were relevant. The other 62.25 % were irrelevant.

Below we check for recall.

Recall
The results for the ten test cases can be seen below (Table 2):

The average recall rate for the 10 documents was higher than the precision rate. We managed to get a rate of 0.736; meaning that a percentage of 73.6 % of the documents that should have been retrieved, were in fact retrieved. The other 26.4 % of the documents which should have been returned were not returned to the user.

Table 1 Precision results for ten test cases with a threshold of 0.15

Test case	Relevant retrieved	Total retrieved	Precision rate
1	7	21	0.33
2	5	17	0.29
3	3	7	0.43
4	1	2	0.5
5	2	8	0.25
6	1	3	0.33
7	1	3	0.33
8	1	2	0.5
9	6	16	0.375
10	4	9	0.44

Table 2 Recall results for ten test cases with a threshold of 0.15

Test case	Relevant retrieved	Relevant should be retrieved	Recall rate
1	7	7	1
2	5	5	1
3	3	3	1
4	1	2	0.5
5	2	4	0.5
6	1	4	0.25
7	1	4	0.25
8	1	1	1
9	6	7	0.86
10	4	4	1

Next we altered the threshold rate to check if there was a difference in the results. The threshold was then set to 0.35.

Precision

The results for the ten test cases can be seen below (Table 3):

The average precision rate in this case was 0.305 (30.5 %). Considering that the precision rate was previously 0.3775, this value has decreased by increasing the threshold. That would mean that many relevant documents are ranked lowly (from between 0.15 and 0.35) and thus when the threshold is increased, these documents are removed.

Recall

The results for the ten test cases can be seen below (Table 4):

The total recall rate with a threshold of 0.35 was 0.445 (44.5 %), a dramatic decrease from the 73.6 % obtained with a threshold of 0.15.

As discussed above, the threshold was changed to a higher value, assuming that the precision rate would benefit, however the opposite effect was obtained. Less relevant documents were returned when the threshold was increased meaning that

Table 3 Precision results for ten test cases with a threshold of 0.35

Test case	Relevant retrieved	Total retrieved	Precision rate
1	4	11	0.36
2	5	17	0.29
3	2	6	0.33
4	1	2	0.5
5	1	7	0.14
6	1	5	0.2
7	0	1	0
8	0	1	0
9	5	9	0.56
10	2	3	0.67

Table 4 Recall results for ten test cases with a threshold of 0.35

Test case	Relevant retrieved	Relevant should be retrieved	Recall rate
1	4	7	0.57
2	5	5	1
3	2	3	0.67
4	1	2	0.5
5	1	4	0.25
6	1	4	0.25
7	0	4	0
8	0	1	0
9	5	7	0.71
10	2	4	0.5

many relevant documents were given weights between 0.15 and 0.35, thus omitted from the list when the threshold is 0.35. To obtain a better effect, the keywords fed to the information retrieval module must be enriched with much more information. It would be ideal if the document itself was parsed to ensure that it does not contain terms which have nothing to do with the dream inputted. This might eliminate irrelevant documents from being retrieved and increase the precision rate. The precision rate might increase if the input is of a longer length. To evaluate this, we submitted a dream found on the dreammoods[6] website:

> I have been having these nightmares about my girlfriend and kids being either attacked by 1 black dog; a shadowy figure, or the newest one from last night, they we're screaming for my help and I was running around looking for them franticly and couldn't find them. My girlfriend kept telling me the youngest one was hurt and they needed my help badly, and then she kept asking me why wouldn't I help them.

[6]http://dreammoods.com/.

For a threshold of 0.15 we have the following figures:

$$\text{Precision Rate } = 6/39 = 0.15$$

$$\text{Recall Rate } = 6/7 = 0.86$$

The precision rate obtained with a threshold of 0.35 was:

$$\text{Precision Rate } = 3/9 = 0.33$$

And the recall rate was calculated as follows:

$$\text{Recall Rate } = 3/7 = 0.43$$

As we can see above, a larger piece of text contributed to a higher recall rate when a smaller threshold was used but a lower threshold in both cases (smaller and larger threshold).

6 Conclusion and Further Work

From this research work we saw that question processing and information retrieval techniques can be used for dream interpretation systems. The accuracy of the system, which is still at a primitive level, depends heavily on the processing techniques.

The system did suffer a lag in time when it came to parsing longer pieces of text. In the case where the text was too long the program simply stopped functioning completely. This would definitely need to be fixed in the future. The aim would be to reduce the time complexity from O(n^2) to O(n), since various nested loops were used during processing.

As part of the further work, it would be interesting to experiment with the idea of query optimization using genetic algorithms to ensure that not only the correct documents are returned but also formulated to return the correct part of the document to the user.

In the future we intend to use databases to store interpretations, each of which will be indexed by its stem and provide additional information such as lexical terms and POS tagging. This will allow for a smoother mapping between query and document and would eliminate the dependency of the system on the dreammoods[7] website.

To summarize, we are satisfied by the result because it highlights the possibilities of using semantic analysis for document similarity and that such a system can exist.

[7]http://dreammoods.com/.

References

1. Azzopardi, J.: Template-Based Fact Finding on the World Wide Web. University of Malta, Msida (2004)
2. Baeza-Yates, R., Riberiro-Neto, B.: Modern Information Retrieval. ACM Press (1999)
3. Grundwls, L., Kwok, K.L.: Sentence Ranking Using Keywords and Meta-keywords. Springer, New York (2008)
4. Jurafsky, D., Martin, J.H.: Speech and Language Processing. Pearson Prentice Hall, New Jersey (2009)
5. Liang, J., Nguyen, T., Koperski, K., Marchisio, G.: Ontology-based natural language query processing for the biological domain. In: Proceedings of the HLT-NAACL BioNLP Workshop on Linking Natural Language and Biology, pp. 9–16. Association for Computational Linguistics, New York (2006)
6. Manning, C.D., Raghaven, P., Schutze, H.: An Introduction to Information Retrieval. Cambridge University Press, Cambridge (2009)
7. Pasca, M.: Open-Domain Question Answering from Large Text Collections. CSLI Publications, United States (2003)
8. Sexton, T.: Dream Analysis of Carl Jung and Sigmund Freud: The Difference Lies in the Unconscious. http://voices.yahoo.com/dream-analysis-carl-jung-sigmund-freud-the-466946.html (2007). Accessed 21 March 2012, from Yahoo Voices

Social Trust in a Familiar Community

Eseosa Oshodin, Francisco Chiclana and Samad Ahmadi

Abstract Most computation of social trust have been used for maintaining communities of individuals, based on their past activities. The behaviour of idle individuals or non-contributors to the communities have been totally ignored in the computation as they might affect the representation of the trust computed for the individuals. If the trust for an individual have been misrepresented, other individuals in the community will erroneously disengage or engage with the individual. In this paper, a new trust metric is proposed which is based on user's pattern of interaction that will be able to assist other users in their decision making on whether to join or leave a community. Different trust features are analysed to evaluate trust values for each user, which is used to determine the trust communities of the users.

1 Introduction

In the context of social network, people tend to have similar preferences with their trustworthy friends and also disclose their personal information to them. According to Nepal [7], trust between people can be described as a "social element" for engagement, acceptance and other decisions. Trust is important to everyone's daily life-activities. For example, before a person accepts the help of another person/group or a group decides to accept a person, they will require to know the trustworthiness of that person or group. This trustworthiness of people form familiar communities where their relationship can be maintained and it can also be referred to for trust information, which can assist in decision making. Examples of the familiar communities exist in Facebook and twitter, where users become familiar after several inter-

E. Oshodin (✉) · F. Chiclana · S. Ahmadi
De Montfort University, Leicester, UK
e-mail: eseosa.oshodin@dmu.ac.uk

F. Chiclana
e-mail: chiclana@dmu.ac.uk

S. Ahmadi
e-mail: sahmadi@dmu.ac.uk

M. Bramer and M. Petridis (eds.), *Research and Development in Intelligent Systems XXXII*, DOI 10.1007/978-3-319-25032-8_8

actions or engagements with themselves. But when trust information is unknown to a person, an inaccurate decision or no decision may be made by the person.

Sociologists, such as Luhmann [5], referred to trust as a means for building familiar communities to reduce complexity when different individuals have diverse views, opinions and goals. But Psychologists tend to define trust as the belief for a person to behave as expected. According to Erikson [2], trust emerges from the first stage of human life (0–2 years) where infants learn how to depend and rely on other people to satisfy their needs but when the infants are neglected by other people, they will learn the mistrust as they view this people to be undependable and unreliable. Both experience and familiarity of the infants with other people will enable them to discern similar situations in the future.

In general, trust is a conditional means to portray individuals' behaviour based on their previous pattern of interaction. In other words, it is the probability that future behaviour of individuals will be faultless and similar to their previous true behaviour in all situations. From this definition, three factors can be drawn out: familiarity, experience and honesty. Familiarity factor reduces the complexity of the dynamic nature of the community [5], the experience factor reflect the similarity in behavioural patterns, while the honesty factor controls the representation of the behavioural patterns in a familiar community. A new trust model will be introduced which relies on these factors.

The rest of the paper is structured as follows: Sect. 2 provides a review on the relevance of trust in communities and the relationship between familiarity, experience and trust. Methods for evaluating familiarity and experience from previous research are also discussed to compare their various features. Section 3 introduces the proposed approach for trust computation that will assist in observing users behaviour. Finally, Sect. 4 summarizes this study with its discoveries.

2 Literature Review

2.1 Trust, Experience and Familiarity

The definition of trust by Rutter [10] (cited in [1]) referred to trust as a "generalized expectancy" from an individual or group given to another individual or group to fulfil tasks. In other words, trust from agent A to agent B is the belief in the ability of agent B to accomplish given tasks to an expected and satisfactory level. However, users may still remain in a state of uncertainty based on their engagements with others.

Hardin [3] described trust between nodes in his game experiment to be asymmetric and referred to it as a one-way trust (see Fig. 1). This type of trust exists when either of the nodes (initiator) decides to act first based on its trust towards the other node (respondent) and the respondent might decide to either respond to or ignore the action of the initiator based on its own trust towards the initiator.

Fig. 1 One-way trust network

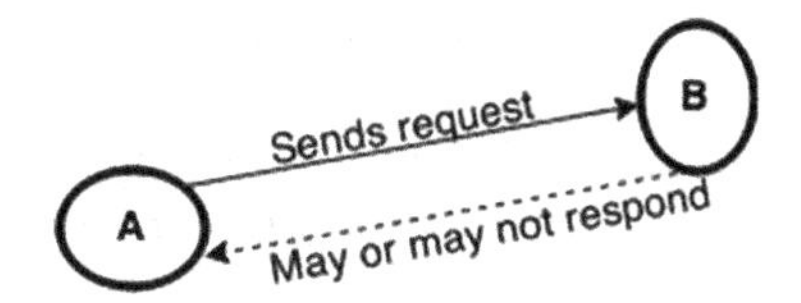

A different view of trust by Marsh [6] states that trust is a useful tool for discernment of the behaviour of other individuals based on previous experience. In [10], it was claimed that trust can be based on "local experience" (experience from engagement between two individuals within a community), interaction and daily routine of individuals which can further improve the participation of others. From these definitions, trust based on experience can be used as a criteria to consider a user into an active community. That is, without its existence, communities will cease to evolve or might completely disappear.

Marsh [6] also stated that in order to determine the outcomes from new diverse situations of interactions, we will need to observe the behaviour of existing ones which were affected by their "importance". He described this "importance" as a measure of the expected benefits to be gained from situations. This importance factor is believed by Marsh [6] to cause an increase in trust based on the fact that as a situation becomes important, the choice for cooperation amongst individuals increases. The importance of individuals should also be considered to determine if the individuals will be able to fulfil the tasks in future situations.

Wu and Chiclana [13] determined the importance degree of an individual by considering the trust degree and consensus level of the individual in a social network. The importance degree was used in evaluating the weights for each individual in order to aggregate the individual preference relation into a collective one. This collective preference relation is then used for giving suggestions or advices to individuals in the same group. If a user doesn't know the preference of another individual who has no or few engagements in the past, the user can use the collective mechanism to decide whether to trust this individual.

According to Luhmann [5], trust can also exist in a familiar community and there may be changes in the behaviour of members in that community. If these changes occur, the existing members might have to either adapt or disengage from the community. External non-members may decide to join the community based on either their high interest in the new characteristics of the community or their familiarity with the exemplar of that community. High interest in sincere characteristics of the community motivates the non-members to initiate communication with members of that community based on their belief in the importance of the community to meet their needs.

In the communication field, Gerck [16] defined trust as an important tool for a communication path but it cannot be transferred from a source to destination through the same path. This means that an individual will rely on information through a channel or path from another individual (source) based on trust inferred from multiple paths relating the source with other individuals in a community. In other words, trust of agent A for agent B can be seen as a rationale for accepting information from agent B based on agent B's previous experiences with others in a community. Gerck [16] still went further to refer trust as a reasonable "collective agreement" which means that there should be equivalent-honest patterns which can be aggregated to make accurate judgement.

2.2 Evaluation of Familiarity and Experience

Previous research work in [15] used both Familiarity and Experience as parameters for modelling trust. The familiarity value was initially measured based on the work by Carter and Ghorbani [1], where similarity values were also considered in the measurement, since two individuals can become friends when they have similar characteristics or behaviour. The similarities between agents were determined using Eq. (1) which is based on Hamming distance of value hierarchies. This hierarchy values represent the importance value from each agent given to other agents in a community. An example of the similarity evaluation will be shown later in the proposed model of this current paper.

$$S_{(a,a_i)} = 1 - \frac{H_{(\tau_a,\tau_{a_i})}}{N} \tag{1}$$

Where: *τ_a and τ_{a_i} are hierarchy values for agent $\mathbf{a}$ and agent $\mathbf{a}_i$; $H_{(\tau_a,\tau_{a_i})}$ is the total number of distinction between the hierarchy values for agent $\mathbf{a}$ and $\mathbf{a}_i$; N = Total number of hierarchy values given to the agents.*

These similarity values were used in [15] to initialize familiarity when an individual (for example, a customer) **u** has not had any encounter before with another individual (selling agent) **a**. Thus, the initial familiarity $F_{o(u,a)}$ was then evaluated by considering other previous engagements the individual **u** had with other individuals (selling agents) $\mathbf{a}_i$, using weighted average functions (as shown in Eq. (2)) [15].

$$F_{o(u,a)} = \sum_{i=1}^{n} \left[F_{(u,a_i)} \cdot \frac{S^2_{(a,a_i)}}{\sqrt{\sum_{i=1}^{n} S^2_{(a,a_i)}}} \right] \tag{2}$$

Where: *$F_{(u,a_i)}$ is the existing familiarity value for user $\mathbf{u}$ with other agents $\mathbf{a}_i$; $S_{(a,a_i)}$ is the similarity value between the target agent $\mathbf{a}$ with other agents $\mathbf{a}_i$.*

It is possible to evaluate $F_{o(u,a)}$ using previous engagement feedbacks, as we need to predict the familiarity value for each users since their behaviours in a social community is dynamic in nature.

Previous research [7] referred to "popularity trust" as a value given to a member of a community by other members of that same community. This can represent the familiarity value for the member since it is based on existing association with other members. This value was evaluated with Eq. (3) which is based on the positive and negative feedbacks given to the member by other members of the community, after this member had previously carried out an active or passive interactions towards these other members. For instance, if a user sends an email towards another user, it can be referred to as an active interaction; while if a user views another user's profile page, it can be referred to as a passive interaction. The positive feedbacks for an active interaction will then be the response by replying to the email previously sent, while that for a passive interaction will be the response by viewing on the other individual's page. The negative feedbacks is the number of non-response to previous emails or viewing actions.

$$F^{x}_{(a)pop} = \frac{\sum_{b=1}^{|M|-1} \frac{|I^{x+}_{ab}|+1}{|I^{x+}_{ab}|+|I^{x-}_{ab}|+2}}{|M|-1} \tag{3}$$

Where: *$F^{x}_{(a)pop}$ is the popularity value given to user $\boldsymbol{a} \in \boldsymbol{U}$ on a situation $\boldsymbol{x}$; $|M|$ is the number of members in the community $\boldsymbol{U}$; $|I^{x+}_{ab}|$ and $|I^{x-}_{ab}|$ represent the number of positive feedbacks and negative feedbacks respectively, given by user $\boldsymbol{b} \in \boldsymbol{U}$ on user $\boldsymbol{a}$'s interactions on situation $\boldsymbol{x}$.*

This formalization of popularity value was based on the research by Josang and Ismail [4] where beta probability distribution function was used since a choice (either positive or negative) on a situation or context **x** is uncertain. This evaluation is related to Laplace rule of succession [14] where both success and failure outcomes are considered. This popularity value relies on the Probability expectation value [4], $Prob_{(E)}$, which is evaluated using two parameters, α and β (as shown in Eq. (4)), related to the number of positive feedback and negative feedback received, respectively.

$$Prob_{(E)} = \frac{\alpha}{\alpha + \beta} \tag{4}$$

From Eq. (3),

$$\alpha = |I^{x+}_{ab}| + 1 \tag{5}$$

$$\beta = |I^{x-}_{ab}| + 1 \tag{6}$$

Josang [4] referred to $|I^{x+}_{ab}|$ and $|I^{x-}_{ab}|$ as the degree of satisfaction and the degree of dissatisfaction respectively, whereby $|I^{x+}_{ab}|$ measures the level of success in user **a**'s actions related to situation **x** while $|I^{x-}_{ab}|$ measures the level of failure in user **a**'s actions related to situation **x**.

Substituting α and β into Eq. (4), the probability of the positive feedback to be true in the future, $Prob_{(E^{+}_{ab})}$ will then be:

$$Prob_{(E^{+}_{ab})} = \frac{|I^{x+}_{ab}| + 1}{|I^{x+}_{ab}| + |I^{x-}_{ab}| + 2} \tag{7}$$

From Eq. (3), the popularity value for each individual can then be evaluated as:

$$F^{x}_{(a)pop} = \frac{\sum_{b=1}^{|M|-1} Prob_{(E^{+}_{ab})}}{|M| - 1} \tag{8}$$

This estimated-popularity value using Eq. (3) reveals how well that user is known to the community based on the existing interactions. If users have few interactions with the community members, their familiarity value could be predicted based on their similar members' previous interactions with other members.

Nepal [7] also went further to evaluate engagement value based on both active and passive engagement interactions, whereby the number of responses to a user's message is consider. The engagement value $E^{x}_{(a)C}$ was also evaluated using the beta probability distribution function (as shown below) which is similar to Eq. (3) since it also depends on the positive feedbacks and negative feedbacks for its evaluation.

$$E^{x}_{(a)C} = \frac{\sum_{b=1}^{|M|-1} \frac{|C^{x+}_{ab}|+1}{|C^{x+}_{ab}|+|C^{x-}_{ab}|+2}}{|M| - 1} \tag{9}$$

Where: *$|C^{x+}_{ab}|$ and $|C^{x-}_{ab}|$ represent the number of positive and negative feedback respectively, given by user **a** on user **b**'s interaction on situation **x**; $|M|$ is the number of members in the community.*

This evaluation depends more on the degree of satisfaction because the degree of satisfaction increases the experience level of a user which obeys the theory in [11] that reveals the performance from an event or situation affecting the experience level. The experience level measures how much a user of a community has satisfied other users of the same community after several interactions. If a user **a** sends number of posts or messages to another user **b**, the number of views (passive interactions) or replies (active interactions) to those messages can be considered as the degree of satisfaction for user **b** who viewed or responded to the messages. This degree value is given to user **b** by user **a** who evaluates the positive feedbacks received on its messages. Whereas, the number of messages that were not viewed or had no response can be considered as the degree of dissatisfaction (C^{x-}_{ab} from Eq. (9)).

From both evaluation of Familiarity and experience with Eqs. (3) and (9) respectively, the sincerity of the individuals was not considered, as the truthfulness of their feedbacks were not evaluated.

3 Modelling Trust from a Community

In this section, the proposed trust metric is presented and tested with the data provided in the **Appendix**. If a new user joins a community of other users or has had few interactions with other users in the community, the social trust information is required to assist in deciding on engagement with the user.

3.1 Familiarity-Experience Based Social Trust

Familiarity between individuals will be established when one of the individual believes that the other individual is important in fulfilling a task. Since the familiarity between two individuals can be uncertain, the similarity can reveal the probability of this two individuals to be familiar with each other in the future [1, 15]. The similarity in behaviour can be used in the comparison between a particular individual and another member in a community, by viewing the way they both consider the importance of other members of the community in fulfilling various tasks. Hence, the familiarity between two individuals can be represented with their similarity value.

As earlier discussed, the similarity values between individuals can be evaluated using Eq. (1), where the difference in hierarchy of importance value for each member is considered. For example, given a hierarchy table which can be used to evaluate the similarity between node $\mathbf{V}_1$ and $\mathbf{V}_3$ as shown in Table 1, where the evaluation of the importance value given to each individual by node V_1 and V_3 will be discussed later in this paper.

Using Eq. (1) on Table 1's data, the similarity value between V_1 and V_3 is:

$S(V_1, V_3) = 1 - \frac{H_{(\tau V_1, \tau V_2)}}{N} = 1 - \frac{3}{4} = 0.25$

The familiarity value for an individual $\mathbf{F}_i$ is therefore evaluated as the mean of similarity values $\mathbf{S}_{ij}$ between the individual and other members of the community (as shown in Eq. (10)) as it reveals the central value of the probability distribution for familiarity. This value denotes the possibility of the individual to be contacted in the future by members of the community. It ranges between 0 and 1, where the value 0 denotes that there is the possibility that an individual will have no or few interaction with other individuals in the future while the value 1 denotes that the individual will definitely be familiar with other members in the network.

$$F_i = \frac{\sum_{i=1}^{j=n} S_{ij}}{n} \tag{10}$$

***Where** n = the number of members **j** from the community that have interacted with the individual **i**.*

The experience value for a user can be defined as the likelihood of the user to understand the behaviour of other members in a community. If user **i** has a task to fulfil for another user **j**, the outcome from the roles might either be successful or unsuccessful to a certain degree. The experience value for user **i** can be evaluated using Eq. (9), whereby all user **i**'s engagement outcomes (feedbacks) with other

Table 1 Hierarchy table with importance value for each node

Node	V_1	V_3	V_5	V_{32}
V_1	0	−3	0	0
V_3	3	1	0	11

members of the community are considered in discerning their behavioural pattern. The overall engagement outcome is then referred to as user **i**'s experience value for the whole community with values ranging [0,1], where the value 0 denotes that the user in the community will have impartial views on other members' behaviour while the value 1 denotes that the user will definitely understand other members' behaviour in the community.

As earlier discussed, both familiarity and experience contribute to trust computation, as they exist within a probabilistic metric space.[1] Thus, the trust value of an individual ST_i can be defined as the output from both familiarity value and experience value inputs. As both inputs are in the range [0,1], the output can be modelled with a conjunction operator on function domain $[0, 1] \times [0, 1]$, which can be generalized using a t-norm operator.[2] A T-norm type, product t-norm T_{prod} is chosen here for the trust evaluation, as it is differentiable so it assures smooth transitions in changes on any of the input values.

Using T-norm function $T : [0, 1] \times [0, 1] \rightarrow [0, 1]$ as a product T-norm for all F and E,

$$T_{prod}(F, E) = F \cdot E \tag{11}$$

then

$$ST_i = F_i \cdot E_i \tag{12}$$

$$ST_i = \frac{1}{n^2} \sum_{i=1}^{j=n} F_{ij} \cdot E_{ij} \tag{13}$$

Where: *F_{ij} and E_{ij} are the respective familiarity and experience between user i and j; F_i and E_i are the overall familiarity and experience values respectively; n is the number of interaction between user i and j.*

Another factor which can be consider in the trust computation is the honesty value H_i which is given to an individual by the community. According to [1], honesty factor encourages trust between pairs of individuals, as it is believed that a honest individual will not defame or misrepresent the character of another individual. We can then define honesty factor as the degree of representation to an item's characteristic or preference. For instance, if user **i** sends 5 messages to user **j** but user **j** only responded to or accepted 4 of the messages, then it is believed that 0.2 of the sent messages denote that user **j** must have misrepresented the preference or characteristic of user i. The overall honesty value for an individual (as shown in Table 5) can be determined by considering the number of times different users' preference is misrepresented by the individual based on various patterns of their engagements.

[1]In mathematics field, Probabilistic metric space is the generalization of metric spaces whereby the distances between points are defined by probability distribution.

[2]T-norm is a binary operation that can be used in generalizing the conjunction within probabilistic metric spaces and it satisfies the commutative, monotonic, associative and identity (where the neutral element is 1) properties.

Therefore, an improved social trust value for an individual αST_i which requires the honesty factor can be defined as the dot product of the individual's familiarity, experience and honesty values (as shown below). Both ST_i and αST_i are in the range [0,1] with 0 denoting a no-trust status for the individual while 1 denotes an absolute trust status for the individual.

$$\alpha ST_i = F_i \cdot E_i \cdot H_i = ST_i \cdot H_i \tag{14}$$

3.2 Collection of Data

From the original dataset in [8], 44 samples of interactions between 15 users (as shown in Table 6 in the Appendix) were randomly selected to carry out experiments to observe the behaviour of the users. The interaction between members of the network is the exchange of messages amongst themselves, where the number of messages sent ($\mathbf{V}_3$) by a member ($\mathbf{V}_1$) to another member ($\mathbf{V}_2$) occurred at different times. V_2's number of responses ($\mathbf{V}_4$) to the number of messages previously sent (V_3) by member V_1 can be assumed to be the number of messages from V_2 to V_1 represented in other situations (as shown in Tables 2 and 3), where the initial V_1 (initial sender of messages) becomes V_2 while the initial V_2 (initial receiver of messages) then becomes V_1.

For instance, from a directed network shown in Fig. 2 (extracted from Table 6), a situation when node **1** sends **32** messages to node **3** denotes the interaction where node 1's 32 messages were sent at different times to node 3 while the situation when node **3** sends **35** messages to node **1** denotes the interaction where node 3 provided 35 responses to node 1's messages at different times. These situations can also be viewed in the reverse way where node 3 is initially considered as the sender of the messages and the node 1 is seen as the receiver of the messages. These information on both the number of sent messages and the number of responses to the messages are then represented in Tables 2 and 3. Other individual nodes' interaction data in the network were also obtained in similar pattern as Tables 2 and 3.

Table 2 Node 1's Interaction with other nodes

$\mathbf{V}_1$	$\mathbf{V}_2$	$\mathbf{V}_3$	$\mathbf{V}_4$
1	2	1	0
1	3	32	35
1	30	4	1
1	32	1	1
1	36	12	6
1	44	5	4

Table 3 Node 3's Interaction with other nodes

V_1	V_2	V_3	V_4
3	1	35	32
3	2	7	0
3	9	1	0
3	26	4	0
3	32	19	8
3	36	9	5
3	44	5	1

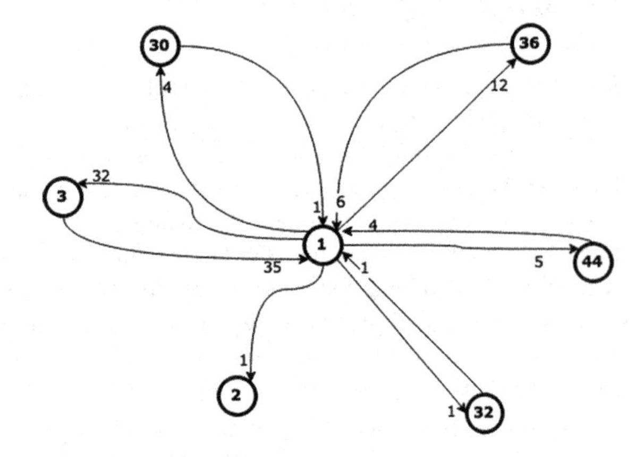

Fig. 2 Directed network for node 1's Interactions

3.3 Analysis with Data

With this data given in Tables 2 and 3 along with other nodes' similar interaction data, both the Familiarity and Experience values for each node were further evaluated using Eqs. (10) and (9) respectively. As earlier discussed, the similarities between pairs of individuals in the network required the importance values for the evaluation of each individual's familiarity value. The importance value $IV_{(ij)}$ for node **i** given by node **j** was determined by calculating the difference between the number of messages sent by node j and the number of messages received by node i.

$$IV_{(ij)} = ms_{(ij)} - res_{(ji)} \tag{15}$$

Where: *$ms_{(ij)}$ is the number of messages sent by node j to node i while $res_{(ji)}$ is the number of messages from node j received by node i.*

The importance value for each individual in the network is then represented in a matrix where the assigners of importance value are the row variables of the matrix while the receivers of importance value are the column variables of the matrix. The number of distinction between the importance values ($H_{(\tau_i,\tau_j)}$) given by i and j which is required for the evaluation of similarity values using Eq. (1) can be determined

Table 4 Trust without honesty factor

Node	F	E	T
1	0.5238	0.5746	0.3009
2	0.6000	0.0000	0.0000
3	0.4238	0.3947	0.1673
5	0.6857	0.3333	0.2285
6	0.6571	0.5000	0.3285
7	0.5762	0.0000	0.0000
8	0.5476	0.5347	0.2928
9	0.6143	0.1899	0.1166
19	0.6143	0.3333	0.2047
26	0.6286	0.3333	0.2095
30	0.6905	0.5555	0.3836
32	0.4809	0.5666	0.2725
36	0.4381	0.3799	0.1664
44	0.5000	0.4147	0.2073
90	0.6667	0.6660	0.4440

from this matrix. Therefore, using Eq. (10), the familiarity values for each individual is evaluated.

Also, with the generated data (Tables 2, 3 and similar relational tables for other nodes' interaction), the honesty values for each nodes were also evaluated (Eq. (16)) as the ratio of the total number of cases with response count (number of responses) being less than number of messages sent, $CN_{res<ms}$ to the total number of interaction N_{int}, in respect to a particular node's (node i) interaction with other members (node j) in a network.

$$H_i = \left(\frac{CN^{ij}_{res<ms}}{N^{ij}_{int}} \right) \tag{16}$$

The social trust value for each individual were evaluated with and without the honesty values using Eqs. (12) and (14) respectively for each individual in the network. The results obtained from these evaluations, generated using the R software environment [9][3] were all represented in separate tables as shown in Tables 4 and 5.

From Fig. 3, the boxplot [12] compares the trust without honesty values and the trust with honesty values. This shows that the honest trust data (Table 5) has the property of a Platykurtic distribution [12], as it has lower excess kurtosis[4] which is in contrast to the other boxplot for non-honest trust data (Table 4) that has a normal

[3]Personalized functions can be created in R environment to operate on data for obtaining accurate results which can be used in further analysis.

[4]A lower excess kurtosis is the peakedness measurement with negative value (lower peak) that is less than zero kurtosis value for a normal distribution.

Table 5 Trust with honesty factor

Node	*F*	*E*	*H*	αT
1	0.5238	0.5746	0.6667	0.2006
2	0.6000	0.0000	0.0000	0.0000
3	0.4238	0.3947	1.0000	0.1673
5	0.6857	0.3333	1.0000	0.2285
6	0.6571	0.5000	0.5000	0.1643
7	0.5761	0.0000	0.0000	0.0000
8	0.5476	0.5347	0.2500	0.0732
9	0.6142	0.1899	1.0000	0.1166
19	0.6142	0.3333	1.0000	0.2047
26	0.6285	0.3333	1.0000	0.2095
30	0.6904	0.5550	0.0000	0.0000
32	0.4809	0.5666	0.4000	0.1089
36	0.4381	0.3799	0.7142	0.1188
44	0.4999	0.4147	0.5000	0.1037
90	0.6667	0.6660	0.0000	0.0000

distribution property. This denotes that the Honest trust data has higher variability in the values than the non-honest trust data, as they are lesser around the mean value. This means that the data with honesty values will enable users to accurately distinguish from any other user they intend to engage with. From Table 5, a node with definite honest ($H = 0$) values for its familiarity and experience will have an impartial trust values (i.e. $\alpha ST = 0$). This situation denote that both the familiarity value of the node given by the members of the network and the experience or inexperience value of the node given to the whole members of the network were accurately represented and assigned to a certain degree, as it can be seen with node 2, 7, 30 and 90

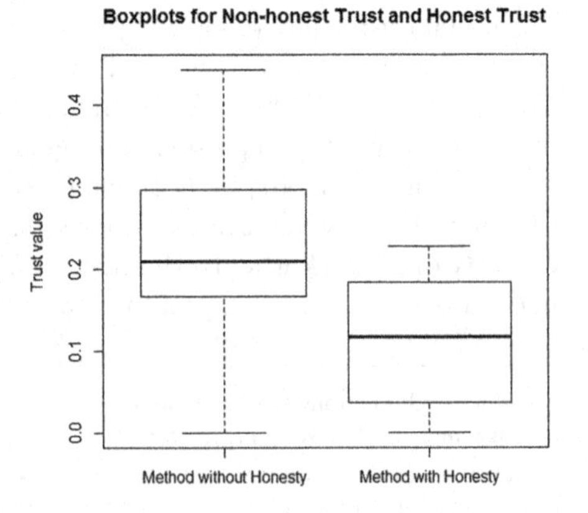

Fig. 3 Boxplots for Trust Data

of the network. This also shows that as these nodes are definitely honest, they will be considered by other nodes for future engagements, irrespective of their trust degree.

Other nodes with certain degrees of dishonesty for both familiarity and experience values also acquired certain degree of trust. The nodes with definite dishonest values will have higher trust values which signifies that these nodes were crafty in their behaviours to earn trust from others.

4 Conclusion

A new trust model based on three factors: familiarity, experience and honesty, was proposed in this paper. These factors which all have degree of uncertainty, are obtained from the interaction that occurs within a community of users. The honesty factor tends to affect the other two factors: familiarity and experience. From the experiments carried out, a better rationale to the inferred trust values were clearly seen from the data with honesty values as compared to that without the honesty value. It was also revealed that honest or accurately assigned values given to individuals based on their importance to roles are likely to be considered for future engagements even though they don't care about the engagement history of other members in the community. Further investigation is required to observe if individuals who were misrepresented or given dishonest values in a community, can influence or motivate other members of the community.

Appendix

Table 6 Sample data for the experiment

V_1	1	1	1	1	1	1	3	3	3	3	3	3	3	5	6	6	8	8	8	8	9	9
V_2	2	3	30	32	36	44	1	2	9	26	32	36	44	2	7	90	7	9	19	32	8	32
V_3	1	32	4	1	12	5	35	7	1	4	19	9	5	1	1	1	1	6	1	8	56	25
V_1	19	26	30	32	32	32	32	32	36	36	36	36	36	36	36	44	44	44	44	44	44	90
V_2	8	7	1	1	3	8	9	26	1	3	6	19	32	44	90	1	3	19	32	36	90	6
V_3	4	1	1	1	8	12	6	1	6	5	5	3	2	2	2	4	1	3	2	1	5	1

References

1. Carter, J., Ghorbani, A.A.: Towards a formalization of value-centric trust in agent societies. Web Intelligence & Agent Systems, pp. 167–183 (1992)
2. Erikson, E.H.: Childhood and Society. Vintage, London (1995)
3. Hardin, R.: Trust. Polity Press, Cambridge (1985)
4. Josang, A., Ismail, R.: The beta reputation system. In: Proceedings of the 15th Bled Economic Commerce Conference, Citeseer, pp. 41–55 (2002)
5. Luhmann, N.(2000) Familiarity, confidence, trust: problems and alternatives. In: Gambetta, D. (ed.) Trust: Making and Breaking Cooperative Relations, vol. 6, pp. 94–107. University of Oxford, Department of Sociology (2000)
6. Marsh, S.P.: Formalising trust as a computational concept. Ph.D. thesis, University of Stirling, www.cs.stir.ac.uk/research/publications/techreps/pdf/TR133.pdf (1994)
7. Nepal, S., Sherchan, W and Paris, C. *Strust: A trust model for social networks*. In Proceedings of the 10th International Conference on Trust, Security and Privacy in Computing and Communications (2011), IEEE, pp. 841–846
8. Opsahl, T.: Datasets. [Online], n.d. Available at http://toreopsahl.com/datasets/
9. R Core Team. *R: A language and environment for statistical computing*. R Foundation for Statistical Computing, Vienna, Austria, 2014
10. Rutter, J.: From the sociology of trust towards a sociology of e-trust. International Journal of New Product & Innovation Management (2001)
11. Söderlund, M. and Gunnarsson, J.: Customer familiarity and its effect on satisfaction and dissatisfaction. SSE/EFI working Paper Series in Business Administration (2000)
12. Stipak, B.: Boxplots. Available at http://web.pdx.edu/~stipakb/download/PA551/boxplot.html, n.d [Accessed 04/08/2015]
13. Wu, J. and Chiclana, F.: A social network analysis trust-consensus based approach to group decision-making problems with interval-valued fuzzy reciprocal preference relations. Knowledge-Based Systems (2014)
14. Zabell, S.: The rule of succession. *Erkenntnis* (1989), vol 31, pp. 283–321. Available at http://www.researchgate.net/publication/226714676
15. Zhang, J., Ghorbani, A.A., Cohen, R.: A familiarity-based trust model for effective selection of sellers in multiagent e-commerce systems. International Journal of Information Security **6**, 333–344 (2007)
16. Towards real-world models of trust: Reliance on received information. [Online], (1998). Available at http://mcwg.org/mcg-mirror/trustdef.htm

AI in Action

Surface Reconstruction from Point Clouds Using a Novel Variational Model

Jinming Duan, Ben Haines, Wil O. C. Ward and Li Bai

Abstract Multi-view reconstruction has been an active research topic in the computer vision community for decades. However, state of the art 3D reconstruction systems have lacked the speed, accuracy, and ease to use properties required by the industry. The work described in this paper is part of the effort to produce a multi-view reconstruction system for a UK company. A novel variational level set method is developed for reconstructing an accurate implicit surface for a set of unorganised points (point cloud). The variational model consists of three energy terms to ensure accurate and smooth surface reconstruction whilst preserving the fine details of the point cloud and increasing speed. The model also completely eliminated the need for reinitialisation associated with the level set method. Implementation details of the variational model using gradient descent optimisation are given, and the roles of its three energy terms are illustrated through numerical experiments. Experiments show that the proposed method outperformed the state of the art surface reconstruction approaches.

1 Introduction

Reconstructing an accurate and smooth 3D surface from a point cloud is a challenging problem as the point cloud consists of unorganised and unconnected points. In general, there are two approaches to represent a surface: explicit or implicit. Explicit representation [1–3] describes the location of points as well as the local geometry of a surface in a explicit manner. It can be accurate, but less robust and less flexible

J. Duan (✉) · B. Haines · W.O.C. Ward · L. Bai
School of Computer Science, University of Nottingham, Nottingham, UK
e-mail: psxjd3@nottingham.ac.uk

B. Haines
e-mail: psxbh@nottingham.ac.uk

W.O.C. Ward
e-mail: psxwowa@nottingham.ac.uk

L. Bai
e-mail: Bai.Li@nottingham.ac.uk

M. Bramer and M. Petridis (eds.), *Research and Development in Intelligent Systems XXXII*, DOI 10.1007/978-3-319-25032-8_9

in handling arbitrary and dynamically changing surface topology. Implicit representation [4–7] usually constructs a surface as an evolving level set function [8]. The approach is topologically flexible and robust in dealing with noisy and non-uniform point clouds [6, 7].

One of the most successful implicit surface reconstruction methods based on level set was proposed by Zhao et al. [5]. The model however has some drawbacks. First, periodical reinitialisation is needed to keep the evolving level set close to a signed distance function to maintain stable surface evolution. This is a tedious and expensive procedure and may even cause the surface to shrink [9]. Second, the model is non-convex, and the reconstruction result is sensitive to the initial condition. In addition, if the point cloud contains fine details or concave regions, the evolving level set surface often gets stuck in a local minimum even if the initial surface is very close to the true surface.

In this paper, a novel variational model is introduced for surface reconstruction from point clouds which overcomes the problems with the existing implicit reconstruction methods. The steps of the reconstruction process are (1) a distance function is derived using a 3D fast sweeping algorithm, which extends the 2D fast sweeping algorithm [10]; (2) the distance function is then used to find an initial surface of the 3D volume enclosed by the point cloud; (3) the proposed variational model is then used to evolve this initial surface. Gradient descent optimisation is then used to find the minimum of the variational model and accurately reconstruct the surface. Experiments show that the proposed method is more accurate than the state of the art methods including the Poisson method [11]. The three reconstruction steps are detailed in the following sections.

2 Preprocessing

2.1 Distance Function for Point Cloud Using Fast Sweeping

Given an unorganised 3D point cloud $\{x_i\}$, its distance function $d(x)$ satisfies the following Eikonal equation

$$|\nabla d(x)| = f(x), x \in \Omega \backslash \{x_i\} \tag{1}$$

where $f(x) = 1$ and $d(x_i) = 0, x \in \{x_i\}$.

To solve partial differential equation (1) computationally in a 3D discrete grid space, the fast sweeping algorithm for 2D problems [10] is extended to 3D in this paper. The discretisation of (1) in 3D using the Godunov upwind difference scheme is as follows, where i, j, k represent the coordinate indexes of the computational grid:

$$\left[(d^n_{i,j,k} - d^n_{xmin})^+\right]^2 + \left[(d^n_{i,j,k} - d^n_{ymin})^+\right]^2 + \left[(d^n_{i,j,k} - d^n_{zmin})^+\right]^2 = f^2_{i,j,k} \tag{2}$$

where $d^n_{xmin} = min(d^n_{i,j+1,k}, d^n_{i,j-1,k})$, $d^n_{ymin} = min(d^n_{i+1,j,k}, d^n_{i-1,j,k})$, and $d^n_{zmin} = min(d^n_{i,j,k+1}, d^n_{i,j,k-1})$. In addition, $x^+ = \begin{cases} x & x > 0 \\ 0 & x \leq 0 \end{cases}$.

To handle boundary conditions, the one-sided upwind difference is used for each of the 6 boundary faces of the grid space. For example, at the left boundary face, one-sided difference along the x direction is computed as follows

$$\left[(d^n_{i,1,k} - d^n_{i,2,k})^+\right]^2 + \left[(d^n_{i,1,k} - d^n_{ymin})^+\right]^2 + \left[(d^n_{i,1,k} - d^n_{zmin})^+\right]^2 = f^2_{i,1,k}$$

d^n_{xmin}, d^n_{ymin} and d^n_{zmin} are then sorted in increasing order and the sorted version is recorded as a_1, a_2 and a_3. So the unique solution to (2) is given as follows:

$$d^{n+1}_{i,j,k} = min(d^n_{i,j,k}, \tilde{d}_{i,j,k}) \tag{3}$$

where $\tilde{d}_{i,j,k}$ is a piecewise function with three components

$$\tilde{d}_{i,j,k} = \begin{cases} \frac{a_1+a_2+a_3+\sqrt{3f^2_{i,j,k}-(a_1-a_2)^2-(a_1-a_3)^2-(a_2-a_3)^2}}{3} \\ \frac{a_1+a_2+\sqrt{2f^2_{i,j,k}-(a_1-a_2)^2}}{2} \\ a_1 + f_{i,j,k} \end{cases}$$

The three components correspond to the following three conditions, respectively

$$f^2_{i,j,k} \geq (a_1 - a_3)^2 + (a_2 - a_3)^2$$

$$(a_1 - a_2)^2 \leq f^2_{i,j,k} < (a_1 - a_3)^2 + (a_2 - a_3)^2$$

$$f^2_{i,j,k} < (a_1 - a_2)^2$$

To solve (3), which is not in analytical form, the fast Gauss-Seidel iteration with alternating sweeping ordering algorithm is employed, and the distance function $d(x)$ in (1) is obtained (Fig. 1).

2.2 The Volume Enclosed by Point Cloud Using the Distance Function

The distance function $d(x)$ is then used to find the volume $u(x)$ enclosed by the point cloud. An annular binary image $I(x)$, shown in the first two images in Fig. 2, can be obtained by thresholding the distance function $d(x)$.

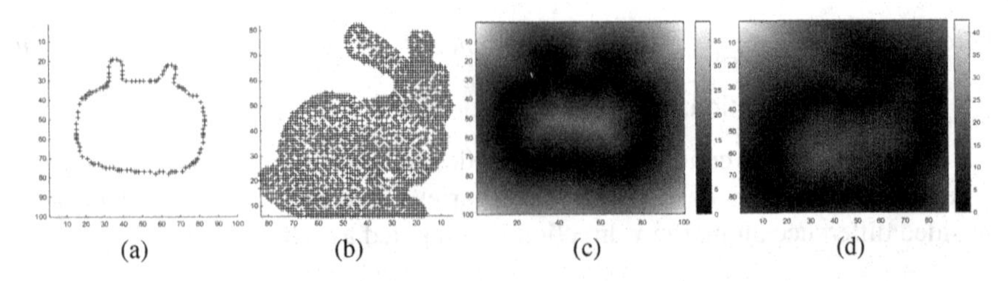

Fig. 1 Calculating the distance function of the original point cloud. **a–b** are 2D and 3D point cloud respectively. **c** is the distance function for (**a**). **d** is a slice of the distance function for (**b**)

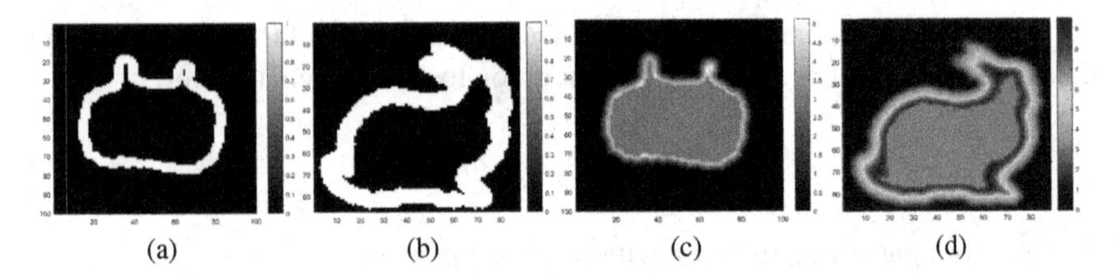

Fig. 2 Finding enclosed volume from annular binary images. **a–b** are two annular binary images. **c–d** are corresponding volumes of (**a**) and (**b**) respectively. (**b**) and (**d**) are slices from the volumes

With the annular binary image $I(x)$, the fast sweeping algorithm is applied to find the volume enclosed by the point cloud. First, $f(x) = I(x)$ replaces $f(x) = 1$ on the right-hand side of (1), rendering $d(x)$ no longer a distance function. Second, in order to solve $d(x)$, the grid points in the 6 boundary faces are set to zero and other grid points are set to very large values for eight directional sweeping Gauss-Seidel iterations. Once the $d(x)$ is found, so is the volume enclosed by the point cloud $u(x) = d(x)$, as shown in the last two images in Fig. 2.

The volume $u(x)$ is then used to find a good initial surface to speed up variational level set evolution in Sect. 3. The volume $u(x)$ is first thresholded to obtain a new binary image $s(x)$ shown in the first two images in Fig. 3. A simple algorithm (such as March Cube) can be employed to find all the points $\{\widetilde{x}_i\}$ on the boundary of the object in the binary image $s(x)$. These new points $\{\widetilde{x}_i\}$ are very close to the original unorganised point cloud $\{x_i\}$. The fast sweeping algorithm is applied again to the

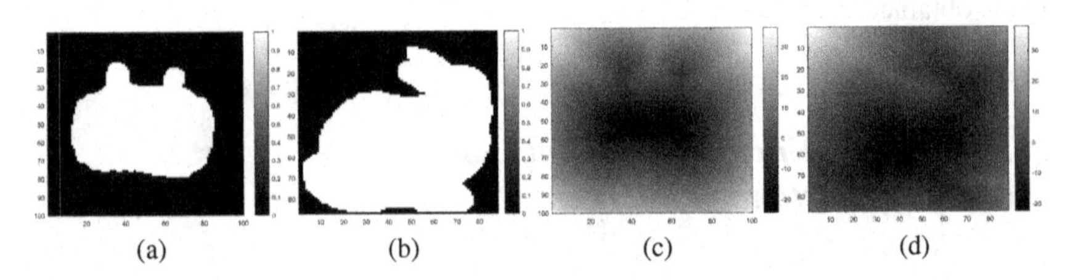

Fig. 3 Calculate a signed distance function from the binary image. **a–b** are two binary images. **c–d** are corresponding signed distance maps of (**a**) and (**b**) respectively. (**b**) and (**d**) stand for one slice of the 3D data

new point cloud $\{\widetilde{x}_i\}$ to obtain a new signed distance function ϕ_0 using the sign information obtained from the binary image $s(x)$ (i.e. inside is negative and outside is positive) as a good initialization for the level set evolution in Sect. 3.

3 The Proposed Variational Model for Optimising Reconstruction

The functional of the proposed variational model consists of three energy terms to optimise the reconstructed surface from point cloud $\{x_i\}$.

$$E(\phi, c_1, c_2) = E_R(\phi) + \lambda E_I(\phi, c_1, c_2) + \beta E_B(\phi) \tag{4}$$

The first term E_R is a regularization term to keep the reconstructed surface smooth while maintaining the level set function ϕ as a signed distance function:

$$E_R(\phi) = \int_\Omega d(x)|\nabla H(\phi)| + \frac{\mu}{2}\int_\Omega (|\nabla\phi| - 1)^2 \tag{5}$$

where $d(x)$ is the distance function calculated from the original point cloud $\{x_i\}$ in Sect. 2.1, and μ is a positive penalty parameter for the deviation of ϕ from a signed distance function. Larger μ leads to more similarity between ϕ and signed distance function. The first term in this functional is equivalent to GAC (geodesic active contour) model [12]. It is the weighted (by $d(x)$) length/area of the boundary using the co-area formula for the TV (total variation) model [13].

The second data fitting term E_I incorporates the information derived from the dataset to ensure that the level set function ϕ remains a signed distance function, thus eliminating the need of reinitialization:

$$E_I(c_1, c_2, \phi) = \int_\Omega Q(c_1, c_2)H(\phi) \tag{6}$$

where c_1 and c_2 represent the mean values inside and outside of the zero level set of ϕ.$Q(c_1, c_2) = (c_1 - u(x))^2 - (c_2 - u(x))^2$, and $u(x)$ is the 2D/3D image computed from Sect. 2.2. This term follows the work of the two-phase piecewise constant Chan-Vese model [14] to maintain regions of similar intensity values. The term can help improve the accuracy of reconstruction by capturing fine features of the object with similar intensity values. The penalty parameter λ on this term in (4) is positive.

The third balloon force term E_B includes the area/volume information inside the zero level set of ϕ to speed up surface evolution as well as segment concave objects:

$$E_B(\phi) = \int_\Omega d(x)H(-\phi) \tag{7}$$

The parameter β for this term can be positive or negative depending on whether inside or outside of the zero level set is defined as positive. In this paper, if the initial boundary is placed outside the object, the coefficient takes a positive values so that the zero level set can shrink during level set evolution. If the initial boundary is placed inside the object, the coefficient takes a negative value to expand the boundary.

The proposed model (4) can be solved by an optimization procedure. First ϕ is fixed to optimise c_1 and c_2 as follows

$$c_1 = \frac{\int_\Omega uH(\phi)}{\int_\Omega H(\phi)};\ c_2 = \frac{\int_\Omega u(1-H(\phi))}{\int_\Omega (1-H(\phi))}$$

Then c_1 and c_2 are fixed using the following gradient descent flow starting with $\phi = \phi_0$ to minimize (4)

$$\frac{\partial \phi}{\partial t} = \left(\nabla \cdot \left(d\frac{\nabla\phi}{|\nabla\phi|}\right) - \lambda Q(c_1, c_2) + \beta d\right)\delta(\phi) + \mu\left(\Delta\phi - \nabla\cdot\left(\frac{\nabla\phi}{|\nabla\phi|}\right)\right) \quad (8)$$

In practice, the Heaviside function $H(\phi)$ and Dirac function $\delta(\phi)$ in (4) and (8) are approximated by their regularized versions with a small positive number ε

$$H_\varepsilon(\phi) = \frac{1}{2} + \frac{1}{\pi} arctan\left(\frac{\phi}{\varepsilon}\right)$$

$$\delta_\varepsilon(\phi) = \frac{1}{\pi}\frac{\varepsilon}{\varepsilon^2 + \phi^2}$$

4 Discretisation

$$\begin{aligned}\nabla\cdot\left(d\frac{\nabla\phi}{|\nabla\phi|_\varepsilon}\right)_{i,j,k} &= d_{i,j+\frac{1}{2},k}\frac{\partial_x^+\phi_{i,j,k}}{\sqrt{\left(\nabla_x^+\phi_{i,j,k}\right)^2 + \left(\nabla_y^0\phi_{i,j+\frac{1}{2},k}\right)^2 + \left(\nabla_z^0\phi_{i,j+\frac{1}{2},k}\right)^2 + \varepsilon^2}}\\ &- d_{i,j-\frac{1}{2},k}\frac{\partial_x^-\phi_{i,j,k}}{\sqrt{\left(\nabla_x^-\phi_{i,j,k}\right)^2 + \left(\nabla_y^0\phi_{i,j-\frac{1}{2},k}\right)^2 + \left(\nabla_z^0\phi_{i,j-\frac{1}{2},k}\right)^2 + \varepsilon^2}}\\ &+ d_{i+\frac{1}{2},j,k}\frac{\partial_y^+\phi_{i,j,k}}{\sqrt{\left(\nabla_y^+\phi_{i,j,k}\right)^2 + \left(\nabla_x^0\phi_{i+\frac{1}{2},j,k}\right)^2 + \left(\nabla_z^0\phi_{i+\frac{1}{2},j,k}\right)^2 + \varepsilon^2}}\end{aligned}$$

$$
\begin{aligned}
&- d_{i-\frac{1}{2},j,k} \frac{\partial_y^- \phi_{i,j,k}}{\sqrt{\left(\nabla_y^- \phi_{i,j,k}\right)^2 + \left(\nabla_x^0 \phi_{i-\frac{1}{2},j,k}\right)^2 + \left(\nabla_z^0 \phi_{i-\frac{1}{2},j,k}\right)^2 + \varepsilon^2}} \\
&+ d_{i,j,k+\frac{1}{2}} \frac{\partial_z^+ \phi_{i,j,k}}{\sqrt{\left(\nabla_z^+ \phi_{i,j,k}\right)^2 + \left(\nabla_x^0 \phi_{i,j,k+\frac{1}{2}}\right)^2 + \left(\nabla_y^0 \phi_{i,j,k+\frac{1}{2}}\right)^2 + \varepsilon^2}} \\
&- d_{i,j,k-\frac{1}{2}} \frac{\partial_z^- \phi_{i,j,k}}{\sqrt{\left(\nabla_z^- \phi_{i,j,k}\right)^2 + \left(\nabla_x^0 \phi_{i,j,k-\frac{1}{2}}\right)^2 + \left(\nabla_y^0 \phi_{i,j,k-\frac{1}{2}}\right)^2 + \varepsilon^2}}
\end{aligned}
\tag{9}
$$

In order to evolve the level set function ϕ in (8), discretisation is needed for $\nabla \cdot \left(d\frac{\nabla\phi}{|\nabla\phi|}\right)$ and $\nabla \cdot \left(\frac{\nabla\phi}{|\nabla\phi|}\right)$ in 3D based on the finite difference scheme. Let $\Omega \to \mathbb{R}^{MNL}$ denote the 3D grid space of size MNL. The second order coupled (with distance function d) curvature term $\nabla \cdot \left(d\frac{\nabla\phi}{|\nabla\phi|}\right)$ at voxel (i,j,k) can be discretised (9), where ε is a small positive number to avoid division by zero. Note that the half-point difference scheme is used here for (9) in order to satisfy rotation-invariant characteristics. The distance function d on half-points between each two voxels (i,j,k) are given as

$$
d_{i,j+\frac{1}{2},k} = \frac{d_{i,j+1,k} + d_{i,j,k}}{2}, \quad d_{i,j-\frac{1}{2},k} = \frac{d_{i,j-1,k} + d_{i,j,k}}{2}
$$

$$
d_{i+\frac{1}{2},j,k} = \frac{d_{i+1,j,k} + d_{i,j,k}}{2}, \quad d_{i-\frac{1}{2},j,k} = \frac{d_{i-1,j,k} + d_{i,j,k}}{2}
$$

$$
d_{i,j,k+\frac{1}{2}} = \frac{d_{i,j,k+1} + d_{i,j,k}}{2}, \quad d_{i,j,k-\frac{1}{2}} = \frac{d_{i,j,k-1} + d_{i,j,k}}{2}
$$

The first order forward ∂_x^+ and backward ∂_x^- discrete derivatives along x, y and z directions on voxel (i,j,k) can be defined as

$$
\partial_x^+ \phi_{i,j,k} = \phi_{i,j+1,k} - \phi_{i,j,k}, \quad \partial_x^- \phi_{i,j,k} = \phi_{i,j,k} - \phi_{i,j-1,k}
$$

$$
\partial_y^+ \phi_{i,j,k} = \phi_{i+1,j,k} - \phi_{i,j,k}, \quad \partial_y^- \phi_{i,j,k} = \phi_{i,j,k} - \phi_{i-1,j,k}
$$

$$
\partial_z^+ \phi_{i,j,k} = \phi_{i,j,k+1} - \phi_{i,j,k}, \quad \partial_z^- \phi_{i,j,k} = \phi_{i,j,k} - \phi_{i,j,k-1}
$$

The central differences are applied to approximate the following discrete first order derivatives on half-points between each two voxels (i,j,k) in (9).

$$\nabla_y^0 \phi_{i,j+\frac{1}{2},k} = \frac{\phi_{i+1,j+1,k} + \phi_{i+1,j,k} - \phi_{i-1,j+1,k} - \phi_{i-1,j,k}}{4}$$

$$\nabla_z^0 \phi_{i,j+\frac{1}{2},k} = \frac{\phi_{i,j+1,k+1} + \phi_{i,j,k+1} - \phi_{i,j+1,k-1} - \phi_{i,j,k-1}}{4}$$

$$\nabla_y^0 \phi_{i,j-\frac{1}{2},k} = \frac{\phi_{i+1,j,k} + \phi_{i+1,j-1,k} - \phi_{i-1,j,k} - \phi_{i-1,j-1,k}}{4}$$

$$\nabla_z^0 \phi_{i,j-\frac{1}{2},k} = \frac{\phi_{i,j,k+1} + \phi_{i,j-1,k+1} - \phi_{i,j,k-1} - \phi_{i,j-1,k-1}}{4}$$

$$\nabla_x^0 \phi_{i+\frac{1}{2},j,k} = \frac{\phi_{i+1,j+1,k} + \phi_{i,j+1,k} - \phi_{i+1,j-1,k} - \phi_{i,j-1,k}}{4}$$

$$\nabla_z^0 \phi_{i+\frac{1}{2},j,k} = \frac{\phi_{i+1,j,k+1} + \phi_{i,j,k+1} - \phi_{i+1,j,k-1} - \phi_{i,j,k-1}}{4}$$

$$\nabla_x^0 \phi_{i-\frac{1}{2},j,k} = \frac{\phi_{i,j+1,k} + \phi_{i-1,j+1,k} - \phi_{i,j-1,k} - \phi_{i-1,j-1,k}}{4}$$

$$\nabla_z^0 \phi_{i-\frac{1}{2},j,k} = \frac{\phi_{i,j,k+1} + \phi_{i-1,j,k+1} - \phi_{i,j,k-1} - \phi_{i-1,j,k-1}}{4}$$

$$\nabla_x^0 \phi_{i,j,k+\frac{1}{2}} = \frac{\phi_{i,j+1,k+1} + \phi_{i,j+1,k} - \phi_{i,j-1,k+1} - \phi_{i,j-1,k}}{4}$$

$$\nabla_y^0 \phi_{i,j,k+\frac{1}{2}} = \frac{\phi_{i+1,j,k+1} + \phi_{i+1,j,k} - \phi_{i-1,j,k+1} - \phi_{i-1,j,k}}{4}$$

$$\nabla_x^0 \phi_{i,j,k-\frac{1}{2}} = \frac{\phi_{i,j+1,k} + \phi_{i,j+1,k-1} - \phi_{i,j-1,k} - \phi_{i,j-1,k-1}}{4}$$

$$\nabla_y^0 \phi_{i,j,k-\frac{1}{2}} = \frac{\phi_{i+1,j,k} + \phi_{i+1,j,k-1} - \phi_{i-1,j,k} - \phi_{i-1,j,k-1}}{4}$$

To discretise the curvature term $\nabla \cdot \left(\frac{\nabla \phi}{|\nabla \phi|} \right)$, we set $d_{i,j+\frac{1}{2},k} = d_{i,j-\frac{1}{2},k} = d_{i+\frac{1}{2},j,k} = d_{i-\frac{1}{2},j,k} = d_{i,j,k+\frac{1}{2}} = d_{i,j,k-\frac{1}{2}} = 1$ in (9) (Fig. 4).

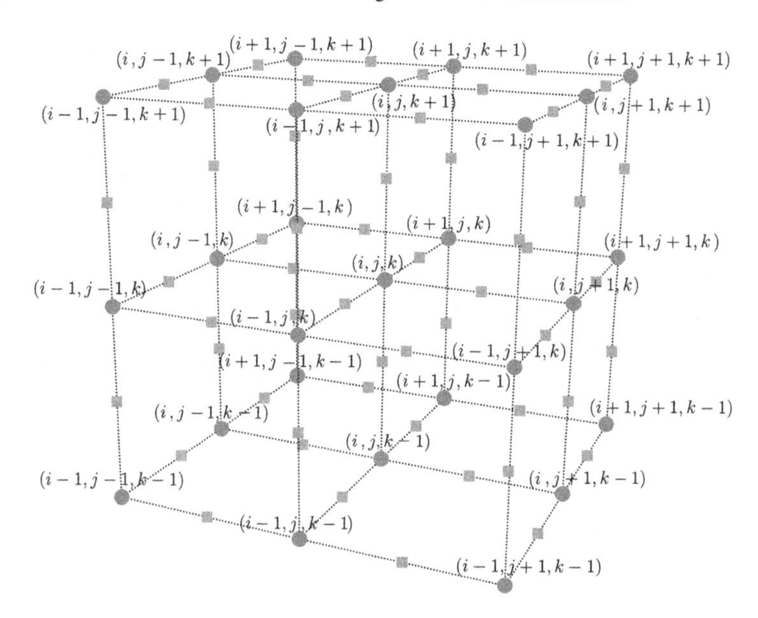

Fig. 4 3D grid space for calculating the discrete differential operators used in Eq. (9). Sphere dots represent voxels. Cube points are half points between two voxels

5 Experiments

Some 2D and 3D reconstruction results using the proposed methods are presented. In Fig. 5, a 2D contour is given of the dataset shown in Fig. 1a. The initial contour obtained in the preprocessing step is very close to the true surface, which can speed up convergent. As the evolution proceeds, the reconstruction by Zhao's method [5] loses the small features of the original data (i.e. the two convex parts). However, the term (6) in the proposed model is able to preserve these features.

Figure 6 shows the reconstructed 3D surface of the Bunny point cloud shown in Fig. 1b. Zhao's method failed to reconstruct the Bunny's ears and feet, while the proposed method succeeded. In addition, we compare our method with Possion reconstruction [11], where the authors show that surface reconstruction from oriented points can be cast as a spatial Poisson problem. They considered the relationship between the gradient of indicator function and an integral of the surface normal field. In detail, the gradient of an indicator function should be zero nearly everywhere except these points on a surface where their gradients are consistent with the inward surface normals. Based on this, Possion reconstruction problem can be simply deemed as calculating indicator function from the gradient field of a surface. As can be seen in this experiment, the reconstruction of Bunny by Poisson loses some texture and looks smoother than that by the proposed method. This experiment demonstrates the effectiveness of the volume data fitting term (6) in the proposed model.

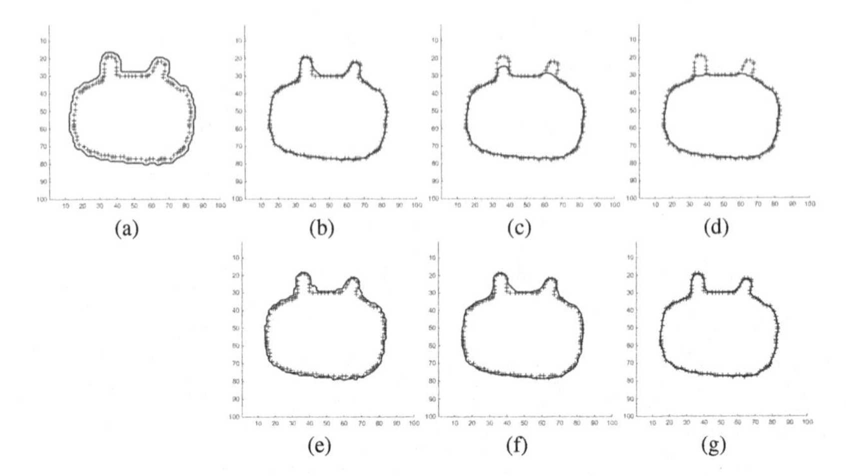

Fig. 5 Comparison of the proposed model (4) with the Zhao's method described in [5]. **a** Same initialisation for both methods; **b–d** Intermediate and final results by Zhao's method; **e–g** Intermediate and final results by the proposed method with $\beta = 0$

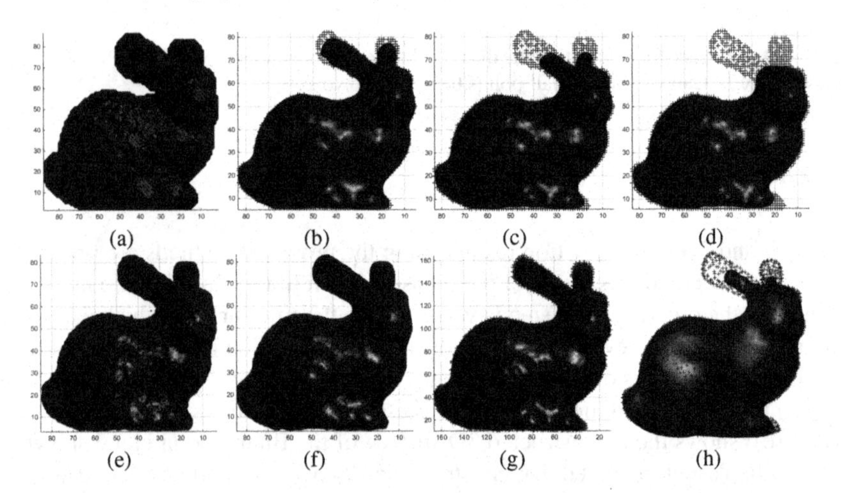

Fig. 6 Comparison with state of the art. **a** Same initialisation for models to be compared; **b–d** Intermediate and final results by Zhao's method; **e–g** Intermediate and final results by the proposed method with $\beta = 0$; **h** Result by Poisson

Figure 7 shows the reconstruction results of a 2D concave object and a 3D human hand that contains fine details (i.e. fingers) and concave regions (i.e. the spaces between fingers). Zhao's method gets stuck in the concave region and also loses the fingers. The proposed model with only the (6) term can preserve the fingers and partially go down the concave region. The proposed model with both terms (6) and (7) succeeds in preserving all features as well as the concave regions in both 2D and 3D cases. This validates the capability of the balloon force term (7) in the proposed model.

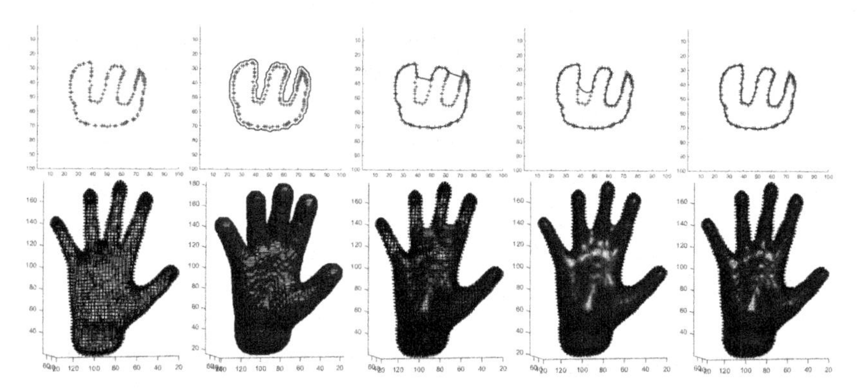

Fig. 7 Effectiveness of the balloon force term (7) in the proposed model using 2D and 3D datasets. *1st column* original data points; *2nd column* initialisation obtained in the first step; *3rd column* results by Zhao's method; *4th column* results by the proposed model without using balloon force term (i.e. $\beta = 0$ in (4)); *5th column* results by the proposed model

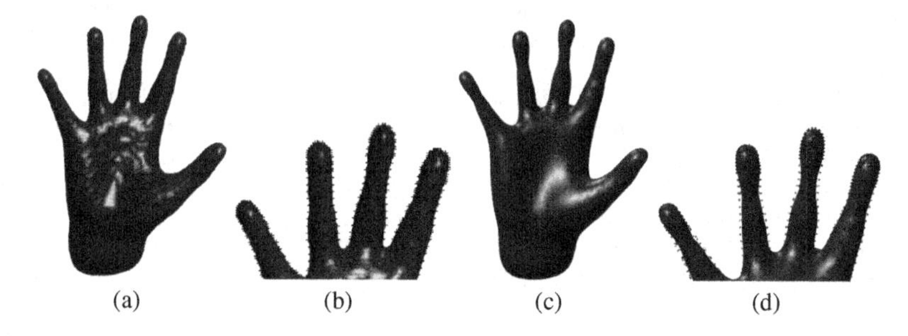

(a) (b) (c) (d)

Fig. 8 Comparison with Poisson. **a** and **b** are reconstructions by the proposed model; **c** and **d** are reconstructions by the Possion method; (**b**) and (**d**) are zoomed-in version of (**a**) and (**c**), with the original data points added

Figure 8 shows that Poisson result is smoother than that of the proposed method, but some fine details/texture, i.e. palm prints, are smeared by the Poisson method. Figure 8b, d show that Poisson result also misses several original data points, and the reconstructed fingers are thinner than those by the proposed method. The proposed method performs better than the Poisson method.

6 Conclusion

In this paper, a novel variational level set method is proposed to reconstruct implicit surfaces from unorganised point clouds. Implementation details of the variational model are given and the functions of its three energy terms are illustrated through numerical experiments. Major advantages of the proposed method over existing approaches include accuracy and detail preservation, without the need for reinitialisation. Experiments show that the proposed model will not get stuck at the local minimum which is commonly associated with implicit reconstruction methods. The components of the proposed variational model are explained and details of the discretisation procedure for implementing the proposed variational model are also given.

References

1. Edelsbrunner, H., Mücke, E.P.: Three-dimensional alpha shapes. ACM Trans. Gr. **13**(1), 43–72 (1994)
2. Amenta, N., Bern, M., Kamvysselis, M.: A new Voronoi-based surface reconstruction algorithm. In: Proceedings of the 25th Annual Conference on Computer Graphics and Interactive Techniques, pp. 415–421 (1998)
3. Boissonnat, J.D.: Geometric structures for three-dimensional shape representation. ACM Trans. Gr. **3**(4), 266–286 (1984)
4. Zhao, H.K., Osher, S., Merriman, B., Kan, M.: Implicit and nonparametric shape reconstruction from unorganized data using a variational level set method. Comput. Vis. Image Underst. **80**(3), 295–314 (2000). Elsevier
5. Zhao, H.K., Osher, S., Fedkiw, R.: Fast surface reconstruction using the level set method. In: Proceedings IEEE Workshop on Variational and Level Set Methods in Computer Vision, 2001, pp. 194–201. IEEE (2001)
6. Ye, J., Bresson, X., Goldtein, T., Osher, S.: A fast variational method for surface reconstruction from sets of scattered points. CAM Report, **10**(01) (2010)
7. Liang, J., Park, F., Zhao, H.K.: Robust and efficient implicit surface reconstruction for point clouds based on convexified image segmentation. J. Sci. Comput. **54**(2–3), 577–602 (2013). Springer
8. Osher, S., Sethian, J.A.: Fronts propagating with curvature-dependent speed: algorithms based on Hamilton-Jacobi formulations. J. Comput. Phys. **79**(1), 12–49 (1988). Elsevier
9. Li, C.M., Xu, C.Y., Gui, C.F., Fox, M.D.: Level set evolution without re-initialization: a new variational formulation. In: IEEE Computer Society Conference on Computer Vision and Pattern Recognition, 2005. CVPR 2005, 1, pp. 430–436. IEEE (2005)
10. Zhao, H.K.: A fast sweeping method for eikonal equations. Math. Comput. **74**(250), 603–627 (2005)
11. Kazhdan, M., Bolitho, M., Hoppe, H.: Poisson surface reconstruction. In: Proceedings of the fourth Eurographics symposium on Geometry processing (2006)
12. Caselles, V., Kimmel, R., Sapiro, G.: Geodesic active contours. Int. J. Comput. Vis. **22**(1), 61–79 (1997). Springer
13. Rudin, L.I., Osher, S., Fatemi, E.: Nonlinear total variation based noise removal algorithms. Physica D: Nonlinear Phenom. **60**(1), 259–268 (1992). Elsevier
14. Chan, T.F., Vese, L.A.: Active contours without edges. IEEE Trans. Image Process. **10**(2), 266–277 (2001). IEEE

3D Spatial Reasoning Using the Clock Model

Joanna Isabelle Olszewska

Abstract A visual, three-dimensional (3D) scene is usually grounded by two-dimensional (2D) views. In order to develop a system able to automatically understand such a 3D scene and to provide high-level specifications of what is this scene, we propose a new computational formalism which allows to perform reasoning simultaneously about the 3D scene and its 2D views. In particular, our approach formalizes both 3D directional relations and 3D far/close spatial relations among objects of interest in the scene. For this purpose, qualitative spatial relations based on the clock model are computed in each of the 2D views capturing the scene and are reconstructed in the 3D space in a semantically meaningful, spherical representation. Our resulting 3D qualitative spatial relations have been successfully tested on real-world dataset and show excellent performance in terms of accurateness and efficiency compatible with real-time applications.

1 Introduction

Automatic understanding and interpretation of three-dimensional (3D) visual scenes is a challenging task which importance has grown with the development of areas such as robotics, involving human-robot interaction via natural language commands [16]; forensics, using crime scene reconstruction [2]; and geographical information systems (GIS), integrating computer graphics and dedicated functions to urban planning [4].

In particular, modelling spatial relations among objects of such visual scenes is a crucial step for these applications. Indeed, the integration of spatial relations into machine vision systems brings an additional level in the task of automatic image understanding, leading to the processing of semantic information besides those provided by visual features.

J.I. Olszewska (✉)
University of Gloucestershire, Cheltenham, UK
e-mail: jolszewska@glos.ac.uk

M. Bramer and M. Petridis (eds.), *Research and Development in Intelligent Systems XXXII*, DOI 10.1007/978-3-319-25032-8_10

Compared to computer vision approaches relying on only quantitative techniques with metric spatial relations (MSR) [6] or with lists of sparse keywords naming scene objects [8], the use of qualitative spatial relations (QSR) allows the action of reasoning on semantically meaningful concepts which is one of the major advantages of QSR [3].

For this purpose, different spatial representations dealing with the most important linguistic spatial concepts [5] related to objects' connectivity, size, and relative directions, respectively, have been introduced in the literature.

Most of the state-of-art spatial relations are two-dimensional (2D), e.g. the topological spatial relations [15] like the RCC-8 model [14], the far/close relations [7], and the directional relations such as left/right relations [17], cardinal relations [18], or clock-based relations [12, 13].

In this paper, we propose a computational formalism that extend the 2D clock model [13] to represent 3D qualitative spatial relations. 3D directional spatial relations are thus obtained based on the 2D semantic clock relations computed in each of the 2D views of the 3D scene. On the other hand, 3D far and close linguistic concepts are extracted from the computation of the related distances among objects in the corresponding 2D views.

The 2D clock-modeled spatial relations mapped into the three dimensions of the scene build then a spherical representation of the 3D qualitative spatial relations. Hence, this approach formalizes 3D qualitative spatial relations in the three dimensions of the space in terms of linguistically meaningful spatial concepts related to clock semantic notions. The resulting high-level specification helps then to set down an abstract description of the 3D scene with respect to the semantic world of clock spatial relations in a way which is coherent between both the two and three dimensions of the scene and which is fully automatic.

In this work, we analyse static scenes, i.e. scenes where objects of interest are not moving. The changes in the scene views are then only due to the variations of viewpoints compared to landmarks.

We implement the 3D QSR relations using Description Logics (DL) [1] which have been widely adopted for knowledge representation in visual systems [9–11].

Thus, our clock-modeled spatial relations define a set of useful, semantic notions to characterize visual 3D scenes involving numerous objects of interest as well as to acquire knowledge about them, and could be incorporated into a complete system for automatic reasoning on 3D spatial relations among objects detected in the 2D views.

The main contribution of this paper is the modeling of 3D qualitative spatial relations based on the clock-face approach.

The paper is structured as follows. In Sect. 2, we present our approach using the clock formalism to model 3D directional spatial relations and 3D far/close semantic relations. These relations have been integrated in a framework for automatic reasoning about visual scenes. The resulting system has been successfully tested on a large dataset as reported and discussed in Sect. 3. Conclusions are drawn up in Sect. 4.

2 Clock-Based 3D Qualitative Spatial Relations

In this work, 3D spatial relations have been defined based on the related 2D clock-modeled, qualitative spatial relations computed for each of the 2D views, namely, top view T, side view S, and front view F of the 3D scene, as displayed in Fig. 1.

Hence, we formalise a clock-based, 3D qualitative spatial relation S as follows:

$$S = R_T \vee R_S \vee R_F, \tag{1}$$

where R_v is a relation R between two objects of interest, namely, the reference object O_{REF} and the relative sought object O_{REL}, present in a 2D view v, $v \in \{T, S, F\}$:

$$R_v = R(O_{REF}\ \ O_{REL})_v, \tag{2}$$

with R a 2D, directional, clock-based spatial relation such as hCK with $h \in \{1, 2, \ldots, 12\}$ and $h \in \mathbb{N}$.

In fact, the clock concept introduced by [13] consists in dividing the image plane in twelve parts around any object of interest in a view. Hence, each portion of the 2D space is then corresponding to an hour. This leads to a semantically meaningful division of the space as a clock face. Figure 1 illustrates that for the different T, S, F views. This concept helps in reducing the uncertainty on the directional relative posi-

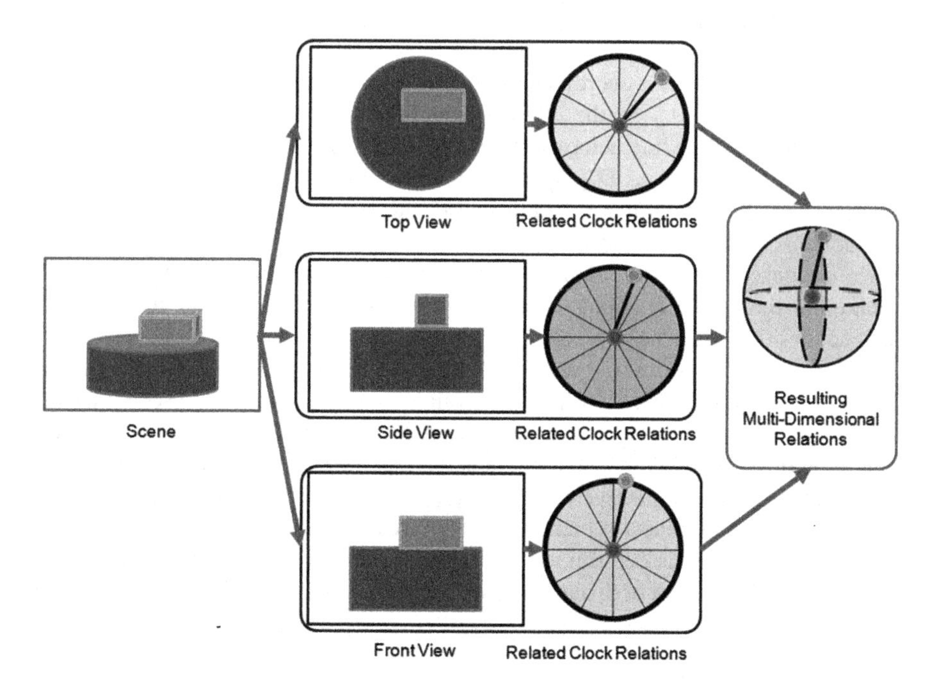

Fig. 1 Illustration of the proposed three-dimensional, directional spatial relations between objects of interest based on the clock model. Best viewed in color

tions between objects in crowded scenes, in which case traditional relations such as *left* or *right* are not enough discriminant as demonstrated in [11].

In this work, the clock notion is used for the formal specification of our 3D relative directional relations, using the 2D JPED relations as follows:

$$\begin{aligned}
1CK &= \{(r_v, \theta_v) | \tfrac{\pi}{6} + 2k\pi < \theta_v \leq \tfrac{\pi}{3} + 2k\pi, k \in \mathbb{N}\},\\
2CK &= \{(r_v, \theta_v) | 2k\pi < \theta_v \leq \tfrac{\pi}{6} + 2k\pi, k \in \mathbb{N}\},\\
3CK &= \{(r_v, \theta_v) | \tfrac{11\pi}{6} + 2k\pi < \theta_v \leq 2k\pi, k \in \mathbb{N}\},\\
4CK &= \{(r_v, \theta_v) | \tfrac{5\pi}{3} + 2k\pi < \theta_v \leq \tfrac{11\pi}{6} + 2k\pi, k \in \mathbb{N}\},\\
5CK &= \{(r_v, \theta_v) | \tfrac{3\pi}{2} + 2k\pi < \theta_v \leq \tfrac{5\pi}{3} + 2k\pi, k \in \mathbb{N}\},\\
6CK &= \{(r_v, \theta_v) | \tfrac{4\pi}{3} + 2k\pi < \theta_v \leq \tfrac{3\pi}{2} + 2k\pi, k \in \mathbb{N}\},\\
7CK &= \{(r_v, \theta_v) | \tfrac{7\pi}{6} + 2k\pi < \theta_v \leq \tfrac{4\pi}{3} + 2k\pi, k \in \mathbb{N}\},\\
8CK &= \{(r_v, \theta_v) | \pi + 2k\pi < \theta_v \leq \tfrac{7\pi}{6} + 2k\pi, k \in \mathbb{N}\},\\
9CK &= \{(r_v, \theta_v) | \tfrac{5\pi}{6} + 2k\pi < \theta_v \leq \pi + 2k\pi, k \in \mathbb{N}\},\\
10CK &= \{(r_v, \theta_v) | \tfrac{2\pi}{3} + 2k\pi < \theta_v \leq \tfrac{5\pi}{6} + 2k\pi, k \in \mathbb{N}\},\\
11CK &= \{(r_v, \theta_v) | \tfrac{\pi}{2} + 2k\pi < \theta_v \leq \tfrac{2\pi}{3} + 2k\pi, k \in \mathbb{N}\},\\
12CK &= \{(r_v, \theta_v) | \tfrac{\pi}{3} + 2k\pi < \theta_v \leq \tfrac{\pi}{2} + 2k\pi, k \in \mathbb{N}\},
\end{aligned} \tag{3}$$

where (r_v, θ_v) are the polar coordinates in a 2D view v, with r_v the radius or distance d_v between O_{REF} and O_{REL} in the 2D view v, and with θ_v the polar angle between the horizontal axis and the line determined by O_{REF} and O_{REL} in the 2D view v.

For example, $1CK(O_{REF}\ O_{REL})_T$ means that the object O_{REL} is *at 1 o'clock* of the reference object O_{REF}, with $h = 1$ in this case and $\frac{\pi}{6} < \theta_T \leq \frac{\pi}{3}$ in the top view T of the studied scene, as shown in Fig. 1.

The proposed 3D formalism allows not only the semantically meaningful interpretation of the spatial relations in both 2D views and in the 3D scene itself, but it also encodes the 3D scene as a coherent spherical model.

Indeed, it gives spherical coordinates (r, ϕ, φ) where r is the radius or the distance between the reference object O_{REF} and the relative one O_{REL} in the 3D scene, while ϕ and φ are the azimuth and elevation angles, respectively. The latter values of the angles could be found via the polar angle in the top view, and the polar angles in the front and side views, respectively, which value is set via the computed r_v radii.

3D farness and closeness semantic concepts are defined intrinsically based on the radius r, which value is determined by its projections in each of the 2D views v, mapping with the r_v radii or distances d_v between O_{REF} and O_{REL} defined as follows:

$$d_v = d(O_{REF}, O_{REL})_v. \tag{4}$$

Algorithm 1 3D Close/Far QSR

```
Given f = 0, c = 0, and v 2D views of the 3D scene
for all i = 1 : v do
    if d_i > Th_i then
        f = f + 1;
    else
        c = c + 1;
    end if
end for
if f ≥ c then
    rl = }far'
else
    rl = }close'
end if
return rl
```

Based on these definitions, the Algorithm 1 automatically computes if an object is close or far from another one in the 3D scene and provides the related semantic value *rl*.

In the Algorithm 1, the threshold values Th_i are set based on the ratio between the width and the height of each of the *i* 2D views of the 3D scene.

3 Experiments and Discussion

In order to evaluate the performance of our representation of the 3D qualitative spatial relations, i.e. the 3D directional clock-based relations and far/close spatial relations, our relations introduced in Sect. 2 have been described next with DL and have been implemented within the STVO ontology [10].

FACT++ has been used as a reasoner for our experiments on reasoning with our 3D relations about a real-world image dataset containing 803 possible spatial relations in between two different objects in top, front, and side 2D views of related 3D scenes. Samples of the datasets and corresponding results have been presented in Figs. 2 and 3.

All the experiments have been run on a computer with Intel Core 2 Duo Pentium T9300, 2.5 GHz, 2Gb RAM, and using Protégé.

The qualitative evaluation of the system is carried out by asking different questions whose answers are boolean. The two main types of possible queries are:

- what are the relation(s) among the given objects O_{REL} and O_{REF}?
- which is/are the object(s) O_{REL} that has the relation *rl* with the given objects O_{REF}?

In the case of the images of Figs. 2 and 3, a sample of the results for questions of type 1 are reported in Tables 1 and 2, respectively.

In the results of the qualitative reasoning on the proposed 3D spatial relations as reported in Tables 1 and 2, we note the excellent concordance between the ground

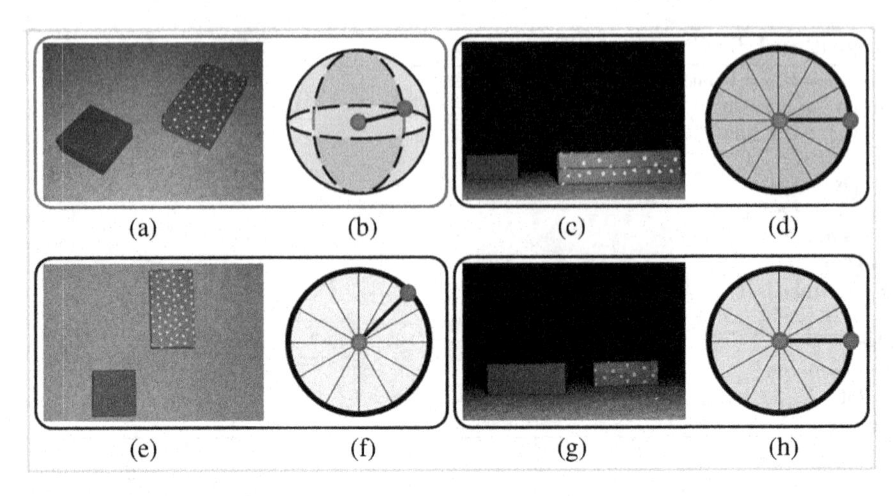

Fig. 2 Example 1 of the multi-dimensional, directional spatial relations of an object of interest (*red one*) related to the reference object (*blue one*) based on the clock model. Best viewed in color. **a** Scene. **b** Multi-dim. relations. **c** Side view. **d** Clock relations. **e** Top view. **f** Clock relations. **g** Front view. **h** Clock relations

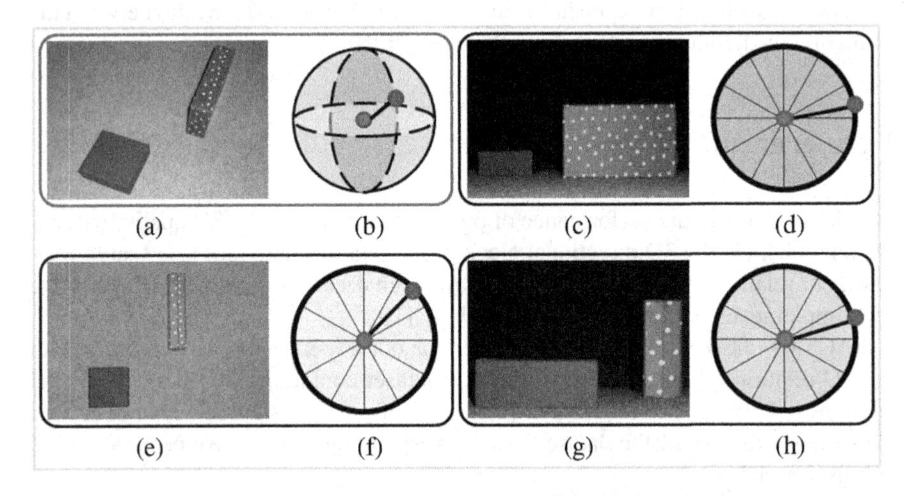

Fig. 3 Example 2 of the multi-dimensional, directional spatial relations of an object of interest (*red one*) related to the reference object (*blue one*) based on the clock model. Best viewed in color. **a** Scene. **b** Multi-dim. relations. **c** Side view. **d** Clock relations. **e** Top view. **f** Clock relations. **g** Front view. **h** Clock relations

truth values set by human users and those computed with our developed system. Quantitative assessment of our relations provides the overall precision of our system tested for type 1 questions on the entire dataset which is of 99.3 ± 0.5 %.

For type 2 questions, it was automatically found that in both cases (Figs. 2 and 3) the object O_{REL} is closed to O_{REF}. These answers were validated compared to ground

Table 1 Sample of results of type 1 questions about scene of Fig. 2

Questions	S	T	F	3D
$1CK(O_{REF}\ O_{REL})_v$	N	Y	N	Y
$2CK(O_{REF}\ O_{REL})_v$	N	N	N	N
$3CK(O_{REF}\ O_{REL})_v$	Y	N	Y	Y

Table 2 Sample of results of type 1 questions about scene of Fig. 3

Questions	S	T	F	3D
$1CK(O_{REF}\ O_{REL})_v$	N	Y	N	Y
$2CK(O_{REF}\ O_{REL})_v$	Y	N	Y	Y
$3CK(O_{REF}\ O_{REL})_v$	N	N	N	N

truth values of the 3D scene. When applied type 2 questions to the entire dataset, the overall accuracy of 3D farness/closeness algorithm was $98.7 \pm 0.5\,\%$.

Hence, the evaluation of the results shows that our clock-based 3D formalism provides an accurate and consistent definition of spatial semantic concepts describing the 3D scene, in real-time.

4 Conclusions

In this paper, we have proposed a new qualitative spatial relation (QSR) formalism which extends the 2D clock-face model to a 3D model, leading to a spherical encoding of the 3D space. The resulting representation allows efficient and accurate reasoning about the 3D scene. The presented method is characterized by providing linguistically meaningful directional spatial relations in all the 3D dimensions of the space as well as the azimuth and elevation angles based on the spatial relations computed in the 2D views grounding the scene. Moreover, the spherical modeling intrinsically links the semantic concepts of farness and closeness to the radius, or distance between the objects in the scene, based on the information extracted from the 2D data. As demonstrated, this conceptualization brings a new insight in the automated analysis of 3D scenes and could be useful for applications embedding navigation systems involving natural language commands such as robotics.

References

1. Baader, F., Calvanese, D., McGuinness, D.L., Nardi, D., Patel-Schneider, P.F.: The Description Logic Handbook: Theory, Implementation and Applications, 2nd edn. Cambridge University Press, Cambridge (2010)
2. Buck, U., Naether, S., Rass, B., Jackowski, C., Thali, M.: Accident or homicide—Virtual crime scene reconstruction using 3D methods. Forensic Sci. Int. **225**, 75–84 (2013)

3. Cohn, A.G., Renz, J.: Qualitative spatial reasoning. Handbook of Knowledge Representation. Elsevier, Amsterdam (2007)
4. Fogliaroni, P., Clementini, E.: Modeling visibility in 3D space: a qualitative frame of reference. In: Proceedings of the International Conference on 3D Geoinformation Science, pp. 243–258 (2015)
5. Freeman, J.: The modeling of spatial relations. Comput. Graph. Image Process. **4**(2), 156–171 (1975)
6. Hijazi, M.H.A., Coenen, F., Zheng, Y.: Classification of retinal images using spatial-histograms and case based reasoning. In: Proceedings of the BCS-SGAI International Conference on Artificial Intelligence (AI) (2010)
7. Kunze, L., Burbridge, C., Alberti, M., Tippur, A., Folkesson, J., Jensfelt, P., Hawes, N.: Combining top-down spatial reasoning and bottom-up object class recognition for scene understanding. In: Proceedings of the IEEE International Conference on Intelligent Robots and Systems (IROS), pp. 2910–2915 (2014)
8. Li, L.J., Socher, R., Fei-Fei, L.: Towards total scene understanding: classification, annotation and segmentation in an automatic framework. In: Proceedings of the IEEE International Conference on Computer Vision and Pattern Recognition (CVPR), pp. 2036–2043 (2009)
9. Neumann, B., Moeller, R.: On scene interpretation with description logics. Image Vis. Comput. **26**(1), 114–126 (2008)
10. Olszewska, J.I.: Spatio-temporal visual ontology. In: Proceedings of the BMVA-EACL Workshop on Vision and Language (2011)
11. Olszewska, J.I.: Multi-target parametric active contours to support ontological domain representation. In: Proceedings of RFIA, pp. 779–784 (2012)
12. Olszewska, J.I.: Clock-modeled ternary spatial relations for visual scene analysis. In: Proceedings of the ACL International Conference on Computational Semantics Workshop, pp. 20–30 (2013)
13. Olszewska, J.I., McCluskey, T.L.: Ontology-coupled active contours for dynamic video scene understanding. In: Proceedings of the IEEE International Conference on Intelligent Engineering Systems, pp. 369–374 (2011)
14. Randell, D.A., Cui, Z., Cohn, A.G.: A spatial logic based on regions and connection. In: Proceedings of the International Conference on Knowledge Representation and Reasoning, pp. 165–176 (1992)
15. Renz, J., Nebel, B.: Spatial reasoning with topological information. Spatial Cognition—An Interdisciplinary Approach to Representation and Processing of Spatial Knowledge. Springer, Berlin (1998)
16. Tan, J., Ju, Z., Liu, H.: Grounding spatial relations in natural language by fuzzy representation for human-robot interaction. In: Proceedings of the IEEE International Conference on Fuzzy Systems (FUZZ), pp. 1743–1750 (2014)
17. Thippur, A., Burbridge, C., Kunze, L., Alberti, M., Folkesson, J., Jensfelt, P., Hawes, N.: A comparison of spatial relation models for scene understanding. In: Proceedings of the AAAI International Conference on Artificial Intelligence (AAAI) (2015)
18. Wallgruen, J.O., Frommberger, L., Wolter, D., Dylla, F., Freksa, C.: Qualitative spatial representation and reasoning in the SparQ-toolbox. In: Proceedings of the International Conference on Spatial Cognition, pp. 39–58 (2007)

Scheduling with Structured Preferences

Roberto Micalizio and Gianluca Torta

Abstract The problem of finding a feasible schedule for a partially ordered set of tasks can be formulated as a Disjunctive Temporal Problem (DTP). While there exist extensions to DTPs that augment them by associating numeric costs to the violation of individual temporal constraints, they still make the restrictive assumption that the costs associated with constraints are independent of one another. In this paper we propose a further extension, which enables the designer to specify (directional) *dependencies* between the preferences associated with the constraints. Such preferences are represented by exploiting Utility Difference Networks (UDNs) that allow for the definition of structured objective functions based on the notion of *conditional difference independence* (CDI). Thanks to such conditional independencies, the specification of the utilities and the computation of the utility of (partial) solutions explored during the search for an optimal solution, turn out to be very similar to how probabilities are handled within a Bayesian Network. The paper presents a branch-and-bound algorithm for solving this new class of problems, analyzes its computational complexity and reports some encouraging experimental results.

1 Introduction

Since the seminal work by Dechter et al. [4], Temporal Constraint Satisfaction Problems (TCSPs) have drawn the attention of several AI researchers, and many problem formulations have been proposed along the time. Notably, the notion of Disjunctive Temporal Problems (DTPs) has been introduced in [17], in order to overcome the limits of Simple Temporal Problems (STPs) [4] by enabling the specification of temporal constraints consisting of disjuncts, each of which represents a temporal interval within which legal solutions can be found. This class of problems is expres-

---

R. Micalizio · G. Torta (✉)
Dipartimento di Informatica, Università di Torino,
C.so Svizzera 185, 10149 Torino, Italy
e-mail: gianluca.torta@unito.it

R. Micalizio
e-mail: roberto.micalizio@unito.it

M. Bramer and M. Petridis (eds.), *Research and Development in Intelligent Systems XXXII*, DOI 10.1007/978-3-319-25032-8_11

sive enough to model scheduling problems [12], as well as other problems of interest in AI (e.g., diagnosis [7, 14]). More recently, the research has been focused on how to address temporal *preferences* (i.e., *soft constraints*). Intuitively, a soft constraint allows one to express preferences on the distance between any two time points. For instance, in a calendar management scenario [10], relevant time points are reasonably the start and end times of the activities to be scheduled. Soft constraints can therefore be used to express the preference that some activities should last as long as possible, or that the distance between the ending of an activity and the starting of the subsequent one should be minimized.

While solving an STP or a DTP usually comes down to verifying the satisfiability of the (hard) constraints specified in the problem, solving a temporal problem with preferences requires to find an assignment of values to the time points that not only satisfies all the hard constraints, but also maximizes a given objective function defined over the soft constraints.

Two main problem formulations taking into account preferences have been proposed in the literature. In the first one, named Disjunctive Temporal Problem with Preferences (DTPP) [5], each constraint is augmented with a function that expresses how well an assignment satisfies the constraint itself. Solving a DTPP requires to find an assignment that maximizes the sum of the preference functions for each involved constraint. In the second formulation, named Valued Disjunctive Temporal Problem [9], each constraint is associated with a value representing the cost "paid" by a solution when that constraint is violated. Thus, in such a case a constraint can actually be violated, but its violation comes at a cost. A solution to a VDTP is therefore a solution whose cost is minimal.[1]

Both formulations, however, assume that the preferences (or costs) associated with the constraints are independent of one another. As a consequence, given a possible solution, its preference value can be computed by a linear function, that aggregates the preference value of each single constraint (i.e., how well the solution satisfies each constraint). Of course, such a function becomes the objective function to be maximized/minimized.

Such an assumption may prove to be too stringent in many applicative domains. Surprisingly, the problem of assessing the preference value of an assignment by taking into account dependencies among constraints has received little attention so far. To the best of our knowledge, only in [10] the authors propose the Multi-Criteria extension to DTPPs (MC-DTPP). Intuitively, the problem formulation includes, besides the disjunctive constraints as usual in DTPPs, also a set of *criteria*. Each criterion is a subset of constraints, which are bundled together as they refer to the same specific feature of the problem at hand. For each criterion (and pair of criteria), the user has to specify a weight denoting how "important" a user considers the satisfaction of that set of constraints.

[1]Note that soft constraints can equivalently be defined in terms of preferences or costs. In this paper we will deal with preferences.

In this paper we propose a different extension, that is suitable to capture a different kind of dependencies. We start by observing that in many practical problems not only there exist dependencies among the preferences, but also that such dependencies are conditional: The best choice for satisfying a constraint might be independent on the choices for the other constraints *given* the choices for a limited set of constraints.

We consider the VDTP formulation as our starting point. To represent causal, directional dependencies, we complement the basic VDTP with a Utility Difference Network (UDN) [1] that allows for the definition of structured objective functions based on the notion of *conditional difference independence* (CDI), after which we name our extended problem formulation *CDI-VDTP*. Thanks to such conditional independencies, the computation of the utility of (partial) solutions explored during the search for an optimal solution turns out to be very similar to how probabilities are computed from a Bayesian network.

The paper is organized as follows. After recalling background information in Sect. 2, we motivate our approach with an example in Sect. 3. In Sect. 4 we formally define CDI-VDTPs, and in Sect. 5 we propose a way to solve them. Section 6 presents experimental results, while Sect. 7 critically discusses related work, before concluding in Sect. 8.

2 Background

2.1 DTPs and VDTPs

A DTP is a pair $\langle \mathbf{X}, C \rangle$ where each element $X_i \in \mathbf{X}$ designates a time point, and each element $C_i \in C$ is a constraint of the form $c_{i,1} \vee \cdots \vee c_{i,n_i}$, and each disjunct $c_{i,j}$ is of the form $a_{i,j} \leq X_{i,j} - X'_{i,j} \leq b_{i,j}$, with $X_{i,j}, X'_{i,j} \in \mathbf{X}$ and $a_{i,j}, b_{i,j} \in \Re$.

A VDTP is a tuple $\langle \mathbf{X}, C, S, \varphi \rangle$ where $\mathbf{X}$, C are as in DTPs, while S and φ are defined as follows. The *valuation structure* S is a tuple $S = \langle E, \circledast, \succ \rangle$ where E is a totally ordered (w.r.t. $\succ$) set of *valuations* that can be combined with $\circledast$, a closed, associative, and commutative binary operator on E. Mapping $\varphi : C \rightarrow E$ assigns a cost $e \in E$ with (the violation of) each constraint $C \in C$. In the *weighted* VDTP, structure S is $\langle \Re^+ \cup \{\infty\}, +, > \rangle$ and the function to optimize is:

$$\sum_i \{\varphi(C_i) | violates(S, C_i)\}.$$

2.2 Utility Difference Networks

Given a set of finite-domain variables $\mathbf{A} = \{A_1, \ldots, A_n\}$ (*attributes*), a *multiattribute utility function* $u(A_1, \ldots, A_n)$ associates a numeric value with each assignment

$\mathbf{a} = a_1 \ldots a_n$ to the attributes. Utility Difference Networks (UDN) [1, 2][2] are a graphical representation of multiattribute utility functions that exhibit strong analogies and properties with the way Bayesian Networks (BN) represent joint probability distributions.

UDNs introduce the notion of a *reference value* $\mathbf{a_i^r}$ for each attribute A_i. The notion of *reference utility function* of a subset of attributes $\mathbf{H} \subseteq \mathbf{A}$ is defined as $u_r(\mathbf{H}) = u(\mathbf{H}\overline{\mathbf{h}}^{\mathbf{r}})$, where $\overline{\mathbf{h}}^{\mathbf{r}}$ is the reference assignment for variables $\overline{\mathbf{H}} = \mathbf{A} \backslash \mathbf{H}$. Based on u_r, the Conditional Independence relation CDI_r and the UDNs are defined then as follows.

Definition 1 [2] Let $\mathbf{H_1}$, $\mathbf{H_2}$, $\mathbf{H_3}$ be subsets of attributes. Set $\mathbf{H_1}$ is said to be Conditionally Independent of $\mathbf{H_2}$ given $\mathbf{H_3}$ (denoted $CDI_r(\mathbf{H_1}, \mathbf{H_2}|\mathbf{H_3})$) if for any assignment $\mathbf{h_3} \in dom(\mathbf{H_3})$, $u_r(\mathbf{H_1}|\mathbf{H_2}\mathbf{h_3}) = u_r(\mathbf{H_1}|\mathbf{h_3})$.

Let $\mathbf{A}$ be a set of attributes. A Utility Difference Network (UDN) is a DAG $\mathcal{G} = (\mathbf{A}, \mathbf{E})$ such that $\forall A \in \mathbf{A} : CDI_r(A, Co(A)|Pa(A))$, where $Pa(A)$ are the parents of A, $Dn(A)$ are the descendants of A, and $Co(A) = \mathbf{A} \backslash (\{A\} \cup Pa(A) \cup Dn(A))$.

UDNs decompose a multiattribute utility function into a sum as BNs decompose a joint probability distribution into a product, namely:

$$u(\mathbf{A}) = \sum_{i=1}^{n} u_r(A_i|Pa(A_i))$$

namely, in order to compute the utility of an assignment $\mathbf{a}$ to the attributes, it is sufficient to sum the values of the reference utility functions of each family of the UDN. A table specifying the values of $u_r(A_i|Pa(A_i))$ is named Conditional Utility Table (CUT).

3 Motivating Example

Let us consider a simplified planetary rover scenario as the one discussed in [3], and let us assume that a mission designer is finalizing the mission that a rover has to carry out. The mission plan has already been outlined, and Fig. 1 shows a portion of interest; edges between actions represent precedence links. The basic idea is that, once the rover has collected a soil sample by means of the `DRILL` action, it analyzes the sample and moves (`DRIVE`) to a position suitable for uploading (`COMM`) the collected data. The analysis and the movement could in principle be carried on simultaneously. The designer has to decide the mode with which the activities in the plan segment have to be completed. Such a decision has to be made balancing the quality and accuracy with which some activities are performed, against the time

[2]Utility Difference Networks are called Marginal Utility Networks (MUT) in [2]. In this paper we shall stick to the former name.

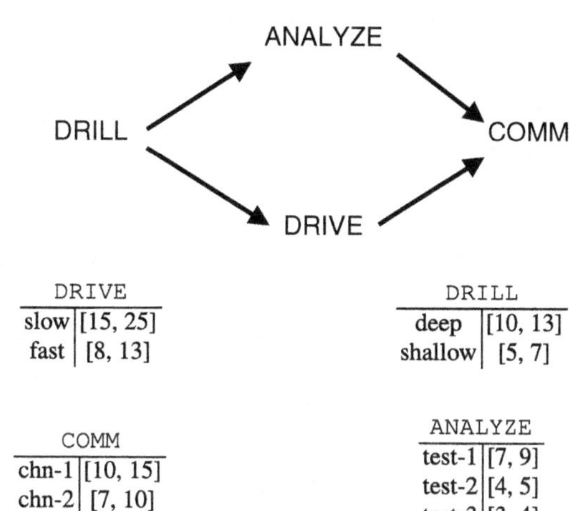

Fig. 1 A segment of a rover plan

Fig. 2 Modes with which activities can be completed and their expected interval durations

DRIVE	
slow	[15, 25]
fast	[8, 13]

DRILL	
deep	[10, 13]
shallow	[5, 7]

COMM	
chn-1	[10, 15]
chn-2	[7, 10]

ANALYZE	
test-1	[7, 9]
test-2	[4, 5]
test-3	[3, 4]

these activities take to be successfully completed. Figure 2 reports, for each action in the plan, the set of action modes and the associated duration intervals. Further inputs for the designer's decision making process are, however, global hard constraints and preferences. The designer has in fact to take into account that the action COMM, must be performed within a communication window, which opens over a precise period. The communication window is a hard constraint since it depends on the position of a satellite functioning as relay, and hence it is outside the control of the mission designer. Moreover, some activity modes are usually more preferred than others. For instance, it is usually preferred, and wiser, to perform a drive action in a slow mode; however, the fast mode can be used, if necessary, to avoid missing the communication window. In the tables of Fig. 2, the modes of each action, considered individually, are ordered from the most preferred down to the least preferred.

The challenge for the designer who has to select a mode for each action arises when we consider actions as being part of a mission. In such a case, the preferred mode for an action might depend on the mode already selected for a previous action. For instance, a scientist would prefer to always drill with modality *deep*, because such a mode usually enables the collection of more interesting samples. On the other hand, when such samples are collected, it is preferable to analyze them with modality *test-1* which is the most accurate one. Both modes, however are very time consuming, moreover the amount of data produced by means of *test-1* mode is usually huge; this impacts the communication, since in that case the 2-channel mode *chn-2* would be preferable, even though the general preference is to use *chn-1* mode.

The problem above could actually be encoded as a VDTP, but the only preferences one could specify would be those informally expressed by the order of the action modes within the tables in Fig. 2. Solving such a problem, thus, would lead to a solution that does not take into account the choices already made. In the following section we first introduce the CDI-VDTP formulation, and then we show how this rover example can be modeled as a CDI-VDTP.

4 Generalizing VDTPs to CDI-VDTPs

A CDI-VDTP is an extension to VDTPs in which the evaluation structure S and mapping φ are substituted by a Utility Difference Network $\mathcal{G}$, and a utility function u over $\mathcal{G}$.

More formally, a CDI-VDTP is a tuple $\langle X, C, \mathcal{G}, u \rangle$, where X and C are as in a standard VDTP; whereas, $\mathcal{G} = \langle \mathbf{A}, \mathbf{E} \rangle$ is a directed acyclic graph representing a Utility Difference Network such that:

- $\mathbf{A}$ is the set of network nodes (attributes). For each constraint $C_i \in C$, there is an attribute $A_i \in \mathbf{A}$ s.t. $dom(A_i)$ consists of the set $\{c_{i,1}, \ldots, c_{i,n_i}\}$ of disjuncts in C_i;
- $\mathbf{E}$ is a set of oriented edges $\langle A, A' \rangle$ such that $A, A' \in \mathbf{A}$. The edges in $\mathbf{E}$ describe the dependencies among the attributes over which one is interested in finding an assignment that maximizes the utility u. For instance, the edge $\langle A_i, A_j \rangle$, means that the selection of a value for A_i (disjunct for constraint C_i) (possibly) affects the utility of the value selection for A_j (i.e., disjunct for C_j) for maximizing the global utility.

Thanks to the properties of UDNs, the utility function u is compactly represented as a set of reference utility functions $u_r(A|Pa(A))$ for each $A \in \mathbf{A}$. In the following, we shall need to compute the maximum utility achievable given an instance $\mathbf{h}$ of a subset $\mathbf{H} \subseteq \mathbf{A}$ of variables. In analogy with the Most Probable Explanation (MPE) for Bayesian Networks, we define the Most Preferred Completion (MPC) of an instance $\mathbf{h}$ as:

$$MPC(\mathbf{h}) = \arg\max_{\overline{\mathbf{h}}}(u(\mathbf{h}, \overline{\mathbf{h}})).$$

Namely, $MPC(\mathbf{h})$ is the instance $\overline{\mathbf{h}}$ that completes $\mathbf{h}$ and yields a maximal utility.

Example 1 Let us consider again the planetary rover scenario, and see how it can be encoded in terms of a CDI-VDTP. The set of temporal variables X consists of a pair of variables for each action in the plan denoting the start and end time of the action itself. For instance, given action DRILL, two variables drl_s and drl_e are included in X. Also the communication window is encoded by means of two variables, cw_s and cw_e. In addition, a variable z is used to encode the time point used as a reference. As for the set C of constraints, we have a soft constraint for each action in the plan, for instance the DRILL action is associated with the following constraint:

$$C_{drl} = \{[10 \leq drl_e - drl_s \leq 13] \vee [5 \leq drl_e - drl_s \leq 7]\}$$

To model the preference value associated with such a constraint, however, we have to consider the dependencies of the constraint. In particular, we can assume that DRILL does not depend on any previous action, but it does influence ANALYZE, which in turn influences COMM. On the other hand, DRIVE can be considered as independent of the other actions. Relying on these observations, in Fig. 3 we sketch the UDN for this problem: Each node corresponds to a constraint in X (including the hard

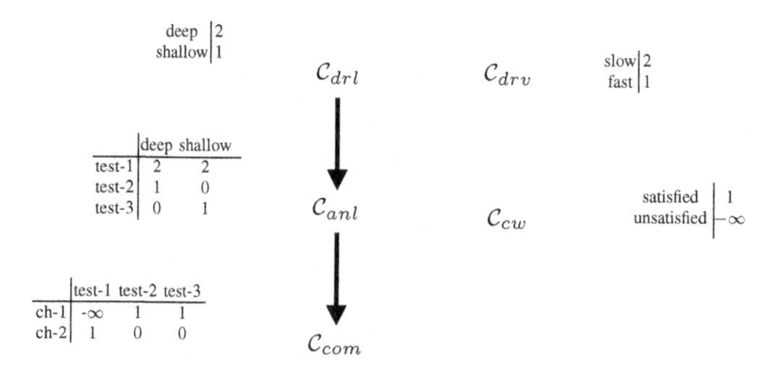

Fig. 3 The utility difference network for the rover example

constraint on the communication window); edges between nodes denote preference dependencies; in addition, in analogy to a Bayesian network, each node is associated with a CUT that defines the preferences for a constraint given its parent nodes.

In this particular case, the utility network has three roots. Two roots are C_{drv} and C_{cw}, representing the constraints associated with the drive action and the communication window, respectively. Being roots, a utility value is directly assigned to each of their disjuncts. For instance, the utility table associated with C_{drv} states that *slow* is generally preferred to *fast*. In addition, since the constraint about the communication window is hard, it is associated with two "fake modes", *satisfied* and *unsatisfied*, this last mode has utility $-\infty$, meaning that any solution that violates the communication window constraint is not acceptable. Note also that these two nodes have no relationships with the other nodes in the network. The third root is C_{drl}, which influences the constraint C_{anl} associated with the analysis action. In this case, the utility associated with each disjunct in C_{anl} depends on the disjuncts that have been selected for its parents (only C_{drl} in this example). The result, thus, is a CUT which looks like a Conditional Probability Table in Bayesian network. The particular table in the figure is to be interpreted as follows; independently of how deep the drill operation is, there is a strong preference in performing *test-1*; however, if the *test-1* is not possible, *test-2* should be preferred when the drill action was *deep*, whereas *test-3* should be preferred when the drill was *shallow*. Similarly, C_{anl} affects C_{com} (i.e., the constraint associated with the communication). Note, in this case, that when the analysis was carried out with mode *test-1*, the usage of mode *ch-1* is practically forbidden. On the other hand, the usage of *ch-1* should be preferred when the analysis was conducted either with *test-2* or *test-3* mode.

It is worth noting that at this stage of development, we assume that the utility values indicated in these tables result from information provided by the problem designer, who takes into account features of the rover that are not explicitly addressed by the temporal problem. For example, the preference on a slow drive could be motivated by safety reasons; whereas the preference of the usage of *ch-1* to *ch-2* could depend on the fact that the second mode is more resource consuming.

5 Solving CDI-VDTPs

Search Process. To solve a CDI-VDTP problem we adopt a strategy similar to the one proposed in [9]. The strategy recursively proceeds in a depth-first manner, and branches are pruned whenever their utility is guaranteed to fall below the cost of the best (i.e., maximal) solution found so far.

Our search strategy is outlined in the algorithm in Fig. 4. The algorithm takes as inputs:

- $\mathbf{h}$: a (partial) assignment of modes to a subset of attributes $\mathbf{H}$, i.e., a (partial) hypothesis;
- $\mathbf{mpc}$: the Most Preferred Completion of $\mathbf{h}$;
- $\overline{\mathbf{H}} = \mathbf{A} \backslash \mathbf{H}$ is the set of attributes whose mode has not been assigned yet;
- lwb: the utility of the best solution found so far;
- Δ: the set of all the best solutions found so far.

It is worth noticing that, while the first three arguments are passed by value, the last two arguments are passed by reference. Thereby, any change made during an invocation of **solve-CDI-VDTP** impacts all instances of the algorithm possibly

solve-CDI-VDTP$(\mathbf{h}, \mathbf{mpc}, \overline{\mathbf{H}}, lwb, \Delta)$
1. $util \leftarrow u(\mathbf{h}, \mathbf{mpc})$
2. **if** $util < lwb$ **then**
3.   **return**
4. **end if**
5. **if** $\overline{\mathbf{H}} = \emptyset$ **then**
6.   **if** $util > lwb$ **then**
7. $\Delta \leftarrow \emptyset$
8. $lwb \leftarrow util$
9.   **end if**
10. $\Delta \leftarrow \Delta \cup \{\mathbf{h}\}$
11.   **return**
12. **end if**
13. $A_i \leftarrow$ **select-attribute**$(\overline{\mathbf{H}})$;
14. $\overline{\mathbf{H}}' \leftarrow \overline{\mathbf{H}} - \{A_i\}$
15. $modes \leftarrow dom(A_i)$
16. **while** $modes \neq \emptyset$ **do**
17. $m \leftarrow$ **select-mode**$(modes)$; $modes \leftarrow modes \backslash \{m\}$
18. $\mathbf{h}' \leftarrow \mathbf{h} \cup \{A_i \leftarrow m\}$
19.   **if consistent**$(\mathbf{h}')$ **then**
20.   **solve-CDI-VDTP**$(\mathbf{h}', MPC(\mathbf{h}'), \overline{\mathbf{H}}', lwb, \Delta)$
21.   **end if**
22. **end while**

Fig. 4 The solve-CDI-VDTP algorithm

active on the stack. In particular, when the search terminates Δ contains the set of best solutions and *lwb* their utility.

At each invocation, the algorithm determines the upper bound of the utility achievable by completing the current (partial) solution **h** (line 1), and checks whether it is lower than the best one so far (line 2); if yes, such a branch is not useful so it is pruned with the return statement. Otherwise, the algorithm checks whether there are still variables to be assigned (line 5): if $\overline{\mathbf{H}}$ is empty, then all attributes have been assigned and **h** is a complete solution. At this stage, the algorithm checks whether the new complete solution is better than any other solution found so far (lines 6–9); in the positive case, *lwb* is updated to be the utility of **h**, and Δ is emptied as all the solutions found so far were not optimal. In any case, **h** is added to Δ (line 10).

In case **h** is still a partial solution, the algorithm tries to get closer to a solution by selecting an attribute A_i from $\overline{\mathbf{H}}$ (line 14). Then the algorithm considers each mode m in $dom(A_i)$ (lines 16–22), in the order determined by function *select-mode* (line 17), and generates new hypotheses from them. In particular, for each $m \in dom(A_i)$, a new hypothesis $\mathbf{h}'$ is obtained by adding the assignment $A_i \leftarrow m$ to **h**. The temporal consistency of the new hypothesis $\mathbf{h}'$ is then verified by means of function *consistent* (line 19), that performs an STP consistency check. Finally, function *solve-CDI-VDTP* is recursively invoked over the new hypothesis $\mathbf{h}'$ and the new set of unassigned variables $\overline{\mathbf{H}}'$ (line 20).

The choice of the next attribute/mode to assign (calls to *select-attribute* and *select-mode*) can benefit from the heuristics established for DTP solving [19], such as conflict-directed backjumping, removal of subsumed variables, semantic branching, and no-good recording. However, in addition to such standard techniques, the choice of the next mode m to try for an attribute A_i can be determined by exploiting **mpc**. In particular, if $\mathbf{mpc} = MPC(\mathbf{h})$ assigns mode m_{mpc} to attribute A_i which is chosen next, that should be the first mode to try for A_i, since it maximizes the utility according to the UDN. Note that, in general, given a hypothesis **h** there may be several completions that maximize the utility, that may assign different modes to A_i. If the MPC computation is able to return all of them, the calls to *select-mode* should return them before the other modes of A_i.

MPC Computation. As pointed out in [2], one of the most desirable characteristics of UDNs is that most inference algorithms developed for BNs can be adapted with small changes to perform useful inferences on UDNs.

In particular, the computation of the MPC of a hypothesis **h** can be performed by adapting algorithms for computing the MPE of some evidence in a BN. We have chosen to use the jointree algorithm (see, e.g., [16]), which is particularly well-suited to the reuse of partial results for the incremental computation of the MPC of a new hypothesis $\mathbf{h}'$.

A jointree $\mathcal{T}$ derived from a UDN $\mathcal{G} = \langle \mathbf{A}, \mathbf{E} \rangle$ is an undirected tree whose nodes (clusters) are subsets $Cl_i \subseteq \mathbf{A}$ s.t. each family $Fam(A_j) = \{A_j\} \cup Pa(A_j)$ $(A_j \in \mathbf{A})$ is associated with a cluster Cl_i that contains $Fam(A_j)$. The computation of MPC follows the same steps of the classic jointree algorithm for BNs, except that the products of probabilities are replaced by sums of reference utilities, and sums of probabilities are replaced by the computation of the max of reference utilities. For example, the *potential* of a cluster Cl_j is:

$$\phi_i = \sum_{Fam(A_j) \subseteq Cl_i} u_r(A_j|Pa(A_j))$$

instead of being the product of the CPTs contained in Cl_i.

After arbitrarily choosing a root, the jointree algorithm consists of an inward and an outward message passing phase, where messages flow respectively from the leaves to the root and vice-versa. In particular, during the inward phase, node Cl_i sends to its parent node Cl_j a message $\mathcal{M}_{i,j}$:

$$\mathcal{M}_{i,j} = \max_{Cl_i \setminus S_{i,j}} \left(\phi_i + \sum_{k \neq j} \mathcal{M}_{k,i} \right) \tag{1}$$

where $S_{i,j} = Cl_i \cap Cl_j$. Assume that message $\mathcal{M}_{i,j}$ has been cached during the computation of $MPC(\mathbf{h})$, and it turns out that it does not change during the computation of $MPC(\mathbf{h}')$, where $\mathbf{h}'$ is derived from $\mathbf{h}$ by adding an attribute assignment. Then, node Cl_i can avoid sending a message to node Cl_j. In turn, if node Cl_j does not receive messages from its children and has an unchanged potential, it can avoid the computation of the message for its parent.

It is easy to see that the replacement of $\sum$ with max in the UDN computations greatly increases the chance that messages can be reused. Indeed, the max operator can "absorb" changes in one or more items leaving $\mathcal{M}_{i,j}$ unchanged.

Computational Complexity. Due to space reasons, we just give some insights about the complexity of the proposed **solve-CDI-VDTP** algorithm; a more detailed analysis of a similar algorithmic approach applied to multi-agent diagnosis can be found in [8]. First of all, we note that the algorithm adopts a recursive strategy for exploring the search space, whose size is bounded by the size of the largest attribute domain, let say D_{max}, and by the number of attributes $|A|$, namely by the upper bound $D_{max}^{|A|}$ (note, however, that the exploration of the whole search space is very unlikely to occur, since this would require that function **consistent** never prunes the domains of the attributes). The two main sub-functions of **solve-CDI-VDTP**, namely **consistent** and *MPC*, can in principle hide further significant computational cost. It is possible to show that the former is polynomial in $|A|$ as the consistency check can be reduced to a number of invocations of checks over a Simple Temporal Network (STN) proportional to $|A|$ (see [8]). On the other hand, *MPC* is more complex, as we have seen, since its computation mirrors the computation of the MPE in Bayesian networks, which, as pointed out by Park and Darwiche [13], can be computed in space and time complexity exponential in the width of a given order of the BN nodes, and such a width is itself $O(|A|)$. In our implementation, we used a jointree algorithm, which in the worst case has complexity $O(D_{max}^{|A|})$.

The computational complexity of the **solve-CDI-VDTP** algorithm is given by multiplying the size of the search space by the complexity of *MPC*, and is therefore exponential in the number of attributes $|A|$: $O({D_{max}}^{|A|} \cdot {D_{max}}^{|A|}) = O({D_{max}}^{2|A|})$.

6 Experimental Results

We have implemented the approach described in this paper as a Perl 5.16 program, exploiting the Boost::Graph module for representing STNs and checking their consistency with the Johnson algorithm, and the Graph module for representing the UDNs. Since the paper presents a new problem (namely, the CDI-VDTP), it is not possible to compare our prototype implementation with previous approaches. Therefore we shall focus on the feasibility of the approach and on the effectiveness of the caching technique discussed above.

The tests have been run on a virtual machine running Linux Ubuntu 12.04, equipped with an i7 M640 CPU at 2.80 GHz, and 4 GB RAM. We have considered three test sets *TS1*, *TS2* and *TS3* of increasing scale, each containing 25 cases. Table 1 reports the following characteristics:

- number *#vars* of variables and number *#constrs* of temporal constraints; *#constrs* is given as the sum of the number of *domain* constraints, shared by all test cases, and the number of constraints that change for each case;
- *#edges* of the UDN describing the dependencies among constraints.

Note that, for all test sets, the UDN networks are non-trivial, since they contain several dependencies among constraint preferences (represented by edges).

In order to appreciate the effectiveness of caching in the jointree algorithm (Sect. 5), we have run the test cases both with and without caching. Table 2 shows the average of the following statistics for the three test sets:

Table 1 Number of constraints, and size of UDNs

	TS1	TS2	TS3
#vars	84	164	244
#constrs	142 + 7	282 + 14	422 + 21
UDN #edges	29	59	89

Table 2 Avg time per sol (sec), and number of sols

Cache	TS1		TS2		TS3	
	yes	no	yes	no	yes	no
time/sol	2.6	6.1	8.8	23.3	15.4	40.9
#sols	3		3.8		3.7	

- *time/sol*: time to compute a solution;
- *#sols*: number of preferred solutions found.

Note that caching reduces the time needed for finding a solution by about 66 % for *TS1*, and about 62 % for *TS2* and *TS3*.

7 Related Work

Since the first formulation of the DTP with preferences (DTPPs) presented in [17], many alternative algorithms and techniques have been discussed in order to efficiently solve the problem. A first class of solutions are based on a semi-ring structure [5], which is used for combining local preference values into a global preference, and for ordering such global preferences so as to compare alternative solutions. Other approaches, such as MAXILITIS [11] are based on SAT algorithms, and ARIO [15] in particular is based on SAT algorithms designed for solving a given DTPP encoded as a Mixed Logical Linear Program (MLLP).

A different formulation of the disjunctive temporal problem with preferences is proposed in [9]. The novel formulation, dubbed Valued Disjunctive Temporal Problems (VDTPs), differs from DTTPs as it associates a single weight to each constraint as a whole, rather than a preference function at the object level as in a DTPP. Such a weight has to be interpreted as a cost a solution gathers when that specific constraint is violated; namely, when the solution does not satisfy any of the disjuncts mentioned in the constraint. In [9], VDTPs are solved by means of a branch-and-bound algorithm exploiting a meta-CSP representation of the temporal problem. In particular, each disjunctive constraint of the temporal problem is associated with a variable of the meta-CSP whose domain corresponds to the set of disjuncts in the constraint itself. The formulation of the CDI-VDTP presented in this paper takes a similar approach in formulating a meta-CSP. Also in CDI-VDTP, in fact, each constraint in the original temporal problem is mapped into a corresponding variable in the meta-CSP; the domain of such a meta-variable coincides with the set of disjuncts mentioned by the constraint itself. A significant difference, however, is that we do not associate a cost to the violation of a constraint as a whole, rather we associate a preference value to each of the disjunct of the constraint (i.e., to each value in the domains of meta-variables).

The approaches and formulations mentioned so far, however, all assume that the preference evaluation of a constraint is independent of the assignments made for the satisfaction of the other constraints. To the best of our knowledge, only the Multi-Criteria approach to DTPPs (MC-DTPPs) [10] takes up the challenge of finding optimal solutions in which the preference value of a constraint does depend on how other constraints are actually satisfied by a given solution. More precisely, in a MC-DTPP, one can define a criterion as a set of constraints; the rationale is that all the constraints related to some particular feature of the problem at hand should be collected within a single criterion. Each criterion is therefore associated with a weight, denoting the importance that criterion has for the user. In addition, a triangular matrix of coefficients is used to represent the magnitude of correlations between any two criteria. The preference value of a solution is therefore computed as a weighted summation of the utilities associated with each criterion. The main difference between MC-DTPP and our CDI-VDTP formulation is that in a CDI-VDTP the dependencies among the constraint are not undirected as in a MC-DTPP. In fact, MC-DTPP criteria define subsets of constraints that are somehow related with each other, but there is no way to express a causal directionality of such relationships. In many practical cases, however, such a directionality exists. (Consider for example business process workflows [6], supply chains and production systems [18], and so on.) The CDI-VDTP formulation takes advantage of the causal directionality, and enables the user to express conditional independences among constraints by relying on the graph-based representation of the UDNs.

8 Conclusions

In this paper we raised the issue of how dealing with preferences that are not completely independent of one another in a disjunctive temporal problem. As far as we know, such a problem has received little attention, and only in [10] a Multi-Criteria DTPP has been proposed.

In this paper we extended the VDTP formulation [9] of temporal problems with the notion of Conditional Difference Independence. The resulting framework, named CDI-VDTP, enables a user to take advantage of the causal dependencies between the preferences associated with the constraints, and to define an objective function shaped over a Utility Difference Network (UDN), in which each node corresponds to a constraint and (oriented) edges between nodes represent causal dependencies. Solving a CDI-VDTP, thus, consists in computing solutions whose utility is optimal; this can be achieved by exploiting algorithms which are similar to those used for computing probabilities in a Bayesian network, but applied to the UDN. In the paper we also presented a branch-and-bound algorithm for solving CDI-VDTPs by exploring the space of possible solutions. Results collected by a preliminary implementation have been discussed, and show that the proposed solution is actually feasible.

As a future work, we intend to further extend the CDI-VDTP formulation with the addition of a set of variables that, although included within the UDN, are not associated with temporal constraints. The rationale would be to explicitly model via these variables aspects of the domain under consideration that might affect the preference values of a subset of constraints. For instance, in the planetary rover scenario, the level of battery power could be represented explicitly within the UDN by means of a specific variable; such a variable could then affect the duration of actions such as drive or communicate depending on the assumed level of power. Problems like planning and diagnosis could therefore exploit such a richer CDI-VDTP to create expectations or verify hypotheses.

References

1. Brafman, R.I., Engel, Y.: Directional decomposition of multiattribute utility functions. Algorithmic Decision Theory, pp. 192–202. Springer, Berlin (2009)
2. Brafman, R.I., Engel, Y.: Decomposed utility functions and graphical models for reasoning about preferences. In: AAAI (2010)
3. Bresina, J.L., Jónsson, A.K., Morris, P.H., Rajan, K.: Activity planning for the mars exploration rovers. In: 15th International Conference on Automated Planning and Scheduling (ICAPS 2005), pp. 40–49 (2005)
4. Dechter, R., Meiri, I., Pearl, J.: Temporal constraint networks. Artif. Intell. **49**, 61–95 (1991)
5. Khatib, L., Morris, P.H., Morris, R.A., Rossi, F.: Temporal constraint reasoning with preferences. In: IJCAI, pp. 322–327 (2001)
6. Mans, R.S., Russell, N.C., van der Aalst, W.M., Moleman, A.J., Bakker, P.J.: Schedule-aware workflow management systems. Transactions on Petri Nets and Other Models of Concurrency IV, pp. 121–143. Springer, Berlin (2010)
7. Micalizio, R., Torta, G.: Diagnosing delays in multi-agent plans execution. In: Proceedings European Conference on Artificial Intelligence (ECAI'12), pp. 594–599 (2012)
8. Micalizio, R., Torta, G.: Explaining interdependent action delays in multiagent plans execution. J. Auton. Agents Multi-Agent Syst. (2015)
9. Moffitt, M.D.: On the modelling and optimization of preferences in constraint-based temporal reasoning. Artif. Intell. **175**(7–8), 1390–1409 (2011)
10. Moffitt, M.D., Peintner, B., Yorke-Smith, N.: Multi-criteria optimization of temporal preferences. In: Proceedings of CP06 Workshop on Preferences and Soft Constraints (Soft06), pp. 79–93. Citeseer (2006)
11. Moffitt, M.D., Pollack, M.E.: Applying local search to disjunctive temporal problems. In: IJCAI, pp. 242–247 (2005)
12. Oddi, A., Cesta, A.: Incremental forward checking for the disjunctive temporal problem. In: Proceedings European Conference on Artificial Intelligence (ECAI'00), pp. 108–112 (2000)
13. Park, J., Darwiche, A.: Complexity results and approximation strategies for map explanations. J. Artif. Intell. Res. **21**, 101–133 (2004)
14. Roos, N., Witteveen, C.: Diagnosis of simple temporal networks. In: Proceedings European Conference on Artificial Intelligence (ECAI'08), pp. 593–597 (2008)
15. Sheini, H.M., Peintner, B., Sakallah, K.A., Pollack, M.E.: On solving soft temporal constraints using sat techniques. In: Principles and Practice of Constraint Programming (CP 2005), pp. 607–621. Springer (2005)
16. Shenoy, P.P., Shafer, G.: Propagating belief functions with local computations. IEEE Expert **1**(3), 43–52 (1986)
17. Stergiou, K., Koubarakis, M.: Backtracking algorithms for disjunctions of temporal constraints. Artif. Intell. **120**(1), 81–117 (2000)
18. Tan, K.C.: A framework of supply chain management literature. Eur. J. Purch. Supply Manag. **7**(1), 39–48 (2001)
19. Tsamardinos, I., Pollack, M.E.: Efficient solution techniques for disjunctive temporal reasoning problems. Artif. Intell. **151**(1), 43–89 (2003)

An Optimisation Algorithm Inspired by Dendritic Cells

N.M.Y. Lee and H.Y.K. Lau

Abstract Clonal Selection Algorithm (CSA) has been widely adopted for solving different types of optimization problems in the field of Artificial Immune Systems (AIS). Apart from the effector functions of *hypermutation* and *proliferation* providing the diversity of solutions in CSA, the metaphors of dendritic cells (DCs) also enthuse a wide-range of effector control functions, for examples, *cytoarchitecture* and *differentiation* for formulating and developing a powerful solution-evolution engine in optimization algorithms. In order to demonstrate the capability of providing a set of good Pareto fronts with regards to the defined objective functions, the interactions and collaboration of the DC-inspired features are studied in this paper. A real-world problem is studied to demonstrate the functionality of the algorithm in terms of its convergence and diversity of solutions. The experimental results also reveal that the framework produces promising solutions in a scheduling problem, particularly the increased problem size in daily real-life operations.

1 Introduction

Artificial Immune Systems (AIS) has widely been adopted for solving optimization problems, including multi-objective optimization [1, 2], combinatorial optimization [3, 4], constrained optimization [5] and functional optimization framework. In the literature, these optimization problems have usually been solved by the *Clonal Selection Algorithm* [6] that is underpinned by the *Clonal Selection Theory* introduced by Burnet [7]. This immunological principle explains that (i) pattern recognition, (ii) differentiation of the antigenic components and (iii) affinity

N.M.Y. Lee (✉) · H.Y.K. Lau
The University of Hong Kong, Pokfulam Road, Kowloon Tong, Hong Kong SAR
e-mail: myleenicole@graduate.hku.hk

H.Y.K. Lau
e-mail: hyklau@hku.hk

M. Bramer and M. Petridis (eds.), *Research and Development in Intelligent Systems XXXII*, DOI 10.1007/978-3-319-25032-8_12

measurement of antigens are incorporated in the regulation the synthesis of antibodies. With respects to the invading antigens, a specific and adaptive immune response is resulted according to the inherent features in the theory for producing optimal solution as summarised in [8], These features include:

- Diversify the population of antibodies by genetic changes in *hypermutation*
- Retain a set of high affinity pattern for *proliferation* and differentiation on the antibodies that have been in contact with the antigens

Foregoing researches studied the paradigm of the clonal selection algorithm and has demonstrated promising results in solving a wide range of optimization problems, particularly in escaping from local optima by the diversification of the population of solutions as mentioned in numerous studies [9]. Apart from the inspiration of the lymphocytes in the adaptive immune system, a novel computational paradigm is introduced that is enthused by the dendritic cells (DCs) in the innate immune system in this study. The proposed DC-inspired computational framework not only encompasses the features from the clonal selection theory as aforementioned, the mechanisms and behaviour of DCs in the metaphor of maturation and migration (as summarised in Table 1) are further enhance the capabilities of decision-making and solution-evolution over the classical DCA proposed by Greensmith [10]. In brief, the intrinsic immuno-features introduced in the proposed DC-inspired framework including,

(i) Threat quantification—that enriches the Discrimination of Self/non-self and Danger Theory that adopted in the DCA and affinity measures in the clonal selection algorithm
(ii) Signal interactions embedded in the cascading pathways/network—the proposed framework is a signal driven optimization algorithm instead of a cell-driven algorithms that have been modeled in the aforementioned AIS algorithms such as clonal selection algorithms [8, 11]

Table 1 Comparison of the proposed framework and the classical Clonal Selection Algorithm in solving optimization problems

Features	The proposed framework	The Clonal Selection Algorithm
Quantification of the solutions (for decision-making)	Threat of the solutions	Affinity measures of the solutions
Communication path(s)	i. Receptors between the dendritic cells and pathogens ii. Between the signal cascading network	i. Antibodies and antigens
Control Operators (for solutions)	i. Differentiation ii. Proliferation iii. Cytoarchitecture iv. Endocytosis	i. Hypermutation ii. Proliferation

(iii) Resulted effector control functions—the controls (e.g. *Cytoarchitecture* and *Endocytosis*) are mediated by mature DCs instead of the antibodies behave in the Clonal Selection Algorithms.

The aforementioned features, namely, the proposed signal cascading network and architecture, are going to enhance the classical DCA. In addition, the orchestral interactions of these characteristics further enhance for the problem solving capabilities in optimisation problems such as diversification of the evolved solutions, instead of limiting to the domains of classification and intrusion detection [12, 13] that are mainly inspired by the danger theory [14]. In the applications of solving detection problems, the sampled data sets (antigens) in the classical DCA are sensed by detectors (DCs) that are further migrated and transformed according to the emitted concentration of cytokines that are regulated by the embedded signals (i.e. PAMPs, safe and danger signals). Such mimic demonstrates a highly adaptive classification paradigm to the sampled data set instead of providing evolution mechanisms to the solutions, as the classical DCA that had been implemented for a decision-making of the unexpected event (as the intrusions) in a dynamic scheduling problem [15]. Whilst the DCA performs the decision-making, the optimal solutions were often evolved based on the AIS mechanisms including clonal selection algorithm and affinity maturation [16, 17]. These AIS-based computing paradigms are efficient in producing high quality of solutions in terms of good acceleration of the solutions convergence and diversity. On the other hands, hybridisation AIS paradigms with the traditional artificial intelligence (AI) are also popular in solving complex engineering problems. For instances of hybridisation of AIS with (i) genetic algorithm (GA) [18, 19] and (ii) particle swarm algorithm (PSA) [20, 21], AIS approaches further address the limitations of the classical approaches such as the efficiency of solution-generation and voiding the premature convergence of evolved solutions respectively, in particular to combinatorial optimisation problems.

Details of the proposed DC-inspired features are given in Sect. 2.2 in the paper. The rest of the paper is organised as follows—the roles of DCs in the human immune system are given in Sect. 2. The competence and performance of the proposed framework is evaluated with a combinatorial optimization problem that is presented in Sect. 3. Discussion and recommendations are presented in the last section.

2 Dendritic Cells and the Optimization Framework

Human DCs are the sentinels [22] in the innate immune system, which capture and process antigens in periphery tissues (such as skin). Subsequently, the cells further migrates to the lymphoid organs and secrete signal molecules (e.g. cytokines and chemokines) for providing a primary inflammatory response, presenting antigens onto the cells and stimulating T-cells in the adaptive immune system. With the

Table 2 The key processes of the DC-induced adaptive immune responses

Process	Description and output of process	Involved immune cells and molecules
Antigen capturing/processing	Discrimination of safe and threat of the invading pathogens	Receptors onto DCs and pathogens, namely, Toll-like receptors (TLRs) and pattern recognition receptors (PRRs) respectively
Maturation and migration	(i) Up-/down-regulation of stimulatory molecules (in maturation) (ii) Intracellular signalling in the signal cascading network (in migration)	Signal molecules and transcription factors in the intracellular pathways and network
Regulation of effector control functions	Regulate and control the fate and migratory properties of DCs and their subpopulation—in terms of (i) the population of the DCs, (ii) phenotypical and functional properties of DCs and (iii) migratory properties	DCs

revealed immunological knowledge of DCs, the paradigm of these cells and molecules provides a powerful computing tool for the evolution of high-quality solutions as given below.

2.1 Human Dendritic Cells

DC-induced adaptive immunity is a three-fold process that involves (i) antigen capturing and processing, (ii) maturation and migration and (iii) initiation of effector control functions, as summarised in Table 2.

Upon DCs encountering the invading pathogens, the magnitude of the microbial threat can be scaled through the formation of synapse [23], including

(i) Microbial viability that refers to the dead or alive of the pathogens. Usually, more vigorous response reacts to viable ("live") microorganisms instead of the "dead".
(ii) Virulence factors, such as toxins, that may harm to the host by causing diseases and illnesses.
(iii) Microbial invasion dictates the magnitude of the immune responses, and the responses are increased from (a) external environment, (b) non-sterile tissues such as skin, (c) sterile tissue and to (d) systematic circulation (e.g. blood).

The resulted threat of pathogens propagates to the downstream signal cascading network that is embedded underneath each activated surface receptors regulates the maturation and the migration properties of DCs. In the process of maturation, the

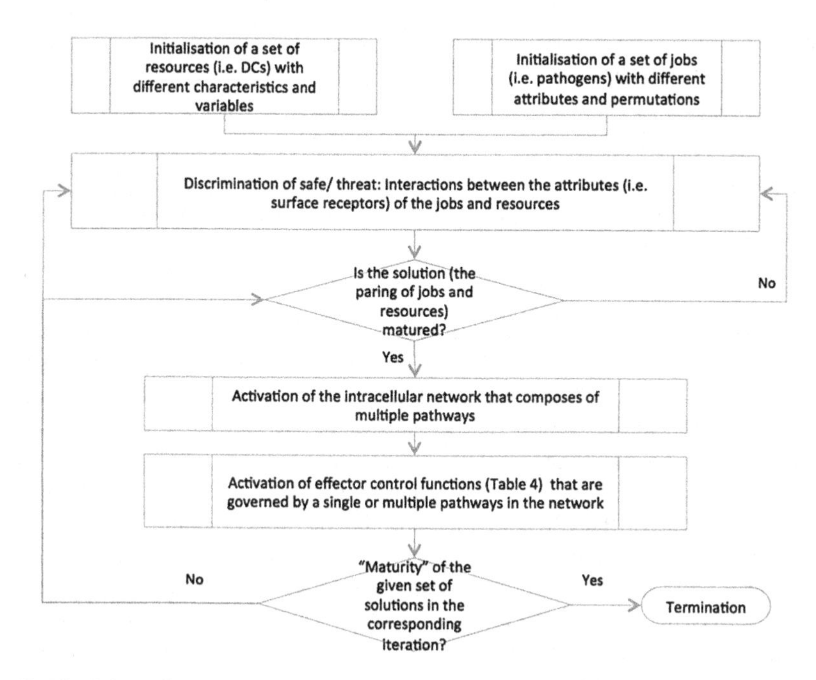

Fig. 1 The DC-mediated framework for solving an optimization problem such as combinatorial optimization problem

phenotype of the surface receptors and their functionalities are changed according to the quantification of the perceived threat. In the context of migration, on the other hand, the migratory prosperities (e.g. kinetic, migratory speed and adhesiveness of the DCs) and immunity control (e.g. antigen up-taking and proliferation of the cells) are regulated upon the activation of T-cell in drained lymph nodes.

The metaphors of threat quantification and the signal cascading network are underpinned the design principles for decision-making and solution-evolution processes in the proposed framework (Fig. 1), are described in the next section.

2.2 The DC-Mediated Framework

The framework (Fig. 1) comprises (i) the macrostructure of the framework and (ii) the microstructure of the network implanted in an individual DC distributed in the framework—a cascading structure facilitates the signal propagation in the solution-evolution process with respects to the quantified threat of the attacks. The macrostructure features the interactions between the artificial components, namely, pathogens and DCs. The analogy is summarised in Table 3.

The framework composes of two key components, namely, DCs and pathogens, which are imitated as resources and jobs in the domain of scheduling problems. Their interactions commenced from (i) the surface receptors (i.e. attributes) of

Table 3 The analogy of the components in the human immune system and the proposed framework for solving combinatorial optimization problems

Human Immune System	The DC-inspired framework
Dendritic cells	Variable(s) of the resources (solutions)
Pathogens	Permutation of the jobs (problem)
Surface receptors/pathogenic recognition molecules (on DCs and pathogens)	Attributes (of the variable and permutation)

Toll-like receptors/Chemokine receptors (TLRs/CCRs) and pathogen-associated molecular patterns (PAMPs) to (i) the signal molecules along the pathways result in producing high quality of solutions as aforementioned, which is distinct from the classical AIS algorithms in particular to the structure of the framework for solving optimisation problems. The framework imitates the DC and its extra- and intra-cellular features, including the surface receptors, signal molecules and its structured pathways—these features leading the signal (i.e. quantified threat) cascading upon the activation of the corresponding effector control functions, which stipulate a robust control to the population of solutions in a given environment.

As illustrated in Fig. 1, the mechanisms (such as the discrimination of threaten solutions) and interactions are adopted as the regulators or control in the decision-making and solution evolution process (Table 4) in the cascading network. Per each iteration, the mechanism begins with a process of decision-making that refers to the differentiation of safe and threat of the solutions according to the factors behaved by DCs [23], which is followed by the activation of the downstream signalling process along the pathways that governs the magnitude of the DC inspired effector control functions. And, these metaphors and features are critical to determine the derived solution, are given in the following sections.

Table 4 Behaviours and immuno-mechanism revealed from dendritic cells and the corresponding analogy

Immuno-mechanisms in Human Immune System	DC-inspired mechanisms in the proposed framework
Discrimination of safe/threat (i) Microbial viability (ii) Virulence factors (iii) Microbial invasion	Differentiate the safe and threat signals of the solutions (i) Feasibility of solutions (ii) Performance of the objective functions (iii) Impacts to the domain
Signal cascading in an intracellular network	Solutions evolution for each paring of variable and permutation
Activation of effector control functions (i) Proliferation and Differentiation (ii) Endocytosis and Cytoarchitecture	(i) The control functions of *proliferation* and *differentiation* regulate the quantities of mature and immature solutions (ii) *Endocytosis* and *Cytoarchitecture* mentioned on the left are used to diversify the generated solutions by changing the attributes of the jobs and resources, such as the permutation of jobs, preference of the jobs, and the availability of the resources

2.2.1 Threat Quantification

In the context of differentiation of safe and threat (signals), signals are emitted from the interacted pathogens can be identified and quantified thru an extracellular signal cascading as the factors given in Sect. 2.1. In other words, the signals of safe and threat associated with the solutions, are estimated based on (i) the feasibility of solutions, (ii) objective functions in the domain (e.g. minimisation of makspan and utilisation of resources in the real-life scheduling problems), and (iii) the impacts of the obtained solutions to the domain. The resulted signal(s) (denoted by $Threat_{DC}$ as presented in Eq. (1)) is estimated the evolved solutions from the perspectives of viability/feasibility of solutions, performance of the objective functions and the corresponding impact to the solutions, that are mimicked the threat components inspired by DCs as aforementioned. The quantified threat further compares with the threshold (δ_{Threat}) that estimated in an iterative manner. More importantly, this may induce the downstream decision-making process if $Threat_{DC} = 1$ in the intracellular network that is implanted in each solution.

$$Threat_{DC} = \begin{cases} 1 & if\, f(feasbility, objectives, impacts) \geq \delta_{Threat} \\ 0 & otherwise \end{cases} \tag{1}$$

2.2.2 Signal Cascading Network and Effector Control

The stimulation and suppression of the proposed effector functions (as summarised in the above table) are regulated by signal molecules in the signal cascading network. The adopted signal molecules are responsible for single or multiple control functions in the network. With regards to the DC-inspired controls, some of them have also been demonstrated in the classical CSA, such as proliferation. In the proposed framework, here, *proliferation* and *differentiation* are taken the roles in adjusting the population and quality of the solutions that are labelled as *safe* and *threat* respectively. Furthermore, the varieties of solutions are regulated by *endocytosis* and *cytoarchitecture* which are the attributes reformation/modification to the variable (of resources) and permutation (of given batch of jobs). The orchestra of various effector control functions is hypothetically a good metaphor for obtaining high quality Pareto font.

3 Experimental Studies

The capabilities of the proposed framework and its performance in terms of convergence and diversity of solutions are investigated in a resource allocation problem that is a class of combinatorial problem. Typically, the optimal resources configuration [24], resources utilization [25] and the shortest dwell time [26] of completing a batch of jobs are the crucial performance indicators in the given domain. In this study, the framework is studied with numerical simulation based on the following objective function,

$$\max f(x) = \sum_{k=1}^{k'} \frac{Occupancy\ of\ resource\ k\ (denoted\ by\ \phi_k)}{Total\ availablity\ of\ resource\ k} \tag{2}$$

where

$$k = 1, 2, \ldots k' \ (k' \text{ is the index of the resource})$$

$$\phi_k = \begin{cases} 1 & \text{Resources is being occupied by a job} \\ 0 & \text{otherwise} \end{cases}$$

Each individual job (denoted by u, where $u = 1, 2, \ldots u'$) is constrained by a workable timeframe between the earliest commencement time and latest completion time depending on the delivery time of the job and the service commitment to the customers.

Experiments are conducted in two scenes according to the size of the problem, (i) 100 jobs and (ii) 300 jobs. As aforementioned, the quality of the solutions is investigated in terms of the convergence and diversity of solutions that are elaborated in the following sections.

3.1 Quality of Solutions

With respects to the objective as defined above, there is a significant improvement of the convergence of the solutions (Fig. 2) with respects to the objective function as defined above. The results also revealed that the percentage of resource utilization has been increased by 10 % in both test environments particularly in the early stage of the evolution process. By comparing the cases of (i) 100 jobs and (ii) 300 jobs, case (ii) produces a set of better solutions in terms of utilization. As observed, some of the solutions have achieved a high utilization rate of up to 43 %.

Despite the limited number of iterations performed, the potential of the proposed framework in evolving good "potential" candidate solution(s) is clearly demonstrated.

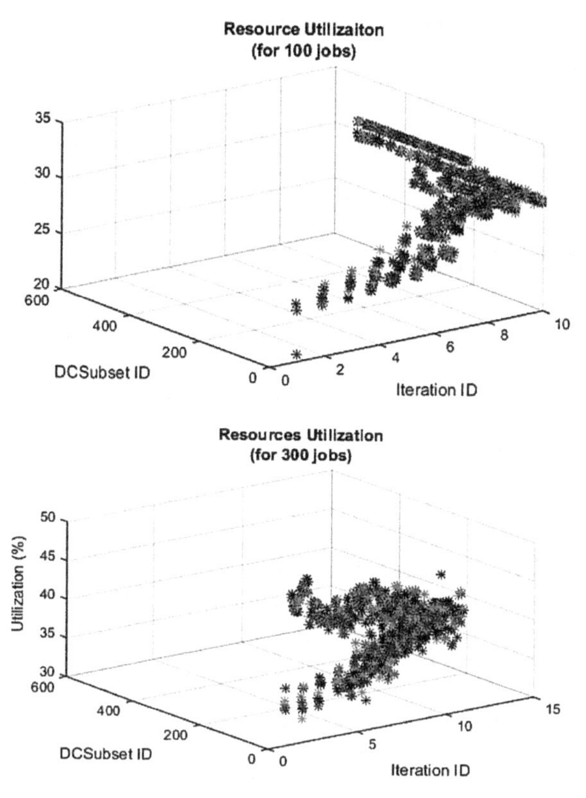

Fig. 2 Convergence of the solutions in resources allocation problems with the sizes of (i) 100 jobs and (ii) 300 jobs

3.2 Diversity of Solutions

Apart from the convergence of solutions, the diversity of the solutions is also investigated. To study the diversity of solutions, the means, maximum and minimum of the objective function (Eq. (2)) are analysed, as depicted in Fig. 3.

The experimental results shown above further illustrate an upward trend in the convergence of the obtained solutions as discussed in last section. In addition, the solutions are evolved and differentiated in a diverse manner. By comparing the solutions obtained in the earliest iterations, the diversity of solutions performed better with a problem with 300 jobs, as compared to a smaller scale problem of 100 jobs under the identical configurations of the number of proliferated and differentiated solutions. On the other hand, the quantities of proliferated and differentiated cells potentially expand the diversity by enlarging the population size of solutions.

Based on the experimental results presented in this section, the framework demonstrates its capability in obtaining optimal solutions especially for a problem with 300 jobs than in a problem with 100 jobs, in terms of convergence and diversity of solutions.

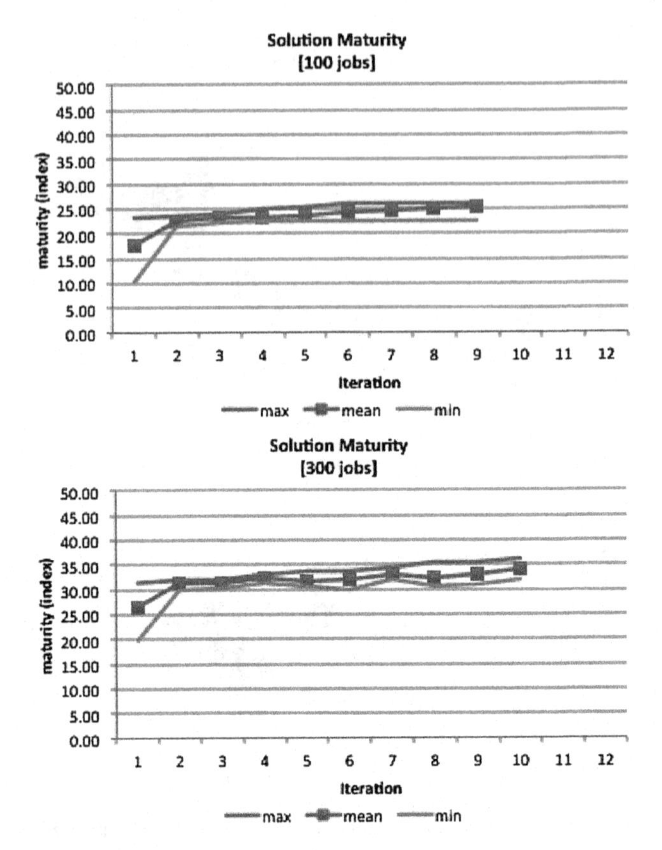

Fig. 3 Diversity of the solutions in resources allocation problems with the scale of (i) 100 jobs and (ii) 300 jobs

4 Conclusions

The key immuo-features and mechanisms of the proposed framework provide a novel approach for the development or evolution of optimal solutions, are inspired by DCs. It is a signal-driven optimization framework, in which, signals are generated from scaling the threat of the solutions, and further regulates the proposed control functions to the generated solutions. In this paper, the functions of the framework have demonstrated problem-solving capabilities, including the convergence and diversity of solutions in solving combinatorial optimization problems. According to the preliminary experimental results, the framework performs well when the number of jobs increased in the combinatorial scheduling problems. In this study, a case with 300 jobs gives satisfactory results in terms of the convergence and diversity of solutions.

As of the proposed framework, forthcoming development of the framework includes the investigation of the abilities of the framework with regards to the

(i) features of the framework and (ii) the behaviour in the search/evolution process. More significantly, benchmarking with other AIS and AI approaches are also considered in the future, in particular to affirming the optimality of the solutions.

References

1. Coello, C.A.C., Cortés, N.C.: Solving multiobjective optimization problems using an artificial immune system. Genet. Program Evolvable Mach. **6**(2), 163–190 (2005)
2. Campelo, F., Guimarães, F.G., Igarashi, H.: Overview of artificial immune systems for multi-objective optimization. In: Evolutionary Multi-criterion Optimization, pp. 937–951. Springer, Berlin (2007)
3. Masutti, T.A., de Castro, L.N.: A self-organizing neural network using ideas from the immune system to solve the traveling salesman problem. Inf. Sci. **179**(10), 1454–1468 (2009)
4. Pasti, R., De Castro, L.N.: A neuro-immune network for solving the traveling salesman problem. In: Neural Networks, 2006. IJCNN'06. International Joint Conference on, pp. 3760–3766, IEEE (2006)
5. Zhang, W., Yen, G.G., He, Z.: Constrained optimization via artificial immune system. Cybern., IEEE Trans. **44**(2), 185–198 (2014)
6. De Castro, L.N., Von Zuben, F.J.: Learning and optimization using the clonal selection principle. Evol. Comput., IEEE Trans. **6**(3), 239–251 (2002)
7. Burnet, S.F.M.: The clonal selection theory of acquired immunity. University Press, Cambridge (1959)
8. De Castro, L.N., Von Zuben, F.J.: The clonal selection algorithm with engineering applications. Proc. GECCO **2000**, 36–39 (2000)
9. Ulutas, B.H., Kulturel-Konak, S.: A review of clonal selection algorithm and its applications. Artif. Intell. Rev. **36**(2), 117–138 (2011)
10. Greensmith, J., Aickelin, U., Cayzer, S.: Introducing dendritic cells as a novel immune-inspired algorithm for anomaly detection. Artif. Immune Syst., Lect. Notes Comput. Sci. **3627**(2005), 153–167 (2005)
11. Hofmeyr, S.A., Forrest, S.: Architecture for an artificial immune system. Evol. Comput. **8**(4), 443–473 (2000)
12. Oates, R., Greensmith, J., Aickelin, U., Garibaldi, J., Kendall, G.: The application of a dendritic cell algorithm to a robotic classifier. In: Artificial Immune Systems (pp. 204–215). Springer, Berlin (2007)
13. Greensmith, J., Aickelin, U., Tedesco, G.: Information fusion for anomaly detection with the dendritic cell algorithm. Inf. Fusion **11**(1), 21–34 (2010)
14. Matzinger, P.: Tolerance, danger and extended family. Annu. Rev. Immunol. **12**, 991–1045 (1994)
15. Qiu, X.N., Lau, H.Y.: An extended deterministic dendritic cell algorithm for dynamic job shop scheduling. In: Research and Development in Intelligent Systems, vol. XXVII, pp. 395–408. Springer, London (2011)
16. Engin, O., Döyen, A.: A new approach to solve flowshop scheduling problems by artificial immune systems. Doğuş Üniversitesi Dergisi **8**(1), 12–27 (2011)
17. Diana, R.O.M., de França Filho, M.F., de Souza, S.R., de Almeida Vitor, J.F.: An immune-inspired algorithm for an unrelated parallel machines' scheduling problem with sequence and machine dependent setup-times for makespan minimisation. Neurocomputing **163**, 94–105 (2015)
18. Hsu, L.F., Hsu, C.C., Lin, T.D.: An intelligent artificial system: artificial immune based hybrid genetic algorithm for the vehicle routing problem. Appl. Math. **8**(3), 1191–1200 (2014)

19. Diabat, A., Kannan, D., Kaliyan, M., Svetinovic, D.: An optimization model for product returns using genetic algorithms and artificial immune system. Resour. Conserv. Recycl. **74**, 156–169 (2013)
20. El-Sherbiny, M.M., Alhamali, R.M.: A hybrid particle swarm algorithm with artificial immune learning for solving the fixed charge transportation problem. Comput. Ind. Eng. **64**(2), 610–620 (2013)
21. Sadrzadeh, A.: Development of both the AIS and PSO for solving the flexible job shop scheduling problem. Arab. J. Sci. Eng. **38**(12), 3593–3604 (2013)
22. Banchereau, J., Steinman, R.M.: Dendritic cells and the control of immunity. Nature **392** (6673), 245–252 (1998)
23. Blander, J.M., Sander, L.E.: Beyond pattern recognition: five immune checkpoints for scaling the microbial threat. Nat. Rev. Immunol. **12**(3), 215–225 (2012)
24. Köchel, P., Kunze, S., Nieländer, U.: Optimal control of a distributed service system with moving resources: application to the fleet sizing and allocation problem. Int. J. Prod. Econ. **81**, 443–459 (2003)
25. Meisel, F., Bierwirth, C.: Heuristics for the integration of crane productivity in the berth allocation problem. Trans. Res. Part E: Logistics Trans. Rev. **45**(1), 196–209 (2009)
26. Koopmans, T.C.: Efficient allocation of resources. Econometrica: J. Econometric Soc. 455–465 (1951)

Short Papers

Graph-Based Multi-Document Summarization: An Initial Investigation

Asma Bader Al-Saleh and Mohamed El Bachir Menai

Abstract With the explosion of information on the Internet, manual summarization has become an unrealistic solution, and the need for automatic summarization–more specifically, multi-document summarization–has grown. Automatic summarization provides the most significant and relevant information while saving time and effort. This paper proposes a graph-based multi-document summarization model to produce extractive multi-document summaries. The model formulates the summarization problem as a capacitated profitable tour problem to optimize the coverage and coherence of the resulting summary. The solution is approximated using swarm intelligence meta-heuristics.

1 Introduction

Summaries exist everywhere and are used by everyone. Browsing a shopping guide, reading a paper's abstract, and viewing news headlines are a few examples of using summaries. Today, the problem of information overload reduces the opportunity for people to reach their targets and increases the time and effort required to do so. Therefore, automatic text summarization–more specifically, multi-document summarization (MDS), whereby a summary is produced from a cluster of related documents–has thus become essential to helping people to identify the relevant and salient aspects of these resources without wasting their time and effort.

Jurafsky and Martin [6] divided the main architecture of the automatic summarizers into three steps: content selection, information-ordering, and sentence-realization. First, during the content selection step, summary contents are extracted

A.B. Al-Saleh (✉)
Department of Computer Science, College of Computer and Information Sciences,
King Saud University, Riyadh, Saudi Arabia
e-mail: asma_alsaleh@ccis.imamu.edu.sa

M.E.B. Menai
Department of Computer Science, College of Computer and Information Sciences,
King Saud University, P.O. Box 51178, Riyadh 11543, Saudi Arabia
e-mail: menai@ksu.edu.sa

M. Bramer and M. Petridis (eds.), *Research and Development in Intelligent Systems XXXII*, DOI 10.1007/978-3-319-25032-8_13

from a cluster of documents. The extracted contents are then ordered during the information-ordering step. Finally, during the sentence-realization step, a cleaning-up process is applied to the ordered contents. Nenkova and McKeown [11] identified three tasks undertaken by summarizers to generate summaries. The first task consists of creating an intermediate representation of the input document(s). The second and third tasks consist of assigning a score to each sentence and selecting summary sentences, respectively.

This paper proposes a (MDS) model that follows the graph-based approach. Its main contribution is proposing a model that formulates the summarization problem as a capacitated profitable tour problem (CPTP) [2] and uses swarm intelligence (SI) meta-heuristics to optimize two objectives in the produced summaries: information coverage and text coherence. In other words, the model will jointly optimize the content selection and the information ordering steps. The remainder of this paper is organized as follows. Section 2 outlines some related work. Section 3 presents the proposed model. Section 4 presents the main conclusion and future work.

2 Related Work

Nenkova and McKeown [11] identified two main approaches to selecting summary sentences: greedy selection and global optimal selection. The greedy selection approach—or the local selection of the best textual unit one at a time—is widely used in text summarization. Although the greedy algorithm is simple and fast, it rarely extracts the best summaries [5]. Therefore, other summarization studies, such as [10, 14, 15], followed a non-greedy direction (i.e., global optimal selection). The main difference between the global optimal selection approach and the greedy approach is that the former searches for the best summary, while the latter searches for the best sentences [5]. Several summarization studies formulated the summarization task as an optimization problem and searched for the best combination of sentences that maximize the summary's overall score. Examples of these problems are knapsack problem [10], set-covering problem [15], and minimum dominating set graph problem [14].

Several objectives to be optimized were considered in text summarization, such as information coverage, diversity (i.e., minimum redundancy), and significance. However, ordering summary sentences to enhance the text coherence is an important objective in (MDS) that has been neglected by most of the summarization studies [12]. Conroy et al. [4] underlined this issue. After generating a summary, this formulated the sentence ordering problem as a traveling salesman problem (TSP). Nevertheless, sentence ordering step could be undertaken jointly with the step of extracting summary content, such as in Nishikawa et al. [12]. However, finding the best (i.e., optimal) summary in all these formulations is an NP-hard problem [10, 14]. Therefore, the use of heuristics or meta-heuristics is required to approximate the solution to (MDS) in a reasonable amount of time. Several approaches were proposed, such as the greedy approach [10, 14], particle swarm optimization (PSO) [1], and dynamic programming [10]. Promising results were produced from these studies. For example, the proposed solution of Lin and Bilmes [9] outperformed all the state-of-the-art methods on the corpora of DUC[1] (DUC 2004 through DUC 2007) in generic and query-driven summarization. The evaluation results of the produced summaries by Shen and Li [14] were close to those of the best methods in several DUC competitions. In addition, some of these studies, such as [13, 15] found that the results of exact and approximate solutions were comparable, although the latter were much faster than the former. Recently, (SI) meta-heuristics have been used in text summarization to select summary sentences. Alguliev, Aliguliyev, and Isazade [1] proposed an optimization model to generate multi-document summaries using discrete (PSO) algorithm to maximize the summary score. Their objective function combined two objectives: coverage and diversity. The experimental results obtained on DUC 2001 and DUC 2002 corpora were promising.

3 The Proposed Model

This study proposes an extractive (MDS) model that jointly optimizes two properties of the produced summary: information coverage and text coherence. The model formulates the summarization problem as a CPTP [2] and optimizes the overall score of the produced summary by using (SI) meta-heuristics. It comprises four main stages: (1) pre-processing, (2) building the intermediate representation of the text, (3) computing the content and the coherence scores, and (4) selecting and ordering summary sentences (see Fig. 1).

3.1 *Pre-processing*

In the pre-processing stage, several steps will be incorporated and applied to the text to prepare it for the summarization process. These steps include text segmentation and tokenization to split the text and identify the sentences and words, respectively. Another step in this stage is stemming that enables treating the different variants of terms equally by reducing them to their stems. In addition, stop word identification step is applied to the text to identify and eliminate from the text the words that appear frequently but have low semantic weights [6]. Examples of these words are ‘and’, ‘the’, and ‘to’.

[1]http://www.nlpir.nist.gov/projects/duc/index.html.

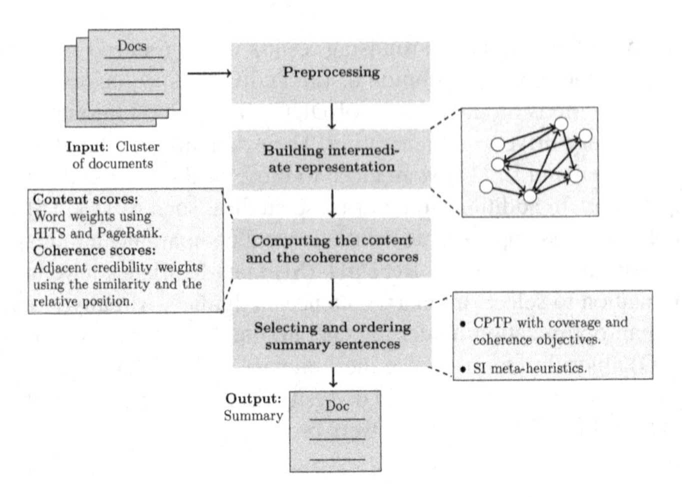

Fig. 1 The main stages of the proposed model

3.2 Building the Intermediate Representation

A directed and weighted graph will be built to represent the text to be summarized. Each sentence will be represented as a graph vertex. The weights of these vertices represent the content scores (i.e., the saliency) of the sentences. Graph edges will be added between sentences that could be adjacent to each other in the summary. The weight of each edge is the adjacent credibility of the pair of sentences it connects (the weights of vertices and edges will be computed in the next stage.). To build this graph, our model will adapt the ordering method of Li et al. [8]. This method orders the sentences of a multi-document summary by employing the indirect relations between them. An edge is added between two sentences if the similarity between one of them and the context of the other one is above a predefined threshold. The context of a sentence is the set of the other sentences in its source document. The authors claimed that their method outperformed the state-of-the-art ordering methods. The main difference between this model and our proposed one is that the former optimizes only the step of ordering summary sentences while the latter will jointly optimize both the summarization and the ordering steps. In addition, our model will give a weight for each vertex.

3.3 Computing the Content and the Coherence Scores

The content score of each sentence will be represented by the scores of words it contains. In order to compute words scores, similar to Wan et al. [16], a bipartite graph between sentences and words will be built. The edges of this graph connect words to sentences in which they occur. In addition, two additional graphs will be built to represent the sentence-to-sentence and word-to-word relationships. To compute the score for each word and sentence, Wan et al. [16] combined the ideas of PageRank [3] and the HITS [7] graph-ranking algorithms. The first algorithm was applied to the sentence-to-sentence and word-to-word relationships while the other was applied to the sentence-to-word relationship (hubs and authorities were represented by sentences and words, respectively). Several differences exist between this model and our proposed one. Our model will generate multi-document summaries, instead of single-document ones, and optimize both the order and the content of summary sentences. Related to the coherence scores, as mentioned above, the weights of edges (i.e., the adjacent credibility) between the connected pairs of sentences will be computed using the proposed formulas in Li et al. [8]. These formulas combine the similarity (computed in the previous phase) and the relative position of the sentences.

3.4 Selecting and Ordering Summary Sentences

This stage represents the core of the proposed model. It generates a summary by considering the content and the order of its sentences. Specifically, the (MDS) problem will be transformed into a CPTP [2] to jointly optimize two objectives: information coverage and text coherence. The coverage is represented by the *profit* and the coherence is represented by the *cost*. The *capacity* represents the summary length. In this model, the coverage will be maximized by selecting sentences that cover more important words, taking into consideration the word weights previously computed. The weight of each word will be counted only once if it is covered by any selected sentence, regardless of the number of times. Therefore, the diversity objective will be maximized implicitly in the summary and the problem of redundancies is reduced, which is an essential issue in any multi-document summarizer.

In addition, the model will optimize the coherence of the produced summaries by considering the order of summary sentences. The model will maximize the overall coherence score of the summary, which is the sum of the credibility weights of the edges it contains. A solution to a CPTP instance will be approximated using (SI) meta-heuristics. Several (SI) meta-heuristics will be investigated, including different variants of ant colony optimization algorithm and bee inspired algorithm. This model differs from the model proposed by Nishikawa et al. [12] on three main points. First, our model will formulate the summarization problem as a CPTP. Second, it will use (SI) meta-heuristics to approximate a solution to CPTP. Finally, it will use different methods to calculate the content and the coherence scores.

4 Conclusion

This paper presented a model for extractive (MDS) in which the text is represented as a graph and the summarization problem is reduced to a CPTP to optimize the information coverage and text coherence in the produced summaries. Its solution will be approximated on using (SI) meta-heuristics. Several (SI) meta-heuristics will be investigated and the performance of the (MDS) system will be tested on standard corpora including DUC corpora.

References

1. Alguliev, R.M., Aliguliyev, R.M., Isazade, N.R.: Formulation of document summarization as a 0–1 nonlinear programming problem. Comput. Ind. Eng. **64**(1), 94–102 (2013)
2. Archetti, C., Feillet, D., Hertz, A., Speranza, M.G.: The capacitated team orienteering and profitable tour problems. J. Oper. Res. Soc. **60**(6), 831–842 (2009)
3. Brin, S., Page, L.: The anatomy of a large-scale hypertextual web search engine. Comput. Netw. ISDN Syst. **30**(1–7), 107–117 (1998)
4. Conroy, J.M., Schlesinger, J.D., Oleary, D.P., Goldstein, J.: Back to basics: classy 2006. In: Proceedings of 2006 Document Understanding Conference (2006)
5. Huang, L., He, Y., Wei, F., Li, W.: Modeling document summarization as multi-objective optimization. In: Intelligent Information Technology and Security Informatics (IITSI), 2010 Third International Symposium on, pp. 382–386, April (2010)
6. Jurafsky, D., Martin, J.H.: Speech and Language Processing: An Introduction to Natural Language Processing, Computational Linguistics, and Speech Recognition, 2nd edn. Prentice-Hall Inc., Upper Saddle River (2009)
7. Kleinberg, J.M.: Authoritative sources in a hyperlinked environment. J. ACM **46**(5), 604–632 (1999)
8. Li, P., Deng, G., Zhu, Q.: Using context inference to improve sentence ordering for multi-document summarization. In: Proceedings of 5th International Joint Conference on Natural Language Processing, pp. 1055–1061, Chiang Mai, Thailand, November 2011. Asian Federation of Natural Language Processing (2011)
9. Lin, H., Bilmes, J.: A class of submodular functions for document summarization. In: Proceedings of the 49th Annual Meeting of the Association for Computational Linguistics: Human Language Technologies, HLT'11, vol. 1, pp. 510–520, Stroudsburg. Association for Computational Linguistics (2011)
10. McDonald, R.: A study of global inference algorithms in multi-document summarization. In: Proceedings of the 29th European Conference on IR Research, ECIR'07, pp. 557–564. Springer, Berlin (2007)
11. Nenkova, A., McKeown, K.: A survey of text summarization techniques. In: Aggarwal, C.C., Zhai, C. (eds.) Mining Text Data, pp. 43–76. Springer, New York (2012)
12. Nishikawa, H., Hasegawa, T., Matsuo, Y., Kikui, G.: Opinion summarization with integer linear programming formulation for sentence extraction and ordering. In: *Proceedings of the 23rd International Conference on Computational Linguistics: Posters*, COLING'10, pp. 910–918, Stroudsburg. Association for Computational Linguistics (2010)
13. Nishino, M., Yasuda, N., Hirao, T., Suzuki, J., Nagata, M.: Text summarization while maximizing multiple objectives with lagrangian relaxation. In: Serdyukov, P., Braslavski, P., Kuznetsov, S., Kamps, J., Rger, S., Agichtein, E., Segalovich, I., Yilmaz, E. (eds.) Advances in Information Retrieval. Lecture Notes in Computer Science, vol. 7814, pp. 772–775. Springer, Berlin (2013)
14. Shen, C., Li, T.: Multi-document summarization via the minimum dominating set. In: Proceedings of the 23rd International Conference on Computational Linguistics, COLING'10, pp. 984–992, Stroudsburg. Association for Computational Linguistics (2010)
15. Takamura, H., Okumura, M.: Text summarization model based on maximum coverage problem and its variant. In: Proceedings of the 12th Conference of the European Chapter of the

Association for Computational Linguistics, EACL'09, pp. 781–789, Stroudsburg. Association for Computational Linguistics (2009)
16. Wan, X., Yang, J., Xiao, J.: Towards an iterative reinforcement approach for simultaneous document summarization and keyword extraction. In: Proceedings of the 45th Annual Meeting of the Association of Computational Linguistics, pp. 552–559, Prague, June 2007. Association for Computational Linguistics (2007)

Towards Expressive Rule Induction on IP Network Event Streams

Chris Wrench, Frederic Stahl, Giuseppe Di Fatta, Vidhyalakshmi Karthikeyan and Detlef Nauck

Abstract In order to gain insights into events and issues that may cause errors and outages in parts of IP networks, intelligent methods that capture and express causal relationships online (in real-time) are needed. Whereas generalised rule induction has been explored for non-streaming data applications, its application and adaptation on streaming data is mostly undeveloped or based on periodic and ad-hoc training with batch algorithms. Some association rule mining approaches for streaming data do exist, however, they can only express binary causal relationships. This paper presents the ongoing work on Online Generalised Rule Induction (OGRI) in order to create expressive and adaptive rule sets in real-time that can be applied to a broad range of applications, including network telemetry data streams.

1 Introduction

Given enough warning, a failure in an IP Network could be avoided or its affect mitigated, either case resulting in reduced disruption. The methodology presented in this paper is constructed for the task of predicting and expressing the causality of the imminent failure in an IP network component by creating rules from frequent patterns in a data stream. Given enough warning a failure could be avoided or its

C. Wrench (✉) · F. Stahl · G. Di Fatta
School of Systems Engineering, University of Reading, PO Box 225, Whiteknights, Reading RG6 6AY, UK
e-mail: C.Wrench@pgr.reading.ac.uk

F. Stahl
e-mail: F.T.Stahl@reading.ac.uk

G. Di Fatta
e-mail: G.DiFatta@reading.ac.uk

V. Karthikeyan · D. Nauck
BT Research and Innovation, Adastral Park IP5 3RE, UK
e-mail: Vidhyalakshmi.Karthikeyan@bt.com

D. Nauck
e-mail: Detlef.Nauck@bt.com

M. Bramer and M. Petridis (eds.), *Research and Development in Intelligent Systems XXXII*, DOI 10.1007/978-3-319-25032-8_14

affect mitigated, either case resulting in reduced disruption. The framework is also developed as an initial step towards a methodology for Generalised Rule Induction (GRI) for data streams which would produce descriptive rules rather than rules limited to the prediction of a single feature. Data Stream Mining is an important topic in Big Data Analytics due to the large amount of streams generated from the World Wide Web. Data Stream algorithms must contend with the rapid pace in which data is generated. A data stream is potentially infinite in size making it impractical to store whilst the rate at which data is generated often limits the algorithm to 'one pass', meaning that the data can only be read once if the algorithm is to keep up with the stream [2]. Algorithms must also contend with concept drifts, i.e. fundamental changes in the underlying data distribution that can occur at any time and can render a previously trained and accurate classifier invalid [14]. When a concept drift has occurred they must either update or retrain the model to keep the accuracy at acceptable levels.

As the network telemetry data is primarily event data it falls under the purview of Complex Event Processing (CEP). Events are loosely defined as anything that happens [5], they are represented in CEP as a tuple typically with a timestamp, an event name and an originating ID, this dataset also contains a duration for the event. In Complex Event Processing the partially ordered data stream can be viewed as an event cloud or a Partially Ordered SET (POSET) [11] and these POSETs take on features depending on the volume and types of events in the set, which can be exploited by statistical means or by the application of machine learning. This paper presents a framework to develop an algorithm for generating generalised rules on data streams of the form:

$$\mathbf{IF}\,(A = 1)\,\mathbf{AND}\,(10 < B < 50)\,\mathbf{THEN}\,(C = \mathrm{true})\,\mathbf{AND}\,(D = 2)$$

These kind of rules are human readable and can be used to predict patterns but also to discover relations between frequently occurring item sets. For example, given the rule above, if event A=1 and (10<B<50) are both true then it is likely that both events C=true and D=2 at the same time. Notable algorithms that can induce such rules are ITRULE [12] and Apriori [3], where Aprioi is a special form of generalised rules that only allows the expression of binary events in the data.

Association Rule Mining (ARM) algorithms based on Apriori have been the subject of some previous adaptation for streams. Work has been done on increasing speed and decreasing the memory usage of itemsets exploration. Algorithms have been developed that attempt frequent itemsets counting in one pass whilst others focus on updating existing rules or inducing rules from the most recent itemsets using time windows. [9] contains details of many of these algorithms. One recent development of online ARM is Transaction-based Rule Changing Mining (TRCM) [1] making use of Rule Matching (RM) [13] which measures the change between rule sets from two consecutive time windows. However, there are no algorithms for streaming data that develop and update expressive sets of rules such as the one above.

The OGRI methodology proposed in this paper makes use of RM and sliding time windows as methods to adapt to concept drift and to keep the rule set current. The rule induction method ITRULE is utilised, which uses information theory to construct a set of *k* best rules from a dataset ascertained by calculating each rule's theoretical information content. The *k* best rules with the highest theoretical information content are maintained in the rule set.

This paper is organised as follows, in Sect. 2 the framework of the OGRI algorithm and the ORGI itself are described. Concluding remarks and ongoing work are found in Sect. 3.

2 Online Generalised Rule Induction Framework

This section outlines our work on expressive rule induction for analysing network telemetry event streams. It first describes the framework for Online Generalised Rule Induction (OGRI) in Sect. 2.1 and then outlines a strategy for inducing expressive generalised rules from the event stream in Sect. 2.2.

2.1 OGRI Framework

Figure 1 is an overview of the OGRI framework and its components.

Step 1 is the pre-processing of the data which includes selecting and normalising features. Events are generated along with a timestamp which is used to determine the intervals between the events arriving. An event is represented as a tuple containing many different data types such as <*timestamp* (time), *location* (nominal), *duration* (integer), *name* (string), *serious* (boolean), ...>. There are over 50 features in the focus telecom dataset, many are pruned and some string manipulation must take place to turn the string values into nominals.

Step 2 gathers events that co-occur. These are determined by their inclusion in a sliding time window which is used so only a given number of the most recent events are processed. Events expire when they leave the window. The events in the window are treated as a POSET. There are issues of order guarantee in a complex

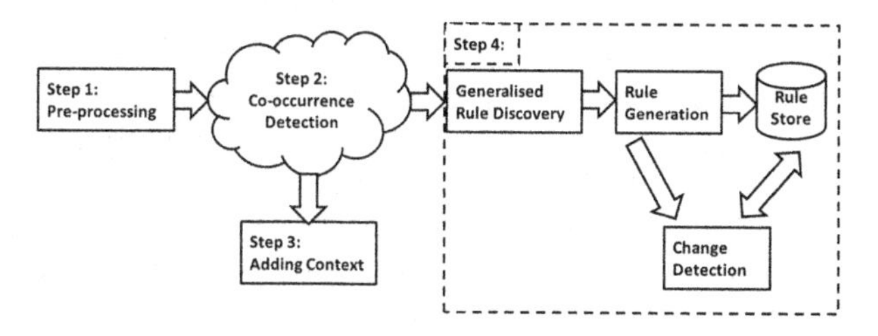

Fig. 1 General framework for online generalised rule induction

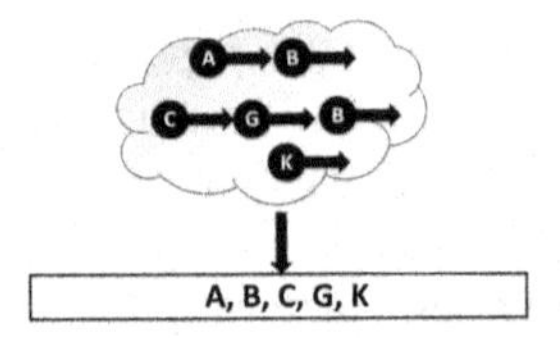

Fig. 2 An POSET of events are converted into a vector for passing to OGRI

system that may interfere with the outcome. An aggressive approach is assumed here and the event cloud is converted into a vector of events without reference to the order in which they arrived, only that they appear within the time window, see Fig. 2. The data contains a number of amalgamated complex events with duration. Duration is an important feature when deciding which events co-occur together, if the sliding window progresses past this event before the event has expired then the set is incomplete. To this end, every event has its duration systematically decreased and only those whose duration reaches 0 are removed. This ensures that events that co-occur with each other, though they may start at different times, still appear in the correct item sets and the association is not lost. In the case of repeated events in the time window the event with the longest remaining duration is saved and the other discarded. It is recognised that this is problematic as the discarded events may have distinctive feature values, the expansion of the system will not suffer from this problem (see 2.2).

Step 3, each event has a station-id representing its geographical location and whose distance to other stations can be looked up in a distance matrix and converted to a similarity for the purpose of weighting events using a metric such as Euclidean distance (other metrics will be explored). Each item has an associated weighting representing the distance to the nearest event in its set of which the highest of these are assigned to the event vector. These vectors are then passed to the pattern detection algorithm.

Step 4 is the induction and maintenance of Generalised Rules as is described in Sect. 2.2.

2.2 Online Generalised Rule Induction

This section establishes the methodology for a Generalised Rule Induction algorithm that only needs a single pass access to the data stream. The methodology is a selection and meaningful combination of descriptive data mining algorithms and adaptive data stream mining techniques. The OGRI algorithm that is currently being developed is based on the ITRULE induction strategy and metric [12], incorporating G-eRules' [8] term generation method for continuous features and RM [13] used in ARM [1, 4]. ITRULE has been chosen because it allows the induction of generalised rules that can have a combination of rule terms on the right hand side representing categorical features, whereas ARM only represents binary features. ITRULE uses the J-Measure to rank and select the *K* best rules from batch data. Rather than producing many rules with subsequent filtering using support and confidence as in ARM, the J-Measure is employed. With the J-Measure rules can already be produced directly ranked according to their theoretical information content [12]. This avoids inducing unnecessary rules that are deleted later and hence is expected to improve computational efficiency. J-Measure is used to measure the information content of a conditional rule such as *IF Y = y THEN X = x*, denoted as: *J(X,Y=y)* [12].

$$J(X; Y = y) = p(y) \cdot j(X; Y = y) \tag{1}$$

As shown in Eq. (1) $J(X; Y = y)$ is essentially a product of $p(y)$, the probability with which the left hand side of the rule will occur, and $j(X; Y = y)$ which is called the j-measure (with a lower case j) and measures the goodness-of-fit of a rule. We plan to extend the rule induction in order to also allow expressing value ranges of continuous features based on the feature evaluation technique used in the G-eRules [8] data stream classifier. This would allow to construct rules of the following form:

IF (10<TEMPERATURE<50) **AND** (DELAY = Low)
THEN (Pressure=Normal) **AND** (TICKET = Logged)

The rule displayed above can be induced by the ITRULE with the exception of term *(10<TEMPERATURE<50)* as it is continuous. Here OGRI algorithm will use G-eRules' Gaussian distribution based method of inducing rule terms by maximising the probability with which the term matches the right hand side of the rule. In the case of G-eRules this is a fixed target feature (class), however, in the case of OGRI this could be a combination of rule terms such as in the example above.

OGRI will be using RM between two consecutive sliding windows in order to detect concept drifts. RM computes two metrics: a maximum similarity score and a difference score [13]. The similarity score *S* determines if a rule is either added or perished and the difference score δ is used to determine if a rule change is emerging or unexpected by comparing the scores with a given threshold *T*. The window size needs to be selected to include events that could directly influence each other whilst keeping the number of events to a manageable size, a dynamic sliding window [7, 10] is currently used here. However, an alternative order guarantee method such as K-Slack [6] will also be considered. An empirical study is planned to compare those methods in the future work. The general *rule induction strategy* is based on a beam search similar to ITRULE, where first only a small number of rules of first order (one rule term on the left and one on the right hand side) are retained and only the best *x* rules are expanded to order 2 (adding an additional rule term), where best refers to the highest theoretical information content according to the J-Measure. Again these rules are then ranked according to their theoretical information content and the *x* best rules are expanded to order 3 and so forth. However, alternative search strategies will be evaluated empirically in the future work.

3 Conclusions

This work represents the ongoing development of the OGRI methodology for generating expressive and generalised rules that can capture significant events in real-time from streaming network telemetry data. The methodology is a combination of the ITRULE and G-eRules algorithm for rule extraction using a sliding time windows approach and RM for detecting concept drifts. This is important in the application domain of network telemetry data as potential faults or outages could be detected early and be described in their causality in real-time. Future work comprises the empirical evaluation of the methodology on telecom network management data.

References

1. Adedoyin-Olowe, M., Gaber, M.M., Stahl, F.: TRCM: a methodology for temporal analysis of evolving concepts in Twitter. In: Artificial Intelligence and Soft Computing, vol. 7895 LNAI, pp. 135–145. Springer (2013)
2. Aggarwal, C.C., Han, J., Wang, J., Yu, P.S.: A framework for clustering evolving data streams. Proc. 29th Int. Conf. Very Large Data Bases **29**, 81–92 (2003)
3. Agrawal, R., Imieliski, T., Swami, A.: Mining association rules between sets of items in large databases. ACM SIGMOD Rec. **22**(2), 207–216 (1993)
4. Chen, M.C., Chiu, A.L., Chang, H.H.: Mining changes in customer behavior in retail marketing. Expert Syst. Appl. **28**(4), 773–781 (2005)
5. Hinze, A., Sachs, K., Buchmann, A.: Event-based applications and enabling technologies. In: Proceedings of the Third ACM International Conference on Distributed Event-Based Systems, pp. 1:1–1:15, New York, July 2009. ACM Press (2009)
6. Hwang, J.H., Balazinska, M., Rasin, A., Çetintemel, U., Stonebraker, M., Zdonik, S.: High-availability algorithms for distributed stream processing. In: Proceedings—International Conference on Data Engineering, pp. 779–790 (2005)
7. Laguna, J.O., Olaya, A.G., Borrajo, D.: A dynamic sliding window approach for activity recognition. In: User Modeling, Adaption and Personalization, pp. 219–230. Springer (2011)
8. Le, T., Stahl, F., Gomes, J.B., Gaber, M.M., Di Fatta, G.: Computationally efficient rule-based classification for continuous streaming data. In: Research and Development in Intelligent Systems XXIV, p. 2014. Springer International Publishing (2008)
9. Lee, V.E., Jin, R., Agrawal, G.: Frequent pattern mining in data streams. In: Aggarwal, C.C., Han, J. (eds.) Frequent Pattern Mining, Chap. 9, p. 199. Springer, New York (2014)
10. Okeyo, G., Chen, L., Wang, H., Sterritt, R.: Dynamic sensor data segmentation for real-time knowledge-driven activity recognition. Pervasive Mobile Comput. **10**, Part B(0), 155–172 (2014)
11. Peer, B., Rajbhoj, P., Chathanur, N.: Complex events processing: unburdening big data complexities. Big Data: Count. Tomorrow's ... **11**(1), 53–65 (2013)
12. Smyth, P., Goodman, R.M.: An Information Theoretic Approach to Rule Induction from Databases (1992)
13. Song, H.S., Kim, J.K., Kim, S.H.: Mining the change of customer behavior in an internet shopping mall. Expert Syst. Appl. **21**, 157–168 (2001)
14. Widmer, G., Kubat, M.: Learning in the presence of concept drift and hidden contexts. Mach. Learn. **23**(3), 69–101 (1996)

A Hybrid Ensemble for Classifying and Repurposing Financial Entities

Peter John Hampton, Hui Wang and William Blackburn

Abstract This preliminary research attempts to address the complexities in classifying the unpredictable textual content of financial reports. The documented experiment extracts a vector of named entities by implementing a hybrid system; a machine learning and logic driven rules engine on an entity per entity basis. We find that recursive pattern matching and manipulation of selected entity classes significantly yields better results for selected named entity types over a standalone Maximum Entropy classifier. We conclude that adopting a hybrid ensemble to store term/key values natively for further lexical analysis is more effective than either approach. We discuss our findings and future research in this promising area.

1 Introduction

The rising volume, variety and velocity of published financial articles has rapidly outpaced mans ability to manage and act upon it. Institutions to private investors are adopting techniques to aid decision-making from automatic knowledge discovery (Columbus [3]). However, various linguistic challenges restrict a computational understanding of the language found in free-form text due to the ontological nature of human life and other cognitive phenomena regarding language. Although there are many documents of interest to analysts, the theme is set on corporate-disclosures, specifically interim accounts due to diverse content. We demonstrate our hybrid ensemble that eventually aims to save time in XBRL (eXtensible Business Reporting Language) annotation of financial statements and related corpora.

P.J. Hampton (✉) · H. Wang · W. Blackburn
Ulster University, BT37 0QB, UK
e-mail: hampton-p1@email.ulster.ac.uk

H. Wang
e-mail: h.wang@ulster.ac.uk

W. Blackburn
e-mail: wt.blackburn@ulster.ac.uk

M. Bramer and M. Petridis (eds.), *Research and Development in Intelligent Systems XXXII*, DOI 10.1007/978-3-319-25032-8_15

2 Related Work

Great effort has been devoted into understanding and acting on the natural language contained within a variety of readily available texts such as analyst studies, social media and news reports. Investors have been overloaded with overwhelming information quantities [1, 7, 10]. The generation of new information has increased, so much so that companies rely on the high processing power of machines for modeling and analysis of the described heterogeneous, often unpredictable data.

It has been noted that the precision of current statistical systems is insufficient for industrial scale adoption in the financial sector [8]. We explore entity classification and propose a hybrid approach to support human or agent based stock market trading based on the analysis of interim statements. Related work in text mining found that although financial statements are historical in nature and focus on past performance, the text-based content contains future focused language. The Management Discussion and Analysis (MD&A)[1] section may include discussion that indicates future financial performance, consumer demand, market outlook and competitive literature within the narrative of past performance [4].

Rule based approaches have been the most popular approach in industry and are largely considered obsolete in academic research. Chiticariu et al. [2] noted that only 6 publications relied solely on rules across a 10-year period reflecting the popularity for machine learning. In contrast Machine Learning (ML) approaches to Named Entity Recognition are currently the most prominent form of research in academia, whereas rule based approaches are favored in industry. The general direction is to let machines self discover and learn from large corpora. However, IBMs industrial survey found that only 1/3 of industry relied exclusively on ML techniques for extraction products [2].

A synthesis of the two could be an attractive alternative for this task, as it is easily adoptable, scalable, and has a relatively transparent debugging features compared to machine learning. Malik et al. [8] complement their black-box, machine learning based approach with rules to handle exceptions whereas Suganthan [9] employ a hybrid approach where rules are used to correct classifier mistakes in order to increase the precision of the overall system. These researchers use rules to cover cases that the system cannot yet handle or prove very difficult to learn. For this initial experiment, we implement a rule-based, recursive pattern matching engine alongside a machine-learning algorithm, in this case, a Maximum Entropy classifier.

[1]The section of a financial statement where management figures discuss numerous aspects of the companys financial position; past, present and future.

3 Experimentation

3.1 Rule Base

Due to the unpredictable and creative use of language within corporate disclosures, the need to manipulate token shape, became apparent at early stages of this study. This component described is a series of objects that manipulate text into a common reusable and storable format. The text subject to our rules-based manipulation module includes monetary values, dates, times, integers and floating point numbers with examples provided in Fig. 1.

We remove the need for a PERCENT named entity type by converting all the percentages found in the text to a decimal format. It was found that performing calculations on the extracted entities was possible if presented in a decimal format, a technique we plan to explore further in future research. We give an example from our rule base, converted from Haskell to Python 3 for readability purposes to convert percentages to a decimal format inline with the original text.

```
from decimal import Decimal
import re

source = '''Revenue has increased 4%.
            However, Cost of Sales has
            inceased 8.94%'''

def morph_percentage(match):
    top = (match.group(0))
    string = Decimal(top.strip('\%')) / 100
    return str(string)

source = re.sub(r'\d+(\.\d+)?\%',
                    morph_percentage, source)

print(source)
```

$$(\$1.1b \rightarrow ('\$', 1100000000))$$
$$(\$4.7million \rightarrow (\$, 4700000))$$
$$(6.3p \rightarrow ('GBP', 0.063))$$
$$(12thMarch2015 \rightarrow ('2015-03-12'))$$
$$(5.2\% \rightarrow (-0.052))$$
$$(66,000,000 \rightarrow (66000000))$$

Fig. 1 Parsed Tokens: examples depicting an input x and output y where (.) represents an input to the annotator objects helper function. (x-(y))

Table 1 The named entity types adopted in this study

Id	Entity type	Example
1	PERSON	Bill Gates, Mrs Smith
2	ORGANIZATION	FCP.L, Fidessa PLC
3	LOCATION	Belfast, Egypt, USA
4	DATE	2014-05-02, 2nd April
5	TIME	13:55, 5pm, Noon
6	MONEY	100000000,1.4 b
7	PERCENT	5 %, 99.9997 %
8	DECIMAL	0.05, 9.11156
9	TEMPORAL	12 Months, 40 Mins

3.2 Machine Learning Base

We implement a somewhat common set of Named Entities shown in Table 1 to achieve semantic granularity.

The NER module is split into two; the classifier and our rule base. We implement a Maximum Entropy Classifier trained on a manually annotated set of corporate disclosures as described by Leong and Hwee [6] where o represents the outcome, h is the context and $Z(h)$ is the normalization function.

$$p(o|h) = \frac{1}{Z(h)} \sum_{j=1}^{k} a_j^{f_j(h|o)} \quad (1)$$

We implement this into the core of our rule-based component. We first segregate the machine learning core and the rule base module to perform entity extraction per entity type. This allows for the comparison of how each approach performed, and what improvements we could make.

4 Results

Simple Random Sampling was employed for testing, and the chosen datasets were interim accounts of British companies on the London Stock Exchange Alternative Investment Market (AIM) dated 2013–2015. This approach was used to test the classifier under real world conditions and serve as a blind test for both cases.

5 Discussion

From Table 2, we can see that the dedicated MaxEnt Classifier produced disappointing results, but out performed the rule-based approach regarding entities 1, 2 and 3. It is yet unclear why the statistical classifier underperformed, but we can assume from studies that a more suitable training set is needed and evidently doesnt reflect the test data and suffers from the labeling bias problem. We found that both approaches underperformed with identifying entity 1 (PERSONS) within a document. Although the testing corpus contained uncommon European names, we believe to improve this entity we would require a far richer corpus for training the classifier.

The rule base, as hypothesized, achieved impressive results for entities 4, 5, 6, 8 and 9, achieving 100 % accuracy for 4 of these entities. The rule base failed to identify a number of temporal entities, which resulted in a diminishing recall. We found that the rules struggled with coreference and context. This could indicate the need for extending the rule engine with grammars. We find that our rule based approach coincided with the advantages described by Gupta and Manning [5]. We combine both approaches where each performed best to achieve the results in Table 3. We found that the machine learning approach is liberal, with higher recall on all points, whereas the rule base is, as expected, conservative by nature with higher precision and lower recall on all entities except 2 and 3.

Although further experimentation is required, we show that complementing a standard maximum entropy classifier for the task of NER with recursive pattern-matching yields significantly better results than by itself. If, or when such research is

Table 2 The results of the maximum entropy classifier and rule base separated by entities

Id	MaxEnt classifier (%)			Rule base (%)		
	Precision	Recall	F1-score	Precision	Recall	F1-score
1	30.55	75.00	43.41	22.12	14.05	17.18
2	31.81	85.87	46.42	10.41	20.23	13.71
3	76.50	90.00	80.70	77.00	82.04	79.44
4	69.87	74.93	72.31	100.00	100.00	100.00
5	55.05	51.66	53.30	100.00	100.00	100.00
6	87.07	92.04	89.48	100.00	100.00	100.00
8	90.87	97.07	93.97	100.00	100.00	100.00
9	72.73	82.72	70.40	100.00	72.43	84.01
T	**64.31**	**81.16**	**71.75**	**76.19**	**73.59**	**74.30**

Entity ids are used to refer to the entities listed in Table 1. Entity 7 has been omitted as discussed in Sect. 3

Table 3 The F1 Score for the all three approaches

	Rule classifier	MaxEnt classifier	Hybrid classifier
F1-score	71.75 %	74.30 %	77.86 %

The hybrid result is the rules and machine learning combined considered mature enough for industrial adoption, the need for perfect precision on entity types such as MONEY, DATES and TIMES are essential for extracting and reusing such data. We demonstrate that rule based approaches are more desirable for entities typically expressed numerically, and would be considered a native data type for most modern database engines.

6 Future Work

Early results show a promising future for cross-entity hybrid techniques and higher precision and recall in financial based entity extraction; a combination of ML and rules for named entity recognition. Extracting relations to repurpose information for tasks such as automated knowledge discovery and automation in financial analysis are a desirable goal, one that could be realized as methodology matures. Future work will focus on extending the Named Entity Types via Ontology Based Annotations inline with XBRL specifications to capture granular semantics for relation extraction. We will explore different machine learning algorithms, specifially sequencing tagging algorithms such as Hidden Markov Models (HMMs), Maximum Entropy Markov Models (MaxEnt) and Conditional Random Fields (CRFs) that have proved to effective in various named entity classification studies. It is important to test them with our amalgamation and establish a future baseline for the enhanced sets of annotations.

References

1. Cao, M., Chychyla, R., Stewart, T.: Big data analytics in financial statement audits. Account. Horiz. (2015)
2. Chiticariu, L., Li, Y., Reiss, F.R.: Rule-based information extraction is dead! Long live rule-based information extraction systems! In: EMNLP, pp. 827–832 (2013)
3. Columbus, C., Viveka, M.T., Murugan, M.N.S.V.: Classifier Based Text Mining Approaches for Data Mining Applications (2015)
4. Kearney, C., Liu, S.: Textual sentiment in finance: a survey of methods and models. Int. Rev. Financ. Anal. **33**, 171–185 (2014)
5. Gupta, S., Manning, C.D.: Spied: Stanford Pattern-based Information Extraction and Diagnostics. Idibon, Sponsor (2014)
6. Leong, C.H., Tou, N.H.: Named entity recognition with a maximum entropy approach. In: Proceedings of the 7th Conference on Natural Language Learning at HLT-NAACL, Edmonton, Canada, vol. 4, pp. 160–163 (2003)
7. Li, X., Xie, H., Chen, L., Wang, J., Deng, X.: News impact on stock price return via sentiment analysis. Knowl.-Based Syst. **69**, 14–23 (2014)
8. Malik, H.H., Bhardwaj, V.S., Fiorletta, H.: Accurate information extraction for quantitative financial events. In: Proceedings of the 20th ACM International Conference on Information and Knowledge Management, pp. 2497–2500. ACM (2011)
9. Suganthan, G.C.P., Sun, C., Gayatri, K.K., Zhang, H., Yang, F., Rampalli, N., Doan, A.: Why Big Data Industrial Systems Need Rules and What We Can Do About It (2015)
10. Vasarhelyi, M.A., Kogan, A., Tuttle, B.: Big data in accounting: an overview. Account. Horiz. (2015)

Data Mining and Knowledge Discovery Approach for Manufacturing in Process Data Optimization

Raed S. Batbooti and R.S. Ransing

Abstract The in-process data offers a rich source for discovering new process knowledge. This is particularly important in manufacturing processes such as casting process where the number of process variables are large (~30–50) and process observations are small (~50–60). A multivariate data analysis technique such as principal component analysis (PCA) is used for discovering knowledge from foundry in-process data. The correlations among factors for a given response are discovered by projecting the data on a reduced dimensional space defined by the principal components. The correlations are discovered among both categorical and continuous data. A new methodology has been introduced that uses scores and loadings in the PCA to define optimal and avoid tolerance limits for factors. Interactions among factors are also considered. The developed approach allows process engineers to adjust system parameters as it discovers factor tolerance limits that contribute most to the overall variance. This information is used to suggest corresponding optimal and avoid limits that would result in the reduction of variance. The workings of the algorithm are demonstrated on a foundry case study.

1 Introduction

PCA is a multivariate technique widely used in different applications to reduce dimension, compress data, simplify data description and extract the most significant information by projecting the data from a higher dimensional space into a sub linear space [1, 2]. The resulted sub space can be used as a dimension reduction space and

R.S. Batbooti (✉) · R.S. Ransing
Swansea University, Swansea SA2 8PP, UK
e-mail: 591619@swansea.ac.uk

R.S. Ransing
e-mail: r.s.ransing@swansea.ac.uk

R.S. Batbooti
University of Basra, Basra, Iraq

M. Bramer and M. Petridis (eds.), *Research and Development in Intelligent Systems XXXII*, DOI 10.1007/978-3-319-25032-8_16

analysis space to infer the relation between variables. This way the PCA transfers the problem from a high dimensional manifold into a linear combination of original parameters known as principal components. This analysis is based on maximum data variance and minimum projection cost.

For manufacturing processes, production goals such as maximization of mechanical properties and/or minimization of mechanical defects are often defined. Such analysis requires using process knowledge to change factor settings in order to optimize the process. This work builds on a new co-linearity index approach based on PCA and penalty matrix approach to reduce production defects [3]. The co-linearity indices are derived from loadings of PCA and used to infer a potential correlation between a response and the associated factors. The work is extended to include categorical variables such as week days by using multiple factor analysis (MFA) as pre-treatment tool [4]. The poster will present the formulation and demonstrate the further development of algorithms to use scores and loadings for finding optimal ranges of factors that are likely to minimize occurrence of defects.

2 Method

In PCA variables represented by loadings and observations are represented by scores in the resulted sub space. However, instead of using loading and score plots, a co-linearity plot and projected scores plots is used to find out the correlation between variables and the contribution of observations on variables respectively. The in-process data consist of m observations and n variables which include one response at least and the remaining is factors. In the case study example shown in the poster, the main aim of the analysis is to reduce the incidence of conchoidal fractured surface area in a steel casting. The fracture surface represents the response, which characterized by 19 continuous variables represent the factors.

2.1 Co-linearity Index (CLI)

The co-linearity index (CLI) is a visual tool to find the relation between specific responses with factors to understand which factors effects a specific response, which is very beneficial to process engineers based on p principal components instead of 2 or 3 principal components in convensional PCA. It can be calculated by using the following steps [3]:

1. Data pre-treatment: centering and standardization are used to preserve the variance and scale the data respectively. Another three sets of transformation that are important in this approach are penalty value transformation for responses (where the response values are transferred to the range [0 1] according to maximum and minimum penalty value thresholds chosen by the analyst),

Multiple Factor Analysis (MFA) to transfer categorical variables into continuous variables (by using an indicator matrix in it each variable is replaced by a set of indicator binary variables taking one if the categorical variable has been observed and zero otherwise) and median-interquartile range transformation for quantitative variables. All above transformation followed by centring and scaling the data by extracting the mean value and divide each column on its Eigen value respectively.

2. Apply PCA on covariance matrix resulted before step (1). $Cov = \frac{1}{n-1} X_T^t$.
3. Estimate the loading matrix based on the following equation $L_s = D_s^{-1} V D_e$ Where: L_s is the standardized loading matrix, V is the matrix of eigenvectors arranged as column vectors in descending order of eigenvalues, D_s is the diagonal matrix of the standard deviations of the columns of X_T and D_e is the diagonal matrix containing the square roots of eigenvalues.
4. Evaluate the correlation matrix from $L_s\, L_s^t$ for p principal components, where the inner product of ith and jth row vectors of L_s represents the correlation between variable i and j. After that co-linearity index can be plotted by plotting angles and length of the loading vectors. Scree plot method is used to choose the optimal number of principal components.
5. Divide the co-linearity index plot into five regions

- The no correlation region between −0.2 to 0.2 co-linearity index.
- The two weak correlation regions between −0.5 to −0.2 and 0.2–0.5 respectively.
- The two strong correlation regions, which include co-linearity index between −1 to −0.5 and 0.5–1 for negative and positive correlation respectively. The CLI plot for steel alloy is displayed in Fig. 1. Where there are five factors correlated with high penalty direction (Pouring temperature (F), %Cr, %P, %S and %Ti) and three factors showed a correlation with low penalty direction (%Zr, Mn/S Ratio and Carbon drop).

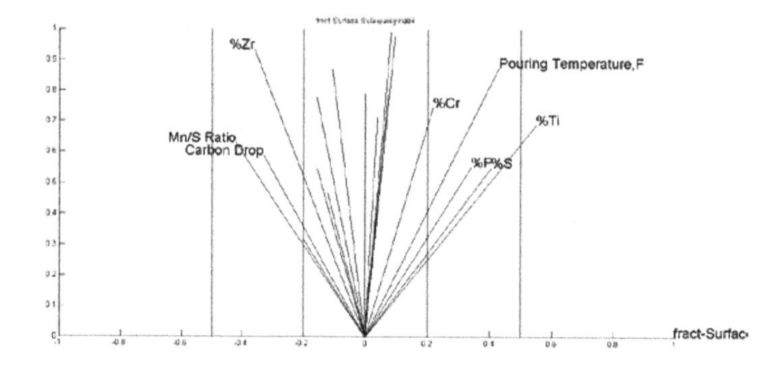

Fig. 1 Co-linearity index plot for the in-process data used in steel alloy

2.2 Scores Projected Space to Predict the Optimal Variables Range and Variables Recommendation

The main steps for calculating optimal system settings for quantitative and categorical variables are summarised below [5]:

1. Find the correlated variables from applying CLI as described in Sect. (2.1).
2. Create a contribution plot for each variable. The horizontal axis of the plot represents the projected scores on the variable, whereas the vertical axis represents the projected scores on the response. The scores lay on the resulted new subspace bounded by positive variable projection axis and corresponding response direction it represents. It should be noted that the response direction is the direction of correlation with the variable. From linear algebra, the projection of score t_i on loading Lj in p dimensions is expressed as:

$$t_i^* = \frac{\sum_{k=1}^{p} L_j(k)^* t_i(k)}{\sqrt{\sum_{k=1}^{p} (L_j(k))^2}} = \frac{L_j \cdot t_i}{\|L_j\|}$$

3. These scores relate to either optimal or avoid ranges with reference to the correlated variable. The observations corresponding to the collected scores is stored in either variable $x^j_{optimal}$ or x^j_{avoid} depending upon whether the correlation is positive or negative and the number of observations stored are counted and stored in a variable n^j_x.
4. Determine the range for factors using the observations stored in $x^j_{optimal}$ or x^j_{avoid}.
5. For categorical variables determine the percentage of occurrences P^j_o:

$$Po^j = \frac{\sum_{i=1}^{n_x^j} x(optimal)_i^j}{n_x^j} \times 100\,\% \text{ or } Po^j = \frac{\sum_{i=1}^{n_x^j} x(avoid)_i^j}{n_x^j} \times 100\,\%$$

where n^j_x = number of elements in x^j.

The categorical variable j is chosen for recommendation as optimal classification if $P^j_o \geq 60\,\%$ and is negatively correlated with penalty values high penalty values. The recommendation will be 'avoid' for a categorical variable i if $P^j_o \geq 60\,\%$ and the variable is positively correlated with high penalty values. Figure 2 shows the main steps of finding the optimal variables range. Score projection and optimal system settings plot displayed in Figs. 3 and 4 respectively.

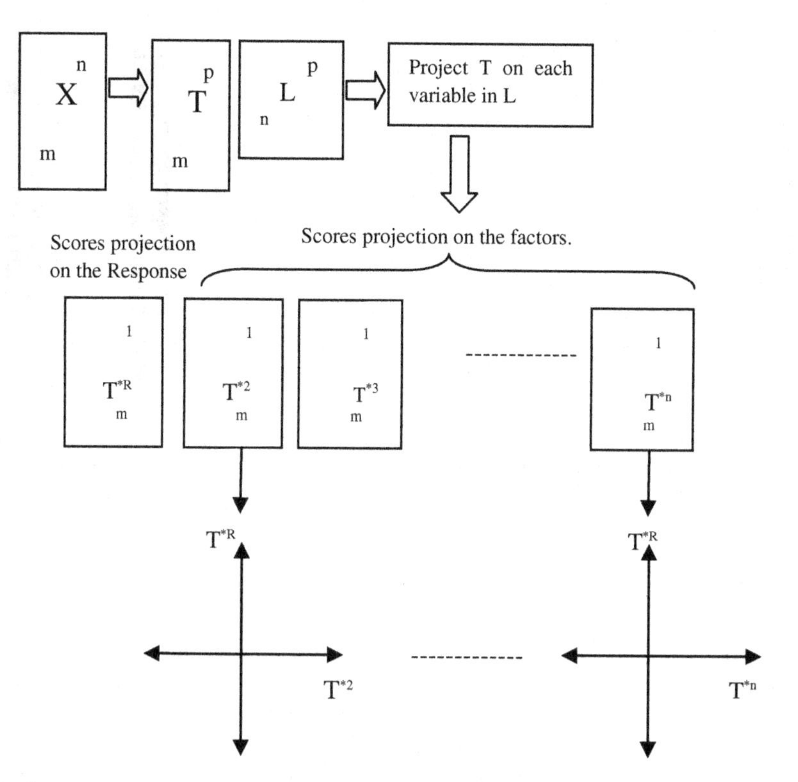

Fig. 2 Scores projection to predict the optimal process settings

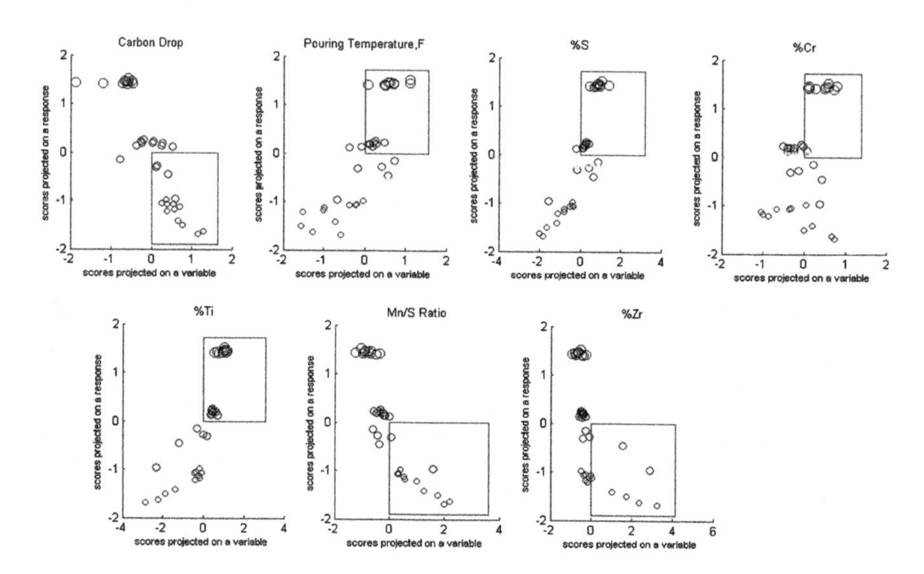

Fig. 3 Scores projection on variables and response of steel alloy

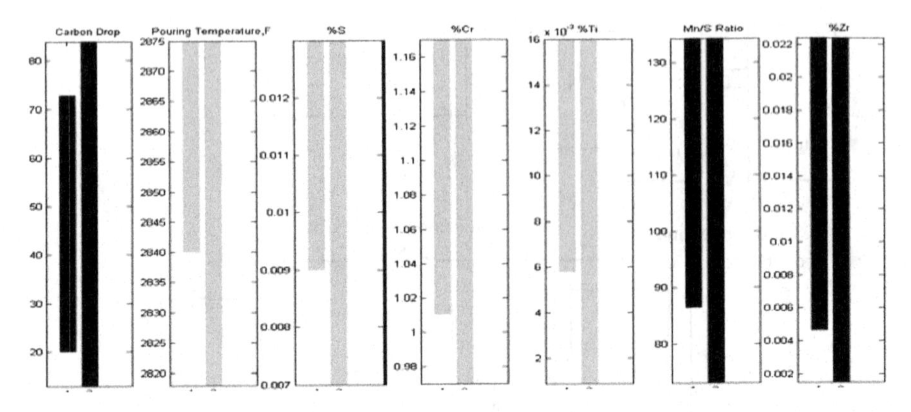

Fig. 4 The optimal range for variables, for each variable the left hand bar represent the optimal (*black bars*) or avoid (the *light bar*) range, obtained range for each variable corresponding to scores bounded by rectangle in Fig. 3 for the variable

3 Conclusion

In the proposed work, an enhanced co-linearity index procedure is used to predict correlations among factors for a given response using foundry in-process. A new approach has been proposed to predict the optimal process settings for correlated variables by using the analogue between loading and scores of principal component analysis as used in the bi-plots. The concept is however extended for p number of principal components. The procedure and results of the algorithm are presented in the context of a foundry case study.

References

1. Bishop, C.: Pattern Recognition and Machine Learning. Springer, New York (2006)
2. Abdi, H., Williams, L.: Principal component analysis. Wiley Interdisciplinary Reviews: Computational Statistics, vol. 2, pp. 433–459 (2010)
3. Ransing, R., Giannetti, C., Ransing, M., James, M.: A coupled penalty matrix approach and principal component based co-linearity index technique to discover product specific foundry process knowledge from in-process data in order to reduce defects. Comput. Ind. **64**(5), 514–523 (2013)
4. Giannetti, C., Ransing, R., Ransing, M., Bould, D., Gethin, D., Sienz, J.: A novel variable selection approach based on co-linearity index to discover optimal process settings by analysing mixed data. Comput. Ind. Eng. **72**, 217–229 (2014)
5. Ransing, R.S., Batbooti, R.S., Giannettia, C., Ransing, M.R.: A quality correlation algorithm to minimise unexplained deviation from expected results in production batches. Comput. Ind. Eng. (2015) (submitted)

Applications and Innovations in Intelligent Systems XXIII

Best Application Paper

Development of Temporal Logic-Based Fuzzy Decision Support System for Diagnosis of Acute Rheumatic Fever/Rheumatic Heart Disease

Sanjib Raj Pandey, Jixin Ma and Chong Hong Lai

Abstract In this paper we describe our research work in developing a Clinical Decision Support System (CDSS) for the diagnosis of Acute Rheumatic Fever (ARF)/Rheumatic Heart Diseases (RHD) in Nepal. This paper expressively emphasizes the three problems which have previously not been addressed, which are: (a) ARF in Nepal has created a lot of confusion in the diagnosis and treatment, due to the lack of standard unique procedures, (b) the adoption of foreign guideline is not effective and does not meet the Nepali environment and lifestyle, (c) using (our proposed method) of hybrid methodologies (knowledge-based, temporal theory and Fuzzy logic) together to design and develop a system to diagnose of ARF case an early stage in the English and Nepali version. The three tier architecture is constructed by integrating the MS Access for backend and C#.net for fronted to deployment of the system.

1 Introduction

This is a collaboration project between the Nepal Heart Foundation (NHF), Nepal and University of Greenwich (UoG). In this research, NHF is providing the data (patient registry), applying diagnosis procedure and practice of ARF, NHF's guidelines, providing other required expert support and information especially in ARF/RHD signs symptoms, management, medication practice etc. Moreover, NHF has continually provided the required support (aspect financial support) from an initial stage to

---

S.R. Pandey (✉) · J. Ma
Computing & Information System, University of Greenwich, London SE10 9LS, UK
e-mail: S.R.Pandey@gre.ac.uk

J. Ma
e-mail: J.Ma@gre.ac.uk

C.H. Lai
Mathematical Sciences Department, University of Greenwich, London SE10 9LS, UK
e-mail: C.H.Lai@gre.ac.uk

M. Bramer and M. Petridis (eds.), *Research and Development in Intelligent Systems XXXII*, DOI 10.1007/978-3-319-25032-8_17

the completion stage of research. NHF is fully participating in the evaluation of developed system and the analysis of the system's result. The training and implementation plan will be set up by NHF in conjunction with the Health Ministry of Nepal and we are giving them an appropriate training and support as required.

Based on the June 2007–Oct 2011 [1] register of National ARF/RHD Prevention and Control Programme from 32 Hospitals of Nepal, it was revealed that 6028 ARF/RHD cases has been registered for secondary prevention. These patients were suffering from ARF and were receiving 3 weekly benzathine penicillin injections or oral antibiotic for secondary prevention of ARF [1].

ARF/RHD is a challenging and life threatening disease for children in the 5–-15 years age groups, especially in the rural area of Nepal. In Nepal, **800 children** per year **die** from ARF/RHD [2]. ARF/RHD can be cured at the initial stage, if we are able to make a proper diagnosis which can not only save children's lives but can also save costs and time. In Nepal, especially in rural areas, lack of awareness and most importantly ignorance of people towards the initial symptoms is a key cause of the damage. Early diagnosis of ARF/RHD will cost the patient NRs. 25.00 but the late diagnosis or negligence of the symptoms can cost them around NRs. 2,00,000–3,00,000 Lakhs [2]. In the context of the Nepalese economy, this amount is very high and the rural population cannot afford this for treatment.

According to National seminar on ARF/RHD Prevention and Control in Nepal, 2011, it has been revealed that *The lack of guideline on ARF/RHD in Nepal has created a lot of confusion in the diagnosis and treatment even among doctors. And* ***the adoption of foreign guideline is not effective as it does not match the Nepali environment and lifestyle*** *[2].* One of the main objective of this research is to design and develop a suitable and affordable CDSS for diagnosis of ARF at an early stage that will fit into the Nepali environment and lifestyle. Furthermore, our development diagnosis system will be a unique procedure and practice for country and individual those who are caring of ARF/RHD patients in Nepal.

The following part of this paper only focuses on the development of application. We have expressively discussed the conceptual concept and theoretical backgrounds of ARF diagnosis process in our previous paper [3].

2 ARF Symptoms and Diagnosis Process

In this section we discuss the ARF signs and symptoms, WHO criteria to diagnose of ARF and the proposed model of ARF diagnosis.

2.1 Signs and Symptoms of ARF

The sign and symptoms have been prepared based on the WHO, WHF and NHF's guidelines/suggestions which are given below [4–6] (Table 1).

Table 1 Sign and symptoms of ARF

Coding	Sign and symptoms	Remarks
A.	**Major manifestations**	
Ar	**Arthritis (large joints pain or tenderness/inflamed)**	
Ar_1	Severe pain (ankles, knees, wrists, elbows, hips, shoulders)	
Ar_2	Pain associated with swelling, hotness, redness, movement restriction	
Ar_3	Migratory/shifting severe pain	
Cr	**Carditis: inflammation of the heart valves**	
Cr_4	Currently present heart murmur	
Cr_5	Chest pain/difficulty in breathing/palpitation	
Ch	**Sydenham's chorea: (St. Vitus Dance)**	
Ch_6	Muscle weakness hands and feet	
Ch_7	Twitchy and jerking movements of hands or feet or facial muscles	
Sn	**Subcutaneous nodules**	
Sn_8	Painless lumps on the outside surface of wrists or elbow or ankles or knees groups 3–4 (up to 12)	Rare in Nepal
Sn_9	Lumps round/firm and freely movable size from 0.5–2.0 cm	
Em	**Erythema marginatum**	
Em_{10}	Painless, flat pink patches on the skin	Rare in Nepal
Em_{11}	Not itchy or painful and has well-defined borders	
B.	**Minor manifestations**	
Fe_{12}	Fever	
Art_{13}	Arthralgia (joints pain)	
Ecg_{14}	Prolonged P-R interval on ECG	
Crp_{15}	Raised or positive CRP	
Esr_{16}	Raised ESR	
C.	**Mandatory/Essential**	
Prt_{17}	Positive rapid strep test	
Ast_{18}	Raised (positive) Anti-Streptolysin O tire (ASOT)	
Gs_{19}	Positive throat culture for GAS infection	

2.2 *Diagnosis Process of ARF*

According to the WHO guidelines [5, 7] the diagnosis process of ARF are as follows:

- The First episode of ARF can be confirmed if
  - There is at least one observable MAJOR symptom and at least two MINOR symptoms manifested plus evidence of GAS infection.

- Recurrent ARF (with no RHD) can be confirmed if
  - There is at least one observable MAJOR symptom and at least two MINOR symptoms manifested plus evidence of GAS infection.
- Recurrent ARF (with existing RHD) can be confirmed if

  Two MINOR symptoms plus evidence of GAS infection.

2.3 *Purposed Model for Diagnosis of ARF*

The purposed model for the diagnosis of ARF in Nepal is given below. This diagnosis model has been prepared based on the expertise guidelines and met the Nepali life style and environment (Fig. 1).

2.4 *Signs and Symptoms for Different Level of the Severity*

Based on the NHF's expertise's guidelines, we have separated the signs and symptoms for different level of the severity that are presented in the below. Based on these signs and symptoms we identify the severity level of ARF for patients (Fig. 2).

3 Methodology

In this section we discuss the applied hybrid methodology's structure and its components. Proposed hybrid methodology is a combination of knowledge-based, temporal and fuzzy logic. It is a three model structure that is integrated into one umbrella and applied for designing and deploying the ARF diagnosis model. The structure of hybrid methodology is given below (Fig. 3):

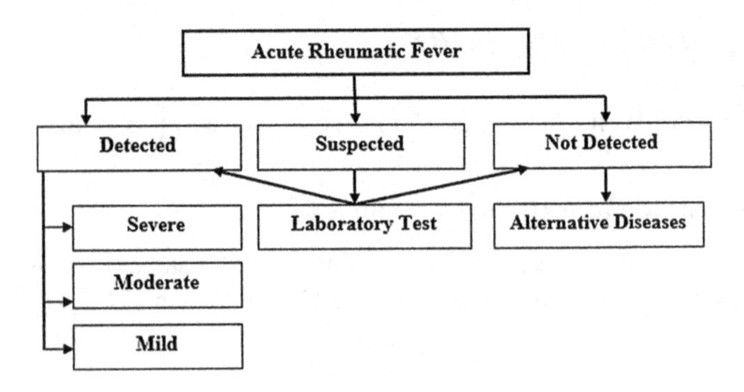

Fig. 1 Purposed model for diagnosis of ARF

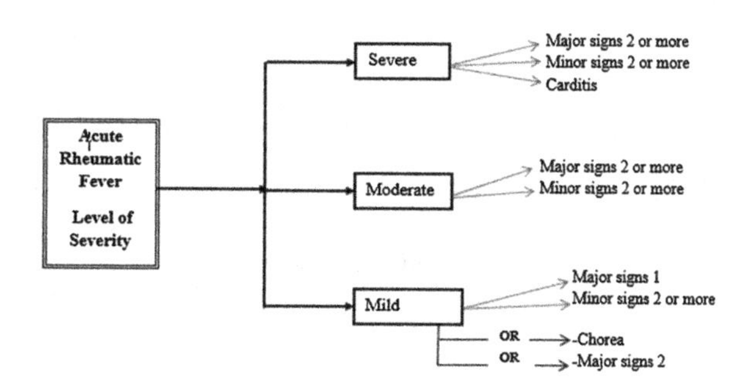

Fig. 2 Signs and symptoms for different level of the severity

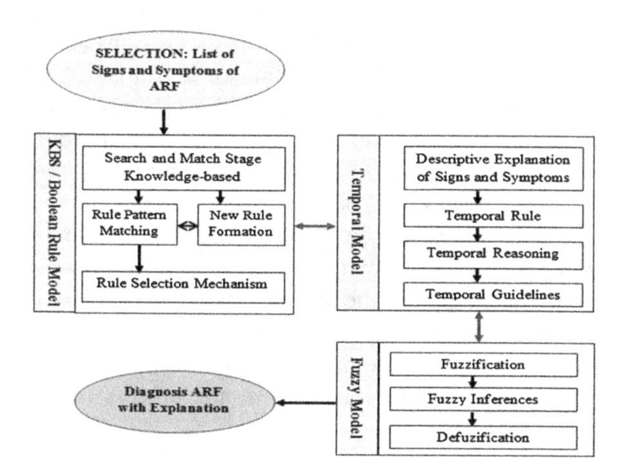

Fig. 3 The structure of hybrid methodology

3.1 Knowledge-Base and Rules to Diagnose of ARF

A knowledge-based model is responsible to identify the ARF status (Detected, Suspected and Not Detected) and the severity level of ARF and construction of the rules. The ARF status and the severity level of ARF has been identified by the presented patients' signs and symptoms. Depending upon the observed traits in a detected ARF patient, the system can diagnose the level of the severity of ARF into one of four categories: suspected, severe, mild or moderate. This categorization (level of the severity) and associated symptoms is shown in Fig. 2. The following Eqs. (1)–(8) have been created to identify the level of the severity of ARF.

$$\textbf{Severe} = \{carditis \wedge essential \geq 1 \wedge major \geq 2 \wedge minor \geq 2\} \tag{1}$$

$$\textbf{Moderate} = \{essential \geq 1 \wedge major \geq 2 \wedge minor \geq 2\} \tag{2}$$

$$\textbf{Mild} = \{essential \geq 1 \wedge major = 1 \wedge minor \geq 1\} \tag{3}$$

$$\mathbf{Mild} = \{chorea\} \tag{4}$$

$$\mathbf{Mild} = \{major \geq 2\} \wedge \{essential \geq 1\} \tag{5}$$

$$\mathbf{Mild} = \{Art_{13} \wedge Fe_{12} \wedge Crp_{15} \wedge essential \geq 1\} \tag{6}$$

$$\mathbf{Suspected} = \{\text{major} = 1 \wedge \text{minor} = 1 \wedge \text{essential} \geq 1\} \tag{7}$$

$$\mathbf{Suspected} = \{Art_{13} \wedge Fe_{12} \wedge essential \geq 1\} \tag{8}$$

We are capturing the patients' sign and symptoms in Boolean expression for example "True" or "False", "0" or "1" etc. We are using "P = present" if signs are presented or observed and "A = absent" if signs are not presented or not observed. The reason for applying Boolean expression is that it can easily capture signs and symptoms for doctors and rural health workers as well as being easy to apply in rule formation. Furthermore, *Boolean principles also underlie logical checking of rule-based decision support systems for inconsistencies, incompleteness and redundancy* [8]. A set of rules have been created for all the severity level of ARF. Each rule has symptoms in the condition part and corresponding severity level of ARF with numbers of positive symptoms are consisting in consequence part.

We have applied the mixed reasoning methods (both backward and forward reasoning/chaining). In the system some of the rules are specially designed and used for forward reasoning and others are used for backward reasoning. The purpose of applying the mixed reasoning process is to maximize the diagnosis efficiency of ARF.

3.1.1 Search and Match Stage (SMS)

SMS is designed and developed to identify the ARF status and the severity level of ARF, based on the observed patients' signs and symptoms and the expertise guidelines. SMS process has 3 different sub components, they are; Rule Pattern Matching (RPM), New Rule Formation (NRF) and Rule Selection Mechanism (RSM). Each component is a sub process of SMS and implemented by a set of algorithms that are explained below. The steps of the algorithm is given below:

Step 1: START → Read and capture all signs and symptoms of patients

Step 2: Define array and store all selected signs and symptoms
Step 3: Create three memory variables to keep the number of major, minor and essential, signs and symptoms from array (which is define in Step 2).
Step 4: Read all signs and symptoms from array
Step 5: Have all signs and symptoms are read from array?

Step 6: IF NO- checks and matches the signs with KBS (ARF manifestation) and identify the satisfied sign then increment value by 1 in appropriate memory variable for example major = major + 1 or minor = minor +1 or essential = essential + 1.
Step 7: Go to next symptoms read (Step 4) and in the Step 5, IF YES then
Step 8: Read the value from three variables (major, minor & essential) and follow the guidelines of KBS, calculate observed sings symptoms and identify the severity level of ARF
Step 9: Display the severity level of ARF with justifications and END

3.1.2 Rule Pattern Matching (RPM)

Pattern matching is the process of matching the new or existing facts against the rule or search for a rule that matches the pattern of data. There are various algorithms used for pattern matching Linear, Rete, Treat, Leaps etc. Here, we used very simplistic pattern matching strategy. The pattern matching task is performed by the inference engine. Pattern matching process will help to fire the exact matching rule by eliminating the unsuitable rules and facts.

After identifying the ARF status, the RPM will automatically activate and identify the exact rule that matches between the pattern and fact (observed signs and symptoms). The pattern of the rule matches the facts in working memory; then the activation of the rule will be fired.

3.1.3 New Rule Formation (NRF)

During the process of RPM, If rule has not been found in the rule-based then NRF is activated which has "Auto Rule Generate Algorithm (ARGA)" that generates a new rule. ARGA is a self-developing algorithm that is applied in this system to formation of new rule automatically. The algorithm is created based on the signs and symptoms for different severity level of ARF and expertise guidelines. This algorithm is very suitable for this system to generate a new rule. However, expertise can revised, delete or add new parameters in the added rule, if required. The benefit of using this algorithm is, to add the new rule, require the domain expertise, knowledge engineer and programmer. So it is time consuming process (depended upon availability of all expertise). Therefore, NRF process act all task (like expertise, knowledge-engineer and programmer) automatically and add the new rule on the rule-based system. The steps of ARGA algorithm is describe below:

Step 1: START → Read and capture all patient's signs and symptoms

Step 2: Search the rule on the rule-based (RPM)
Step 3: If rule found then display the satisfied one with providing the no of positive symptoms and STOP the process
Step 4: If rule not found then "AGRA" activate automatically with displaying all the presented symptoms with providing the options of Add or Cancel rule.
Step 5: "Cancel"—STOP program
Step 6: If "Add"—read all presented symptoms, match the symptoms with each ARF severity's signs and symptoms (Severe, Moderate, Mild and Suspected)
Step 7: Apply the guideline of KBS and add the rule on the rule-based.
Step 8: Display the rule addition message and STOP the program.

3.1.4 Rule Selection Mechanism (RSM)

In this system, RSM concept is simple but very important to the diagnosis of ARF. During the RPM process, it will show only matching rule and based on that make a decision, which will not be accurate enough for the system. Therefore, the main reason to develop and implement this RSM is to diagnose of ARF more effectively. RSM is responsible to search for and select the appropriate rules from every level of the severity—severe, moderate, mild and suspected based on the given total number of positive signs.

3.2 Temporal Logic and Temporal Rule (TR)

The temporal dimension for each signs and symptoms development time is divided up to 24 weeks. Temporal knowledge-based is designed as a temporal lookup table [3]. The sign "GAS infection on the throat" will be a starting time point or origin of time for individual patient. Other captured signs and symptoms presented or development time will be calculated from origin of time and we find out where the particular symptoms consist based on the temporal lookup table. The following equation is used to show the temporal dimension and create a temporal knowledge-based (Table 2).

The sample equation of temporal relation between ARF's signs for absolutely positive are given below:

Table 2 The time span of ARF's signs and symptoms (in days)

Signs and symptom	Time period	Absolutely positive	Very positive	Relatively positive	Suspected
Arthritis	$(P_{11}, P_{12}]$ $x = P_{12} - P_{11}$	$x > 14 \wedge x <= 35$	$(x >= 0 \wedge x <= 14) \vee (x > 35 \wedge x <= 49)$	$x > 49 \wedge x <= 63$	$x > 63 \wedge x <= 168$
Carditis	$(P_{21}, P_{22}]$ $x = P_{22} - P_{21}$	$x > 21 \wedge x <= 42$	$(x >= 0 \wedge x <= 21) \vee (x > 42 \wedge x <= 56)$	$x > 56 \wedge x <= 77$	$x > 77 \wedge x <= 168$
EM	$(P_{31},P_{32}]$ $x = P_{32} - P_{31}$	$x > 14 \wedge x <= 35$	$(x >= 0 \wedge x <= 14) \vee (x > 35 \wedge x <= 140)$	$x > 140 \wedge x <= 168$	$x > 168$
Chorea	$(P_{41}, P_{42}]$ $x = P_{42} - P_{41}$	$x > 35 \wedge x <= 98$	$(x > 21 \wedge x <= 35) \vee (x > 98 \wedge x <= 168)$	$x > 7 \wedge x <= 21$	$x >= 0 \wedge x <= 7$
Subcutaneous nodules	$(P_{51}, P_{52}]$ $x = P_{52} - P_{51}$	$x > 49 \wedge x <= 98$	$(x > 35 \wedge x <= 49) \vee (x > 98 \wedge x <= 119)$	$(x > 21 \wedge x <= 35) \vee (x > 119 \wedge x <= 147)$	$(x >= 0 \wedge x <= 21) \vee (x > 147 \wedge x <= 168)$
Minor symptoms	$(P_{61}, P_{62}]$ $x = P_{62} - P_{61}$	$x > 7 \wedge x <= 49$	$(x >= 0 \wedge x <= 7) \vee (x > 49 \wedge x <= 70)$	$x > 70 \wedge x <= 91$	$X > 91 \wedge x <= 168$

Table 3 Input variables, range and fuzzy set

Input variables	Ranges	Fuzzy sets
All ARF signs and symptoms that presented in above Table 3	0	None
	0–1	Fairly positive
	1–2	Moderately positive
	2–3	Absolutely positive

3.2.1 Relation Between Arthritis and Other Symptoms: Absolutely Positive Case

- Arthritis and Carditis

$$Occurs(arthritis, p_{11}, p_{12}) \wedge Occurs(carditis, p_{21}, p_{22}) \wedge p_{21} - p_{11} = 7 \wedge p_{22} - p_{12} = 7 \wedge p_{12} - p_{21} = 14 \wedge p_{22} - p_{11} = 28, \text{ where} \begin{cases} 14 < p_{12} - p_{11} \leq 35 \\ 21 < p_{22} - p_{21} \leq 42 \end{cases} \quad (9)$$

- Arthritis and Erythema Marginatum (em)

$$Occurs(arthritis, p_{11}, p_{12}) \wedge Occurs(em, p_{31}, p_{32}) \wedge p_{11} = p_{31} \wedge p_{12} = p_{32} \wedge p_{12} - p_{31} = 21 \wedge p_{32} - p_{11} = 21, where \begin{cases} 14 < p_{12} - p_{11} \leq 35 \\ 14 < p_{32} - p_{31} \leq 35 \end{cases} \quad (10)$$

- Arthritis and Chorea

$$Occurs(arthritis, p_{11}, p_{12}) \wedge Occours(chorea, p_{41}, p_{42}) \wedge p_{41} - p_{11} = 21 \wedge p_{12} = p_{41} \wedge p_{42} - p_{12} = 63, where \begin{cases} 14 < p_{12} - p_{11} \leq 35 \\ 35 < p_{42} - p_{41} \leq 98 \end{cases} \quad (11)$$

- Arthritis and Subcutaneous Nodules (sn)

$$Occurs(arthritis, p_{11}, p_{12}) \wedge Occours(sn, p_{51}, p_{52}) \wedge p_{51} - p_{12} = 14 \wedge p_{51} - p_{11} = 35 \wedge p_{52} - p_{12} = 63, where \begin{cases} 14 < p_{12} - p_{11} \leq 35 \\ 49 < p_{52} - p_{51} \leq 98 \end{cases} \quad (12)$$

- Arthritis and Minor Symptoms (ms)

$$Occurs(arthritis, p_{11}, p_{12}) \wedge Occurs(ms, p_{41}, p_{42}) \wedge p_{11} - p_{61} = 7 \wedge p_{62} - p_{12} = 14 \wedge p_{12} - p_{61} = 28, where \begin{cases} 14 < p_{12} - p_{11} \leq 35 \\ 7 < p_{62} - p_{61} \leq 49 \end{cases} \quad (13)$$

3.2.2 Temporal Reasoning (TR) and Temporal Guideline (TG)

TR is to determine and analyse the relation between symptoms as well as provide the linguistic information how closely related a particular symptom with ARF. Temporal abstraction will be applied to analyse the relation between the symptoms and absolute-time granularity, will capture the temporal dimension of symptoms and its relationship with ARF is analysed. We created point-based time patterns of symptoms and, based on these time dimension, we introduce how precisely related of particular signs with ARF [3, 9, 10].

The guideline information is designed in linguistic variable for instances “Absolutely Positive for ARF”, “Very Positive for ARF”, “Relatively Positive for ARF” and “Suspected for ARF”. This linguistic base information will provide support for rural health workers to understand the relation between symptoms and ARF. The linguistic variable and time frame is designed based on the expertise support from Nepal. Temporal rule and temporal reasoning will interpret each patient’s symptoms and how to precisely relate with ARF.

3.3 *Fuzzy Logic*

The main concept of fuzzy logic is that a statement cannot be restricted to being only true or false; it must have some freedom of partial truth. Moreover, Fuzzy logic has a great potential to use linguistic variables, for example few, slow, fast, large, heavy, severe, mild, low, medium, high, short, average, tall etc. [11] Monotonically Increasing Linear Membership Function (MILMF), Root Sum Square, and Centre of Gravity methods are applied for fuzzification, fuzzy inference and defuziffication process respectively. MILMF converted the crisp value into fuzzy value for this system. The monotonically increasing linear membership function and equation is given below [12] (Fig. 4).

3.3.1 Fuzzy Inferences

MILMF is applied to determine the degree of symptoms to ascertain the doctor’s belief of the symptoms’ severity. The fuzzy inference mechanism will map the entire set of rules with membership degrees. Then the output of each rule is aggregated and Mamdani’s Centre of Gravity (CoG) method is applied for the defuzzification process. Root Sum Square (RSS) will apply for the getting the values from firing rules. To achieve this following equation is applied.

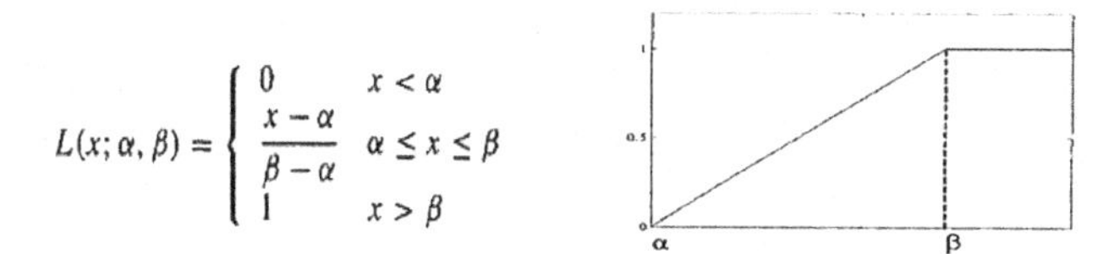

Fig. 4 Monotonically increasing linear membership function and equation

$$\begin{aligned} &\mathrm{Rv_i \ldots Rv_n = Value\ of\ firing\ Rules} \\ &\mathrm{Ov_i \ldots Ov_n = Output\ value\ of\ RSS} \\ &\mathrm{Ov_i} = \sqrt{Rv_i^2} \end{aligned} \quad (14)$$

The equation of defuzzification process given below:

$$\mathrm{COG\ (Y)} = \frac{\sum \mu_y(x_i) x_i}{\sum \mu_y(x_i)} \quad (15)$$

where,

$\mu_y(x_i)$ = Membership value in the membership function
x_i = centre value of membership function

4 System Development

This application is developed by applying three-tier concept. The three-tier architecture enables to increase performance, easy to maintain code, flexibility and reusability. This application has been deployed using C#, window forms on the front-end and MS access as back-end. The front-end is developed to be very simple and easy to use and consists of various forms for example login, patient registration, check-up, patient history, help/guidelines etc. The development environment of system is Visual Studio.Net 2012 and Framework is .Net Framework 4.0. The user interface is Windows and the programming language is C#.net. MS Access 2010 is used for database and ADO.NET is using for data access. The components of layered architecture are given below:

Presentation layer/Graphical User Interface (GUI): Presentation layer contains the user screen of various programs or forms associated with GUI components and code-behind it which handles the assign instruction appropriately. This layer is delivers the information to the user and to the system.

Data Layer: This layer has other sub components; the *business service methods*, *data access logic*, *data model* and all required *SQL coding*. Business service is a method to communication with the data access. This service is calling the data access methods and delivering the result presentation layer. Data model is a set of entities (table) where we stored the records. SQL code is a set of statements that is used to manipulate the data.

Data Access: Data access task is to accessing the data on the database via SQL code.

4.1 User Interface for Diagnosis of ARF Application

This application has various input and output forms. User logon, patient's registration, view patients history/report, diagnosis, patient report, view/update rules etc. Only registered user can use the application. Expertise can update, modify, delete, and add rules etc. Moreover, expertise is responsible to manage all the rules and help/guideline information. The sample interfaces of the application are given below (Figs. 5, 6, 7, 8 and 9).

5 System Testing and Conclusion

The developed system has been tested with NHF's patients' registry data (585 records) sets and it reflects that 93.6 % were matches. Therefore, we proved that this hybrid methodology can handle both temporal abstraction and uncertainty in an integrated way and shown to be effective in the diagnosis of ARF at an early stage.

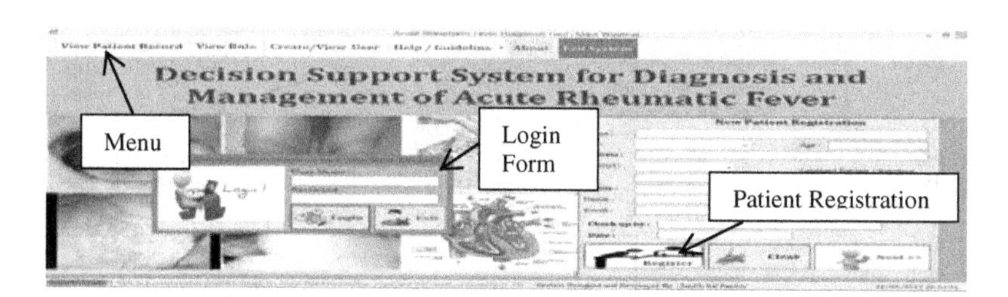

Fig. 5 User login, application's main window and new patient registration

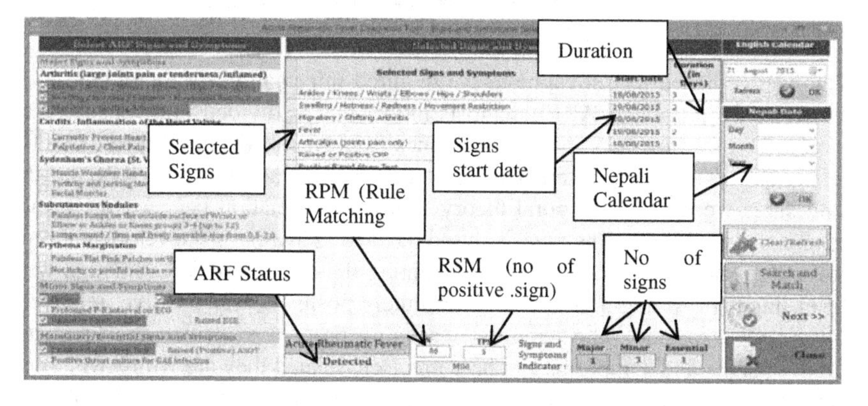

Fig. 6 Patient's signs and symptoms selection form

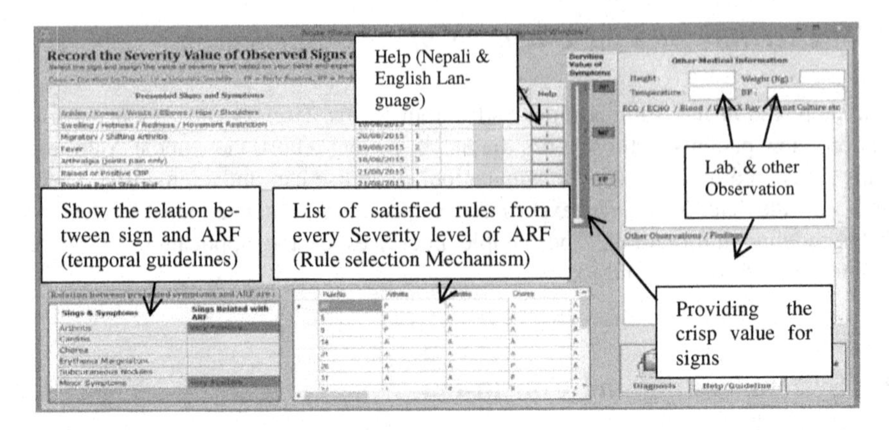

Fig. 7 Signs and symptoms severity level

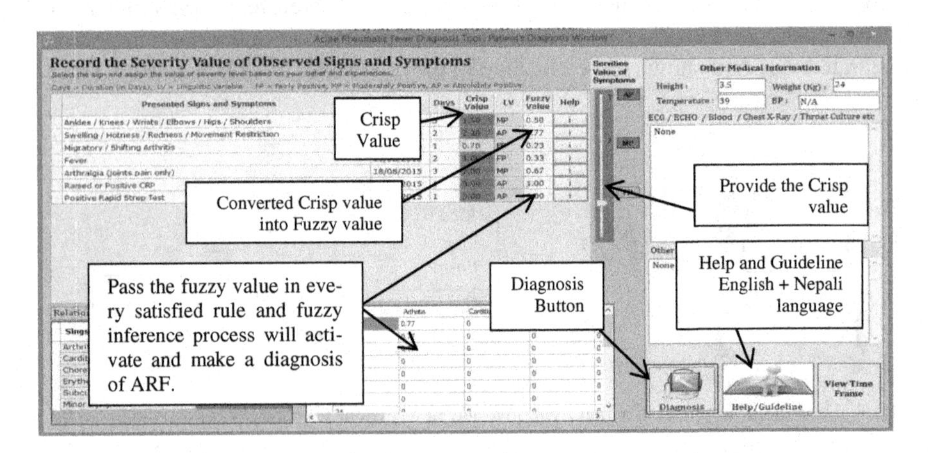

Fig. 8 Recording the fuzzy value and other (e.g. laboratory result, other observations)

Knowledge-based model captured the required information for the diagnosis of ARF. Then we presented ARF diagnosis model that is divided into four severity level of ARF based on the expertise guidelines from Nepal. We developed a rule for the diagnosis of ARF in different level of the severity according to KBS. In the second stage we applied temporal theory to show the relation between the symptoms and analysed them. Then we analysed the ARF's symptoms development time frame and created a temporal rule by explaining the relation of the each sign with ARF in linguistic strength. These are: absolutely positive, very positive, relatively positive and suspected. The temporal reasoning process explained the linguistic strength of signs and symptoms and how precisely related they are with ARF. Then we applied fuzzy logic to handle the uncertainty as well as make a decision of ARF based on the qualitative precise temporal guidelines.

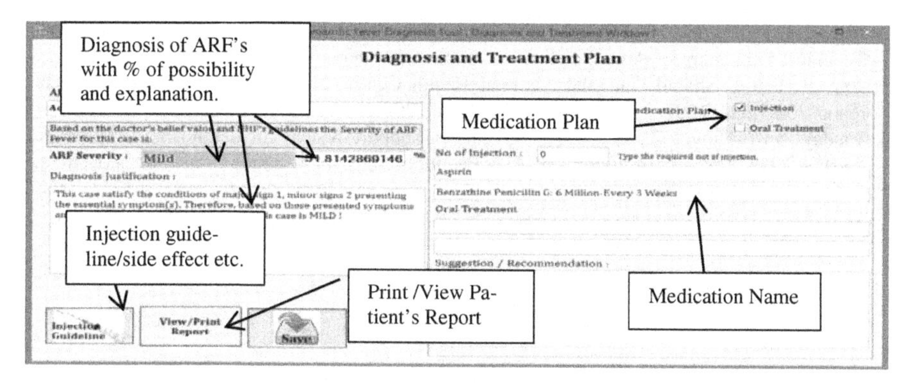

Fig. 9 Recording the fuzzy value and other (e.g. laboratory result, other observations)

Our hybrid methodologies proved to be an appropriate solution to the development of a decision support system for ARF. It is very useful to diagnose ARF an early stage where manpower and technologies are not guaranteed. It is cost effective, easy to use, flexible, and suitable for the Nepali environment and lifestyle and introduces the unique procedure for the diagnosis of ARF in Nepal.

References

1. Regmi, P.R., et al.: Prevalence of Sydenham's chorea in patients with acute rheumatic fever in Nepal. www.nepjol.info/index.php/nhj/article/download/8345/6794. Accessed 03 Apr 2015
2. National Seminar on RF/RHD Prevention and Control in Nepal, July 2011
3. Pandey, S., Ma, J., Lai, C-H., Regmi, P.R.: A Conceptual framework to diagnosis of acute rheumatic fever based on the temporal and fuzzy logic approach. Can. Int. J. Sci. Technol. ISBN: 8472356X (accepted)
4. Rheumatic fever and rheumatic heart disease. Report of a WHO Expert Consultations, WHO Technical Report Series, Geneva, 29 Oct–1 Nov 2001
5. World Heart Federation: Diagnosis and management of acute rheumatic fever rheumatic heart disease. World Heart Federation. http://www.world-heart-federation.org/ (2007). Accessed 24 Feb 2015
6. NHS Choice (Rheumatic Fever): http://www.nhs.uk/conditions/Rheumatic-fever/Pages/Introduction.aspx. Assessed 9 Feb 2015
7. Rheumatic fever and rheumatic heart disease. WHO Technical Report Series 923, Report of a WHO Expert Consultations, Geneva. http://www.who.int/cardiovascular_diseases/resources/en/cvd_trs923.pdf. Accessed 24 Feb 2015
8. Encyclopedia of Medical Decision Making. Contribution by J. Hilden, j.hilden@biostat.ku.dk (2009)
9. Allen, J.F.: Maintaining knowledge about temporal intervals. CACM **26**(11), 832–843 (1983)
10. Ma, J., Knight, B.: A general temporal theory. Comput. J. **37**(2), 114–123 (1994)
11. Pandey, S., Ma, J., Lai C.-H.: Development of decision support system for the diagnosis of arthritis pain for rheumatic fever patients: based on the fuzzy approach. J. Algorithms Comput. Technol. [in press for printing]

12. Pappis, C.P., Siettos, C.: Chapter 15 Fuzzy Reasoning. University of Piraeus, National Technical University of Athens School of Applied Mathematics and Physics, Athens
13. Annual Report, Shahid Gangalal National Health Centre, Nepal. http://www.sgnhc.org.np/report-2007/main%20Page%201%20to%2066.pdf (2007). Accessed 27 Feb 2015
14. Pandey, S.R., Njovu, C., Lai, C-H.: **A** decision support system for diagnosis of rheumatic fever in Nepal. In: Proceeding of 7th Asia Pacific Medical Informatics Conferences, Beijing, China, 22–25 Oct 2012

Applications of Genetic Algorithms

Optimising Skill Matching in the Service Industry for Large Multi-skilled Workforces

R.T. Ainslie, S. Shakya, J. McCall and G. Owusu

Abstract The continued drive to improve efficiency within the service operations sector is motivating the development of more sophisticated service chain planning tools to aid in longer term planning decisions. This involves optimising resource against expected demand and is critical for successful operations of service industries with large multi-skilled workforces, such as telecoms, utility companies and logistic companies. To effectively plan over longer durations a key requirement is the ability to simulate the effects any long term decisions have on the shorter term planning processes. For this purpose, a mathematical model encapsulating all the factors of the shorter term planning, such as skills, geographical constraints, and other business objectives was defined. Attempting to use conventional methods to optimise over this model highlighted poor scalability as the complexity increased. This has motivated the development of a heuristic method to provide near optimal solutions to the model in a shorter timescale. The specific problem we look at is that of matching resource to demand across the skill dimension. We design a genetic algorithm to solve this problem and show that it produces better solutions than a current planning approach, providing a powerful means to automate that process. We also show it reaching near optimal solutions in all cases, proving it is a feasible replacement for the poorly scaling linear model approach.

R.T. Ainslie (✉) · J. McCall
Robert Gordon University, Aberdeen AB10 7QB, UK
e-mail: r.t.ainslie@rgu.ac.uk

J. McCall
e-mail: j.mccall@rgu.ac.uk

S. Shakya · G. Owusu
British Telecommunications Plc, IP5 3RE London, UK
e-mail: sid.shakya@bt.com

G. Owusu
e-mail: gilbert.owusu@bt.com

M. Bramer and M. Petridis (eds.), *Research and Development in Intelligent Systems XXXII*, DOI 10.1007/978-3-319-25032-8_18

1 Introduction

In recent decades, with the advent of privatisation and the increase in competition, there has been a drive in the service operations sector to improve the sophistication of supporting applications up to the level of the far more mature supply sector [1]. Recent research has shown that this is increasingly involving the automation of current processes [2].

One of the key areas for development lies within the field of service chain planning. Planning for a service industry involves the matching of resources' available time to the jobs requiring completion. This matching occurs along the dimensions of time, area and skill. Each job requires a skill to complete, in a certain geographical area, within a certain period of the plan. The planner attempts to match this with the skills and areas a resource can cover within each period, by using some simple rules [3].

Further decisions available to the planner come in the form of budget constraints and business objectives [4]. Here budget constraints impose limits on the overall plan, such as hiring, overtime etc., while business objectives denote the actual utilisation of that budget, such as when and what type of resource to hire. This can add further complexity to the planning problem.

In order to overcome this, a mixed integer linear programming [5] model was formulated to capture all of the factors in the planning process. The aims were to use the model to both provide an improvement over a manual planning process as well as to provide a means to simulate the effects of any long term decisions.

During this process it was found that as more complex decisions were added to the linear model, such as the decision to recruit new resources, the required solution time rapidly increased. This quickly reached infeasible durations for the purposes of dynamic planning and scenario modelling.

This discovery has motivated the investigation of heuristic methods [6–10] capable of providing near optimal solutions to planning problems within a more reasonable amount of time.

In this paper, we present a Genetic Algorithm (GA) [11, 12] to perform resource-demand matching to optimally use the available skills. We investigate whether this genetic algorithm provides a suitable automation of the planning process, whilst also discovering if it could be a good replacement for the badly scaling linear programming approach.

In the remainder of this paper, we first describe the skill matching problem and define it mathematically in Sect. 2. The solution methods we used are covered in Sect. 3, including the newly formulated GA representation, along with the method used to simulate the manual process and linear solver used for comparisons. We then outline the experiments performed to compare their performance in Sect. 4 before presenting the results with their analysis in Sect. 5. Finally we present our conclusions in Sect. 6.

2 Skill Matching Problem

In this section we define the problem of matching resources and their skills to demand, as faced by planners in service industry organisations with large multi-skilled workforces.

2.1 Problem Description

A large multi-skilled workforce contains a number of resources, with each resource having a number of skills. Further to this, each resource also has a preference level allocated to the skill they perform. This value is used to indicate which skills they are better practiced at performing and is a factor in producing an optimal plan (the planner should prefer to use a resource's higher preference skills where possible). Each resource has a total time available to it which requires allocation to the skills it can perform.

The organisation also has a demand, consisting of a set of jobs requiring completion in a particular period. Each job has a skill that is required to complete them, along with an amount of time that the job needs applied to it. The final factor is the priority of the job. This priority is used to ensure that the planner attempts to fulfil the more important jobs first.

The job of the planner, therefore, is to attempt to best match the skills of resources time to the jobs making up the demand, subject to the amount of each resource available in the time period.

For a manual planner however, individually matching between resources and jobs is not practical. In this case, a solution is to plan using the skill variances. The variance for a skill is defined as the total time supplied for that skill from the resources, minus the total time required for that skill from the jobs. A positive variance indicates there is over-allocation to that skill (surplus); a negative variance shows there is not enough time allocated to that skill (deficit). The planner then attempts to move time from skills with a surplus to those with deficits to bring each variance value as close to zero as possible.

This can be defined as the constrained optimisation problem.

$$min\colon \sum_{i,s \in S_i} \omega_{is} x_{is} + \sum_j \varphi_j (\rho_j - y_j) + E \sum_s z_s \tag{1}$$

subject to:

1. $\sum_s x_{is} = \sigma_i \forall i$
2. $y_j \leq \rho_j \forall j$
3. $\sum_i x_{is} - \sum_{s_j = s} y_j - z_s = 0 \forall s$

4. $x_{is}, y_j, z_s \geq 0$
5. $x_{is}, y_j, z_s \in \mathbb{R}$

Equation 1 defines the objective function for resource-demand matching in the single planning dimension of resource skills. The equation can be broken down into three key components, denoted by each summation operator.

The first component calculates the cost of the use of resources in the current plan. The decision variable here, x_{is}, denotes the amount of time from resource i allocated to skill s. This is multiplied by ω_{is} which is the associated cost for this resource using that skill, set based on the resource's preference for the skill. The set of skills a resource has is denoted by S_i, thus this component of the equation sums the allocation multiplied by the cost for each skill each resource has.

The second component calculates the cost of any unmet demand. Here y_j is the decision variable, indicating the amount of time allocated to job j. The other variables in this section denote the total time required for the job ρ_j, and the penalty cost for not allocating enough time φ_j. Thus the unmet-demand cost is calculated by summing the difference between the required and the allocated time to each job multiplied by their penalty cost.

The final component is an additional penalty cost for over-allocation to any skills. The decision variable here, z_s, denotes extra time allocated to skill s that hasn't been assigned to complete a job. The cost E is the penalty factor for this over-allocation. For the purposes of this comparison the cost was set to zero.

Valid solutions to the equation are defined by the constraints. The first constraint (I) states that the total time allocated to a resource's skills should equal the total time available to that resource (σ_i). Similarly, the second constraint (II) limits the amount of time applied to a job to be less than or equal to the time required by that job. The third constraint (III) links the components of the objective function, stating that the total time for a skill supplied from the resource side must equal the total time used by the demand side of the resource-demand matching. The final constraints (IV and V) limit the decision variables to positive real values.

3 Solution Methods

This section outlines the methods we will use to solve the planning problem outlined in Sect. 2. The cost of the solution provided by each of these methods will be calculated using Eq. (1) to provide an equal comparison.

3.1 GA

To solve this problem using a GA, first we design the solution representation. Then define the operators for use with this solution before describing the fitness function.

Finally we briefly outline the pseudo code for the algorithm before covering the implementation we used for the comparison experiments.

3.1.1 Solution Representation

In order to represent an individual resource's available time distribution a list of real values is used. Each value on this list equates to the amount of time that resource spends on each skill they can perform (analogous to xis in Eq. (1)). The sum of these values is constrained to equal the time that resource has available (constraint I).

This list of real values is an individual gene of the solution. The entire solution is made up of a list of these genes, one for each resource in the problem.

Take, for example, a problem containing four resources. The first can perform three different skills and has 8 h available, the second can perform two different skills and has 10 h available, the third can perform four skills and has 12 h available and the fourth can perform two skills and has 8 h available. A valid solution for this problem could be:

$$[6.0, 2.0, 0.0], [5.0, 5.0], [3.0, 2.0, 0.0, 7.0], [1.0, 7.0]$$

Here, the first gene represents the time allocation for the first resource. 6 h of their time is allocated to their first skill, two to their second and zero to their third.

Similarly the second gene represents the time allocation for the second resource, with their time equally split with 5 h for each resource they can perform and so on for the third and fourth.

3.1.2 Operators

The operators chosen for use were a uniform crossover [13] with the tournament selection operator [14].

Due to the nature of the individual genes however, something other than the standard mutation operators had to be defined. Three operators were implemented for this GA. The first completely re-randomises a gene (and is the same operator used when initially generating the gene values for the solutions in the initial population). It randomly allocates the resource's available time, and ensures that the generated gene is valid.

The second involves some minor modifications to the mutated distribution. A random amount is subtracted from one randomly selected value in the gene and added to the other. If the amount to be subtracted is greater than the value selected, then the value is set to zero and its previous value is added to another. For example, given a gene:

$$[4.0, 2.0, 0.0, 2.0]$$

The random modification amount is chosen as 3.0, and the second point is chosen as the source with the first as the destination.

As the second point is 2.0, subtracting 3.0 would cause it to become negative and thus an invalid solution. Instead 2.0 is subtracted from the second point and added to the first, giving the resulting gene:

$$[6.0, 0.0, 0.0, 2.0]$$

In order to retain the effectiveness of this mutation in problems with longer genes the number of changes occurring for any given mutation is chosen randomly (with an upper limit of half the genes length).

The third mutation operator is a hybrid of the first two, completely re-randomising for the first few generations (the number tuneable by the operator) before swapping to just performing minor modifications for the rest of the evolution.

3.1.3 Fitness Function

The fitness function for this genetic algorithm calculates the cost of the solution in a similar method to Eq. 1. As mentioned in the solution representation, the GA solution provides the values for x_{is}. To speed calculation of this portion of the equation, the fitness function contains an index containing the associated cost for each value in the solution (ω_{is}). The fitness function need only multiply these together to calculate the resource portion of the cost function.

Unlike the linear model however, the GA does not explicitly allocate the decision variables for the demand side of the equation.

To calculate the cost of the unmet demand the fitness function splits the jobs into groups by their skill required and orders them in priority order. The time allocated to each skill by the resources is applied to complete the jobs of the same skill, fulfilling the highest priority job first. This provides the values for y_j in Eq. 1.

The values for z_s are calculated as any excess time left over after all jobs of a skill are fulfilled. As mention in Sect. 2 however, for the purposes of these experiments the additional cost was set to zero.

3.1.4 Pseudo Code

The final part of the GA required for implementation is the decision around the stopping condition. As the solution complexity depends on the number of resources and possible skills in a data-set, then the precise number of generations required to reach a near optimal solution would vary per problem. Something dynamic was required to allow the GA to run longer for larger problems or stop sooner on smaller ones. The solution implemented here was to stop the evolution upon reaching stagnation. This involves reaching termination when the best fitness has not improved for a tuneable number of generations.

Bringing all the components together we get the GA algorithm as:

```
1.  Generate Initial Population;
2.  Evaluate population;
3.  Set best solution = solution with best fitness;
4.  Set Stagnation Generations = 0;
5.  While (stagnation generations < termination limit) {
6.        Select population for breeding;
7.        Perform crossover on selected population;
8.        Perform mutation on selected population;
9.        Evaluate population;
10.       If (current best solution has better fitness than
    best solution) {
11.               Stagnation Generations = 0;
12.               Best solution = current best;
13.       }
14.       Else {
15.               Stagnation Generations++;
16.       }
17. }
18. Return best solution;
```

3.1.5 GA Implementation

The previously described GA was implemented in Java making use of the Watchmaker framework [15].

After some simple test runs the crossover probability was set to 0.6 with a mutation probability of 0.02. The mutation was set to perform re-randomisation for the first 500 generations then use the gene modification method thereafter. Population size was chosen as 250 with an elitism value of 3. The algorithm was set to terminate after reaching stagnation for 100 generations.

3.2 Simple Planner Heuristic

To provide some real world comparisons, a method to simulate the manual planning process is also used.

The first step assigns all of a resource's available time to their primary (most preferred) skill. This is analogous with how the data is usually first represented in a planning application, before any manual (or otherwise) planning is performed.

The second step applies a simple rule based optimisation to attempt to move surplus time allocated to one skill to any skills with deficits where possible. The rule attempts to move time to the highest priority skills first. This is essentially the heuristic used when performing manual planning.

3.3 Linear Program

The constrained optimisation problem defined in Sect. 2 was also solved using CPLEX [16]. The model was implemented in java using SCPSolver [17] to interface between Java and the CPLEX solver. This method is known to show poor scalability as the problem complexity increases therefore is not a candidate for use in a planning application. However, it does provide the optimal solution for our experiments and thus provides a good comparison to judge the performance of the solutions provided by the GA.

4 Experiments

In this section we define the experiments to perform a comparison between the methods outlined in Sect. 3. First we outline the motivation for the experiments before explaining the method used.

4.1 Motivation

The purpose of the experiments conducted was to provide an effective comparison between the formulated GA and the simple planner's heuristic used by current planners. The objective being to show that the GA provides an improvement over the current methods by performing a more thorough search of the solution space, thus making a better candidate for use to automate this planning process.

Further to this, the model is also compared against the results obtained by solving the linear model using the CPLEX solver. Although this provides the optimal result, we know that the linear program doesn't scale well to more complex problems. With a drive of this paper being the production of an alternate method with better scalability, the goal at this stage is to show the GA producing near optimal solutions. Solution time is not tested as this is not an issue for the skill matching portion of the model, this will be explored in future papers once more complex parts of the full model have been implemented.

To analyse the relative effectiveness of each method, we look both at some skill variance graphs to analyse the allocation of time to each skill as well as looking at the overall cost of the solutions produced.

4.2 Experimental Method

A logic was built to generate randomised data for a specified number of skills, resources and work-stacks (jobs). Each generated resource was allocated a random number of skills (each with an associated preference) and a random amount of time available. Each work-stack was generated to require a random skill, require a random amount of time and was allocated a random priority level.

To ensure the generated data resembled a possible real-world situation, the range for possible values of time available to a resource and for time required by jobs were set such that selecting the mean in all cases would produce a data-set where the total time available equalled the total time required. This creates a situation where there is a workforce that generally matches the demand, the job of the planner is simply to perform the skill matching between the two.

This logic was used to generate 12 separate data-sets, each with 50 resources, 10 skills and 200 work-stacks, to simulate a typical planning situation for a service company [18].

Each of the methods outlined in Sect. 3 was used to solve the planning problems arising from the resulting data-sets. As the genetic algorithm is non-deterministic the model was run for 100 repetitions on each of those 12 problems and the mean results noted. The alternate methods were all deterministic and thus only run once per problem.

As well as the optimisation methods, a 4th dataset was produced to show the results of simply allocating all of the time to each resources' primary skills. This was to provide a baseline solution.

5 Results

Here we present the results of the comparison experiments. First we look at the graph showing the costs of the solutions produced by each method in each problem. After that we pick out a few skill variance graphs to illustrate how each handles the allocation of resource skills to the required demand.

5.1 Solution Cost

Figure 1 shows the cost values achieved by using Eq. (1) to evaluate the solutions. In this case the lower the value the better the solution. The graph shows that the GA approach is producing better results than the simple planner in all cases. The GA is also producing a cost close to the optimal in all the problems.

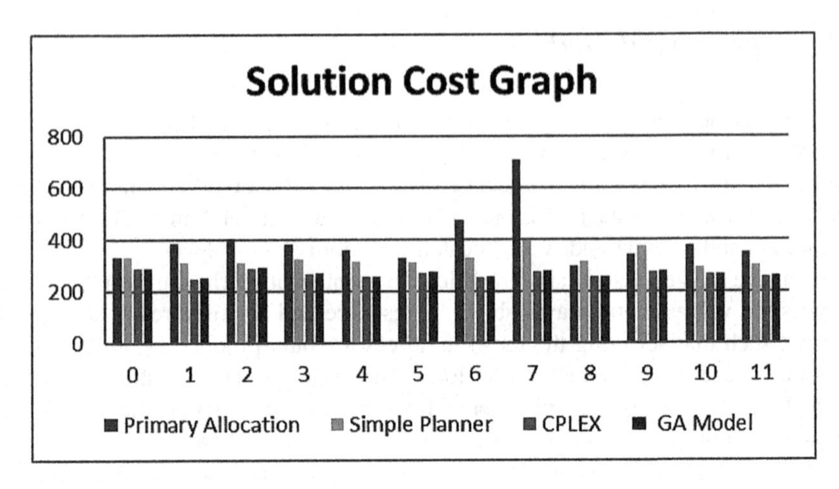

Fig. 1 Solution costs

5.2 Skill Variances

With 10 skills in the data-set there were too many graphs to list them all here, for the purposes of analysis a few representative result sets were selected.

Figure 2 shows a typical looking variance curve; in this case we are looking at the values for skill 8 across the 12 problems. In all cases the optimisation methods reach values the same as or closer to zero than the base solution. Most methods have managed to get close to zero in all the problems; the simple planner heuristic however shows some under-allocation in problem 6.

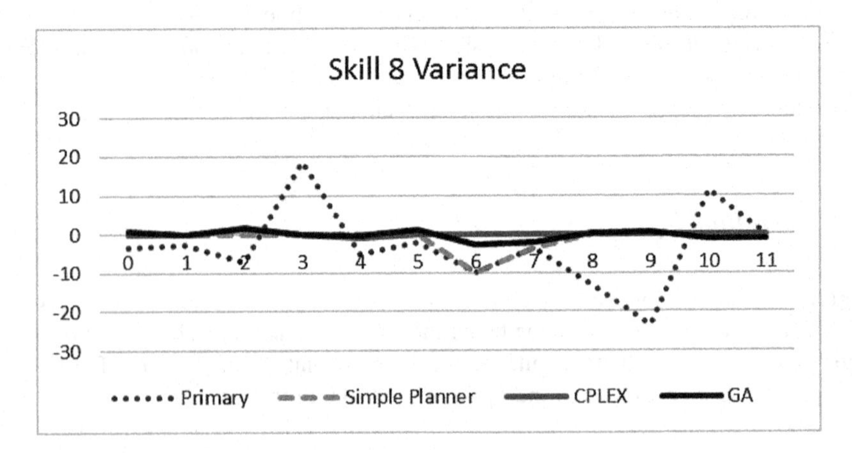

Fig. 2 Typical variance improvement

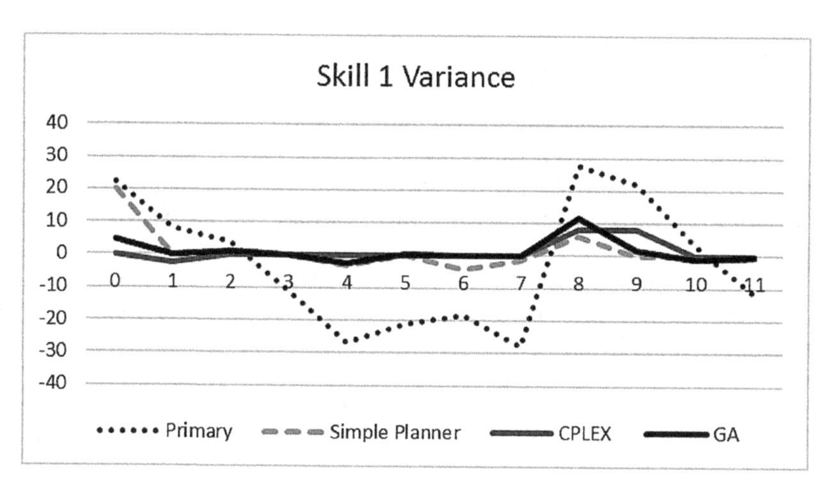

Fig. 3 Higher variance than simple planner in problem 8

Figure 3 shows a less typical variance graph. This graph is chosen as it highlights a case where the GA has produced a higher variance than the simple planner heuristic in problem 8. The rest of the problems produced typical results however.

Figure 4 shows the balancing value for Fig. 3 in problem 8. Here in problem 8 it is the simple planner heuristic that has over-allocated while the GA has reached a balanced allocation. The rest of the graph shows typical values, with the GA generally producing similar or better variance than the simple planner heuristic.

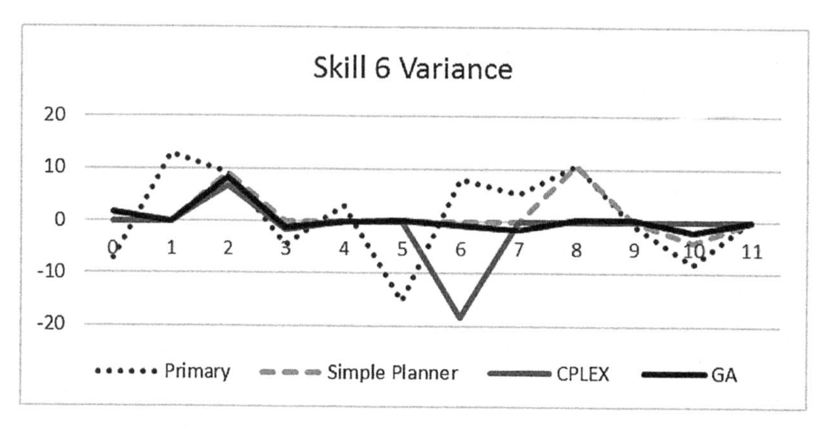

Fig. 4 Lower variance than simple planner in problem 8

5.3 Analysis

The results of the tests show a significant improvement in using the GA model over current simple planning practices, as can be seen in Fig. 1. The GA produces a lower cost solution in all cases. Further illustrated by the same graph is that the GA also performs comparably to the linear model, reaching optimal or near optimal in all cases.

Analysing the variance graphs, we can see that the GA also outperforms the simple planner in most cases here, reaching the same or better variance values for the skills in most problems.

The one case where the GA produces a worse variance, comparatively to the simple planner heuristic, for skill 1 in problem 8 (Fig. 3), is balanced by a better variance for skill 6 (Fig. 4). The differing decisions producing a lower cost for the GA solution, as seen in Fig. 1.

The likely reason for the cost reduction is that the simple planner has a tendency to become stuck within local optima. This is a side effect of starting from an initial point before making moves to improve the solution. A GA, in comparison, performs a more robust exploration of the solution space.

6 Conclusions

In this paper we have used a GA approach to solve the skill matching problem faced by members of the service industry, mainly those with large multi-skilled mobile field workforces.

We have shown that this GA produces superior results to the simple planner heuristic, providing a lower cost solution in all problems they were applied to. As this simulates the current manual planning approach we have proven that the GA would provide an improvement over the current planning process. This provides a strong case for utilising the GA to automate the planning process.

We also demonstrated that the GA we have presented produces near optimal solutions in all the test problems. This proves the GA is a strong alternative to the badly scaling linear programming approach. This provides a platform to perform future work in this area, expanding the GA to cover further planning dimensions, such as that of area and period.

References

1. Voudouris, C.: Defining and understanding service chain management. Service Chain Management, pp. 1–17. Springer (2008)
2. Owusu, G., O'Brien, P., McCall, J., Doherty, N.F.: Transforming Field and Service Operations. Springer, Stellenbosch (2013)

3. Shakya, S., Kassem, S., Mohamed, A., Hagras, H., Owusu, G.: Enhancing field service operations via fuzzy automation of tactical supply plan. Transforming Field and Service Operations, pp. 101–114. Springer (2013)
4. Owusu, G,. Anim-Ansah, G., Kern, M.: Strategic resource planning. Service Chain Management, pp. 35–49. Springer (2008)
5. Floudas, C.A., Lin, X.: Mixed integer linear programming in process scheduling: modeling, algorithms, and applications. Annals of Operations Research, pp. 131–162. Springer (2005)
6. Ashlock, D.: Evolutionary Computation for Modeling and Optimization, Springer, Heidelberg (2006)
7. Kordon, K.: Applying Computational Intelligence: How To Create Value, Springer, Berlin (2010)
8. Shakya, S., Santana, R.: A review of estimation of distribution algorithms and Markov networks. Markov Networks in Evolutionary Algorithms. Adaptation, Learning and Optimization. Series vol. 14, Springer, pp. 21–37 (2012)
9. Kiranyaz, S., Ince, T., Gabbouj, M.: Multidimensional Particle Swarm Optimization for Machine Learning and Pattern Recognition. Springer, Berlin (2014)
10. Dorigo, M., Stutzle, T., Ant Colony Optimization. MIT Press, Cambridge (2004)
11. Goldberg, D.: Genetic Algorithms in Search, Optimization amd Machine Learning. Addison-Wesley, Boston (1989)
12. Haupt, R.L., Haupt, S., Practical Genetic Algorithms, 2nd edn. Wiley, New York (2004)
13. Spears, W.M., De Jong, K.D.: On the virtues of parameterized uniform crossover. Naval Research Laboratory, Washington (1995)
14. Goldberg, D.E., Deb, K.: A comparative analysis of selection schemes used in genetic algorithms. Foundations of Genetic Algorithms 1, pp. 69–93. Morgan Kaufmann Publishers, Inc. (1991)
15. Watchmaker framework for evolutionary computation. http://watchmaker.uncommons.org/. Accessed May 2015
16. CPLEX optimizer. http://www-01.ibm.com/software/commerce/optimization/cplex-optimizer/. Accessed Jan 2015
17. SCPSolver—an easy to use java linear programming interface. http://scpsolver.org/. Accessed Jan 2015
18. Voudouris, C., Owusu, G., Dorne, R., Lesaint, D.: Forecasting and demand planning. Service Chain Management, pp. 51–64. Springer (2008)

Hybrid Optimization Approach for Multi-Layer FTTH Network Design

Kaltham Al Romaithi, Kin Fai Poon, Anis Ouali, Peng-Yong Kong and Sid Shakya

Abstract The growth in demand of bandwidth-intensive services is ever increasing. Telecommunication companies have moved to employing next generation access network architectures to meet the requirements of customers for over a decade. FTTH networks, in particular, have proven to be a cost effective solution for the last mile of access networks. There are many design factors to put into consideration, making the optical network planning design a challenging process. This drives the need for an intelligent network planning system that can minimize both the time and cost for creating a network. In this paper, we consider the design of FTTH GPON network in a greenfield area and we specifically tackle the problem of fiber duct sharing. A hybrid approach, which utilizes Genetic Algorithm and Binary Integer Programming, to solve the problem is detailed. Results are provided in this paper demonstrating the effectiveness and feasibility of such an approach.

1 Introduction

In the last 10–15 years, there has been an increase in subscriber demands for bandwidth due to the use of multimedia applications that are typically bandwidth-intensive services. Due to the limited capacity of copper technologies based on

K. Al Romaithi · K.F. Poon(✉) · A. Ouali
Etisalat British Telecom Innovation Centre, Abu Dhabi, Uae
e-mail: kin.poon@kustar.ac.ae

K. Al Romaithi
e-mail: kaltham.alromaithi@kustar.ac.ae

A. Ouali
e-mail: anis.ouali@kustar.ac.ae

P.-Y. Kong
Khalifa University of Science, Technology And Research, Abu Dhabi, Uae
e-mail: pengyong.kong@kustar.ac.ae

S. Shakya
British Telecom, Ipswich, UK
e-mail: sid.shakya@bt.com

M. Bramer and M. Petridis (eds.), *Research and Development in Intelligent Systems XXXII*, DOI 10.1007/978-3-319-25032-8_19

xDSL, telecommunication companies have been looking to Gigabit Passive Optical Networks (GPON) that can provide the necessary bandwidth [6]. There are several known topologies referred to as FTTx, depending on where the termination of the optical fiber takes place [7]. The next generation access network architectures have the ability to overcome the last-mile access bottleneck by providing high-bandwidth and low delay simultaneously [2]. The closer the fiber to the customer premises, the faster the connection is. Therefore, Fiber to the Home (FTTH) networks have become an attractive solution to operators.

Typically, there are two types of architectures that can be used to implement FTTH networks: Point-to-multipoint (P2MP) and Point-to-Point (P2P). For the latter, a dedicated fiber runs all the way from the Central Office (CO) (i.e., Exchange) to a customer premises. It provides very high bandwidth services, but incurs higher civil engineering costs in general due to the trenches that need to be dug up for cable deployment. On the other hand, the P2MP network based on GPON technology provides an attractive solution to reduce the overall cost. With the P2MP GPONs, no active electronic components are required between a CO and customer premises as shown in Fig. 1. Only optical splitters are used to connect an Optical Line Terminal (OLT) equipment at the CO to a cluster of premises by splitting the fiber in different ratios (1:4 up to 1:64). An Optical Network Unit (ONU) is then used to convert the optical signal back to an electrical signal at the customer premises. Furthermore, FTTH GPON networks have many advantages: they are impervious to electric interference and are more reliable than the copper equivalent [6]. It is considered to be the most OPEX-friendly as there are only two active elements in the access network [4].

Migrating to fiber-based networks, however, requires many practical considerations. The copper wire and equipment from the DSL-based architectures cannot be reused [2]. As fibers are installed in subterranean ducts [1], trenches would need to be dug up for the new infrastructure. Therefore, optimization of those networks becomes paramount. Deployment of FTTH networks usually goes through a design process of cabling, ducting and identifying locations of equipment. This process is typically achieved manually [4]. In addition, there are various planning rules that have to be met while maintaining a certain budget. Due to the process being manual, a high deployment cost is usually incurred. The design process can also take up to

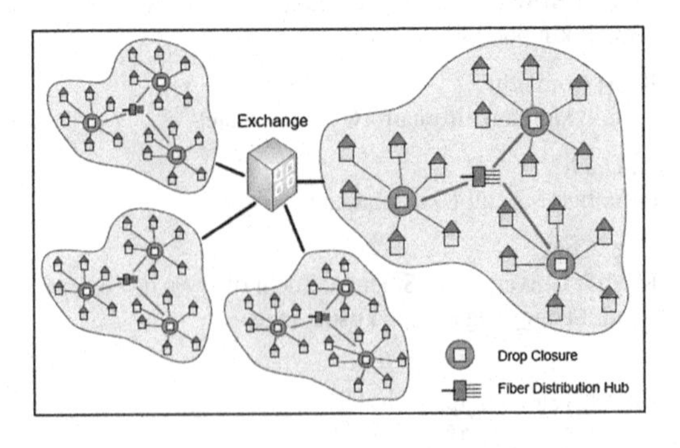

Fig. 1 A typical FTTH network

weeks or months depending on the size of the network. Ideally, the design process should take a short calculation time while minimizing the costs [4]. This is why automation of topological design for the physical layer network is crucial.

Figure 1 illustrates a typical two-level FTTH tree network. The root of the tree is an Exchange, in which an OLT is situated. Through the feeder fiber, it connects to the fiber distribution hub (FDH), where the optical splitters are installed. Each FDH is connected to drop closures (DCs) where the fiber splicing is performed by the use of distribution fibers. They are connected to the optical network units (ONUs) at the customer premises through drop cables. ONUs are required to interface the access network and the end-users network by converting the optical signal back to the electrical signal [6]. In this paper, we consider an individual FDH catchment area similar to the areas highlighted in blue. The approach described in the paper can be applied to other catchment areas.

In the Greenfield scenario, the infrastructure of fiber networks include the civils, duct and cable layers, which are closely-related. The civils layer, also called the trenching layer, specifies which roads in a given area need to be dug up. The duct layer indicates where the ducts would be laid to accommodate optical cables. The cable layer specifies the location of the network equipment (e.g., DCs, FDHs), the connectivity among network equipment and the routes of different cables to be installed. Even though both layers are related, they have conflicting criteria, which makes is difficult to optimize the overall deployment cost across different layers. In this paper, we consider a two-layer optimization problem: the duct and cable layers. We propose a method that minimizes the cost by utilizing a combination of a metaheuristic, namely Genetic Algorithm (GA), and exact techniques.

This paper is organized as follows: Sect. 2 provides a literature review of related problems. Section 3 includes a problem statement. Our approach to this problem, as well as a description of the AI-method used, is detailed in Sect. 4. Results and conclusions are provided in Sects. 5 and 6 respectively.

2 Related Work

The FTTH deployment is increasing worldwide with a huge expected growth in the number of new subscribers [6]. However, the deployment pace is hampered by the CAPEX that telcos need to allocate. Therefore, the problem of minimizing the deployment cost is critical and many attempts have been made based on different approaches. They can be classified into two major categories: exact techniques based on mathematical programming and metaheuristic approaches such as Genetic Algorithm (GA) and Ant Colony Optimization (ACO). Both approaches have their pros and cons. If the network to be optimized is of a medium size, exact techniques, in general, can outperform metaheuristics and lead to the global optimal solution. However, as the number of parameters and constraints increases, the computational time increases considerably and metaheuristics usually perform better [9]. In addition, mathematical programming is problem specific making it harder to adapt new changes. A new mathematical model is usually needed for the new customers requirements whilst only a new objective function is required for metaheuristics.

Based on the former class of approaches, the authors in [1] present a two-level FTTH network planning problem focusing on the fiber duct sharing problem by taking into account both trenching and cabling layers. The authors formulate a Mixed Integer Linear Programming (MILP) model stemmed from network flow optimization. In order to reduce the complexity of the problem, heuristics were presented. The approach described in the paper was capable of solving the problem and arriving at near optimal solutions. However, with dense and highly-clustered networks, the approach was shown to be unsuitable.

Another work in the literature that formulates a MILP model to solve the problem is proposed in [8]. The authors present a model that creates the optimal cable layer. Given the civils layer, the locations of the customers and a set of possible locations of network equipment, the model will select the best locations and the assignment of the customers to them. The model is capable of generating global optimal results for small to medium-sized networks. For big networks, it does not arrive at the optimal solution.

In [11], the authors present an approach that combined mathematical programming and heuristics to solve a Greenfield GPON planning problem and to minimize the deployment cost. The authors apply a simulated annealing-like approach through which the assignment of customers to network equipment and the locations of the splitters were recursively re-allocated until convergence.

On the other hand, the authors in [9] use Ant Colony Optimization as a metaheuristic approach to tackle a two-level FTTH planning problem. For given locations of customers and possible locations of splitter nodes and drop closures, the algorithm would choose the best types of network equipment and generate the cable routing information. The proposed algorithm can handle different types of equipment with different capacities which is in-line with practical FTTH network deployment. The authors describe two ACO-based algorithms and compare their performance. The approach shown yielded good results and it was demonstrated that the approach can be applied in a real-life practical scenario.

The authors in [10] present a GA-based approach to solve the multi-criteria optimization problem of PON deployment. They take into account the power budget, existing network resources and obstacles. The authors describe a semi-automated deployment tool. The algorithm detailed in the paper is split into three main parts. In the first part, the authors make use of a geographical map to set paths between any two points in the network in order to implement obstacle avoidance. The number of separate PON networks needed for an area is then found by a k-means clustering algorithm [14]. GA is then used to add a connection path of ONUs to network equipment one at a time while adding used resources to a pool of existing resources in order to make use of them. In most cases, the results were found to have outperformed the manual solutions.

Many approaches exist in literature that try to tackle the FTTH-PON planning problem, however, most neglect to address the different layers due to the complexity of the problem. In addition, there are many practical considerations that need to be taken into account when it comes to FTTH planning. In this paper, we propose a FTTH design approach that combines both mathematical programming and meta-heuristic techniques to tackle a practical problem aiming at cost minimization for both duct and cable layers.

3 Problem Statement

In this paper, we focus on Greenfield FTTH design scenario where only customer locations and the road network are known. Usually, a greenfield area would cover several thousands of customers and would need to be divided into several FDH catchment areas. We consider each FDH catchment area separately. The objective is to generate the duct and cable layer automatically with minimum cost. The duct layer is composed of three main parts: the main ducts to be laid on the road network, the joint boxes (JBs) to be installed to connect ducts together and special ducts known as lead-in ducts to connect customer plots to the main duct. Minimizing the cost of the duct layer is important due to the fact that wherever there is a duct, a trench would need to be dug. Trenching costs have significant impact on the total deployment cost. Therefore, reducing the cost of the duct layer would subsequently reduce the cost of trenching. On the other hand, the cable layer consists of drop closures (DCs) and optical cables running from each customer premise to a DC and from each DC to the FDH.

Figure 2 illustrates an example of a road network that serves as an input to our design approach. The bigger square represents the entry point to the network, the smaller square is the FDH and the solid circles indicate the locations of customer premises. The road network consists of edges indicated by the grey lines whilst road

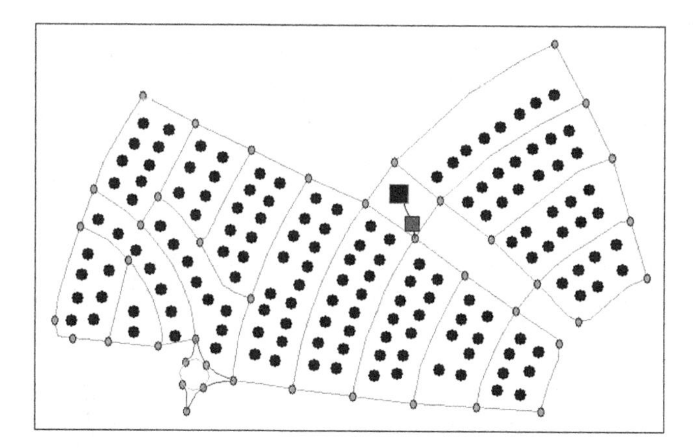

Fig. 2 A given road network

junctions (RJs) are denoted as small grey circles. Given such a road network, the objective of our system is to minimize the network design cost by:

- Generating ducts and the locations of JBs
- Identifying the optimal DC locations
- Finding the best cabling assignment of the customers to the selected DCs and from the DCs to the FDH based on the generated duct layer.

In the next section, we provide a detailed description of our approach for multi-layer optimization.

4 Hybrid Optimization Approach

In this section, we propose an approach that performs multi-layer optimization using mathematical programming, namely binary integer programming (BIP), and Genetic Algorithm (GA). This approach consists of the following steps:

(1) Finding JB locations based on the road network and the customer locations using a BIP model

(2) Generating the duct layer based on JBs and RJs using GA

(3) Creating the cable layer based on the generated duct layer using another BIP model

4.1 Finding JB Locations

As previously mentioned, customer premises are connected to the main duct by means of a lead-in duct. They are joined together by means of a JB. With this planning rule, a BIP model was executed on each road segment. The set of potential locations for the JBs is taken to be the customer projections on the respective road segments. The model determines the best location for placing JBs on each road segment. The mathematical model is detailed as follows:

Sets

- U—A set of all locations of ONUs, where index i refers to the ith ONU.
- D—A set of all possible locations for JBs in the model, where index j refers to the jth JB.

Variables

- y_j—Binary variable indicating the usage of the jth JB where $j \in$ D. It is set to 1 if the jth JB is deployed and 0 if it is not.
- x_{ij}—Binary variable indicating the connection of the ith ONU to the jth JB, $i \in$ U and $j \in$ D. It is set to 1 if the connection exists, and 0 otherwise.

Parameters

- λ—JB capacity.
- c_{jb}—Cost of a JB being deployed.
- c_{leadin}—Average cost per meter of lead-in duct used to connect an ONU to a JB.
- l_{ij}—The length in meter of the edge connecting the ith ONU to the jth JB. This is obtained by running the shortest path algorithm between each pair.

Mathematical Model

Minimize

$$OBJ = \sum_{j \in D} y_j(c_{jb}) + \sum_{i \in U} \sum_{j \in D} x_{ij}\, c_{leadin} l_{ij}$$

Subject to

$$\sum_{j \in D} x_{ij} = 1, \forall i \in U \quad (1)$$

$$\sum_{i \in U} x_{ij} \leq \lambda y_j, \forall j \in D \quad (2)$$

Constraint (1) ensures that each customer is only connected to one JB and (2) makes sure the JB capacity is not exceeded.

4.2 Generating the Duct Layer

Based on the identified JB locations, we generate a cost-efficient duct tree connecting all JBs (and therefore customers) to the FDH. This step is performed by the GA. We briefly describe the standard GA and more details about how the duct tree being created will follow.

4.2.1 Genetic Algorithms

Genetic Algorithms are a family of intelligent search methods inspired by the evolutionary processes in natural systems [5]. By simultaneously evaluating a population of solutions for a particular problem, GAs are considered to be a parallel search optimization method and often perform with remarkable efficiency. The major components within GAs are variables and individuals that are analogous to genes and chromosomes. The individuals within the population are evaluated by means of a fitness function to be optimized. Based on the fitness values, individuals are selected to undergo reproduction operators, namely crossover and mutation. The crossover operator will cross over the individuals, with a defined probability, to form new offspring. With a certain mutation probability, the new offspring will experience mutation at each locus. After undergoing the selection, crossover and mutation processes, a new population is generated to replace the previous generation. The algorithm terminates at a predefined number of generations or when no further improvement of the fitness value is observed for a certain number of generations [12].

4.2.2 Creating Anchor Points for the Duct Tree

The choice of GA instead of the exact method to generate the duct layer is motivated by its flexibility in modeling the problem. Given the road network, a GA with the binary encoding is used to create the duct layer that minimizes the ducting cost. The advantage of using the binary encoding is that the standard crossover and mutation operations can be employed. The length of each individual is equal to the total number of road junctions (RJs) in the network.

Each gene represents a possible anchor point to form a tree structure. A value of 1, in the individual for a particular index of RJ, indicates that a particular RJ has been selected as an anchor point. For example, if there are 50 RJs in the road network, out of which 15 are set to one, then those 15 RJs are regarded as anchor points to form the duct network with a tree structure. Using a pre-calculated distance matrix (based on road lengths) and the tree structure, a ducting cost based on our spanning tree algorithm can be calculated and is used as the fitness value for the GA.

The objective function to calculate the fitness of a given solution is shown in Eq. (3). A large number, *MaxInt*, is used to convert the minimization of a tree structure into a maximization problem. *E* is the set of all the edges that make up a tree for a particular solution.

$$Fitness = MaxInt - \sum_{i \in E} E_i \tag{3}$$

4.2.3 Steps of Generating a Tree Structure

Step 1. Choose a FDH as the head of the tree as shown in Fig. 3.

Step 2. Generate a sorted list of the vertices (v_b to v_f) in ascending order based on their shortest distances from FDH (i.e. FDH, v_b, v_c, v_f, v_e, v_d). The shortest distance between any two vertices are provided by the distance matrix. It should be noted that v_i are the possible anchor points identified by the previous BIP model for finding the JB locations. The list is examined from left to right.

Step 3. Include the first element of the list v_a, labeled as FDH in Fig. 3, in Set T. $\therefore$ T becomes { FDH } and G−T = { v_b , v_c , v_f , v_e , v_d } Connect the first element of the sorted list in set G−T to the element in T based on the shortest distance.

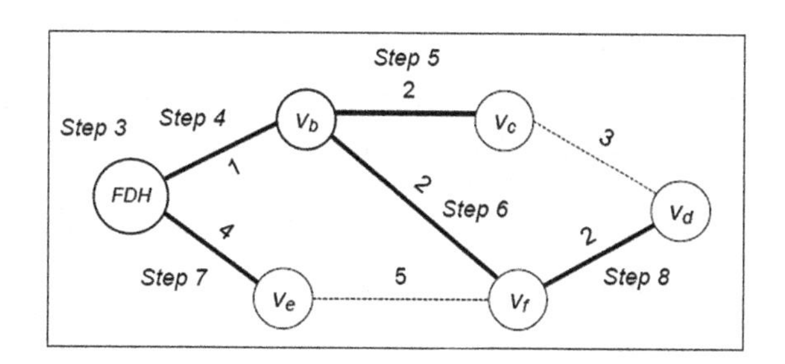

Fig. 3 A customized spanning tree

Step 4. Connect v_b in G−T to v_a in T ∴ T becomes { FDH, v_b } and G−T = { v_c, v_f, v_e, v_d }

Step 5. Connect v_c to v_b ∴ T becomes { FDH, v_b, v_c } and G−T = { v_f, v_e, v_d }

Step 6. Connect v_f to v_b as the distance between v_f and v_b is the shortest. ∴ T becomes { FDH, v_b, v_c, v_f } and G−T = { v_e, v_d }

Step 7. Connect v_e to v_a as the distance between v_e and v_a is the shortest. ∴ T becomes { FDH, v_b, v_c, v_f, v_e } and G−T = { v_d }

Step 8. Finally, connect v_d to v_f. ∴ T becomes { FDH, v_b, v_c, v_f, v_e, v_d } and G−T = ∅

The use of customized spanning tree (CST) algorithm compared with standard MST can reduce the time to create a spanning tree by making use of the distance matrix which has already been created from the shortest path algorithm. In other words, the adoption of CST can shorten the time to evaluate the fitness function within the GA, hence improve the overall performance. In addition, the standard MST includes all the road junctions of a given road network to form a tree structure. However, we do not want to lay a duct to connect every road junction especially if no customer plot is close to it.

4.3 Creating the Cable Layer

Based on the ducting layer generated in step 2, another BIP model is used to create the cable layer, which includes generating the locations of DCs and assigning cables from the customer premises to the selected DCs. The mathematical model is detailed below:

Sets

- U—A set of all locations of ONUs, where index *i* refers to the *i*th ONU.
- D—A set of all possible locations for DCs in the model, where index *j* refers to the *j*th DC.

Variables

- y_j—Binary variable indicating the usage of the *j*th DC where $j \in$ D. It is set to 1 if the *j*th DC is deployed and 0 if it's unused.
- x_{ij}—Binary variable indicating the connection of the *i*th ONU to the *j*th DC, $i \in$ U and $j \in$ D. If the connection exists, it is set to 1, 0 otherwise.

Parameters

- κ—FDH capacity.
- λ—DC capacity.
- l_{dc}^{max}—The maximum distance between a DC and the FDH.
- l_{onu}^{max}—The maximum distance between the ONU and the DC.
- c_{dc}—Cost of a DC being deployed.
- c_{drop}—Average cost per meter of fiber used to connect an ONU to a DC.
- c_{dist}—Average cost per meter of fiber used to connect a DC to an FDH.
- l_j—The length in meter of the path connecting the *j*th DC to the FDH.
- l_{ij}—The length in meter of the path connecting the *i*th ONU to the *j*th DC.

Mathematical Model

Minimize

$$OBJ = \sum_{j \in D} y_j (c_{dc} + c_{dist} l_j) + \sum_{i \in U} \sum_{j \in D} x_{ij} \, c_{drop} l_{ij}$$

Subject to

$$\sum_{j \in D} x_{ij} = 1, \forall i \in U \tag{4}$$

$$l_{ij} x_{ij} \leq l_{onu}^{max}, \forall i \in U, \forall j \in D \tag{5}$$

$$l_j y_j \leq l_{dc}^{max}, \forall j \in D \tag{6}$$

$$\sum_{j \in D} y_j \leq \kappa \tag{7}$$

$$\sum_{i \in U} x_{ij} \leq \lambda y_j, \forall j \in D \tag{8}$$

Constraint (4) ensures that a customer is only connected to one DC. Constraints (5) and (6) are distance constraints whilst (7) and (8) represent the FDH and the DC physical capacities respectively.

5 Results

In this section, we present the results of automatically generated duct and cable layers based on the approaches described in Sect. 4. Each step is discussed in the following sub-sections. An input network as shown in Fig. 2 was used with the assumption that each customer requires only one connection. The planning parameters and the indicative costs of the network equipment are shown in Table 1. A summary of the planning results is shown in Table 2.

5.1 Locations of Joint Boxes

In this step, a BIP model described in Sect. 4.1 is executed. The BIP is implemented using the CPLEX solver [3]. As mentioned, customer premises' locations are first projected to the road segments perpendicularly. Then, the locations of the JBs are generated by the model. In Fig. 4, the smaller grey circles inside the polygons represent the chosen JB locations. The polygons on the network indicate associated customers to the deployed JBs. A post-process is used to generate the lead-in ducts to connect the customer plot points to JBs. For the example network, the model selected 56 JB locations. The chosen locations are centric for each cluster of customers as shown in Fig. 4, and this is done in order to reduce the total lead-in duct length.

5.2 Creation of Duct Layer

In this step, the GA described in Sect. 4 was run on the output of the previous step. The population size and the number of generations were set to 50 and 1000 respec-

Table 1 Design parameters and indicative costs

Planning constraints/cost	
FDH capacity	30
DC capacity	16
JB capacity	4
Max fiber length from FDH to DC	1500 m
Max fiber length from DC to ONU	300 m
DC cost	1968 AED
JB cost	2000 AED
Duct cost	20.00 AED
Drop cable cost	6.47 AED
Distribution cable cost	8.83 AED

Table 2 Planning results

Planning results	
Time to run /s	25
Total network cost /AED	543,011
Number of deployed JBs	84
Number of deployed DCs	14
Total length of main ducts used (m)	5,963
Total length of lead-in ducts (m)	6,070
Distribution cable length (m)	4,700
Drop cable length (m)	9,326

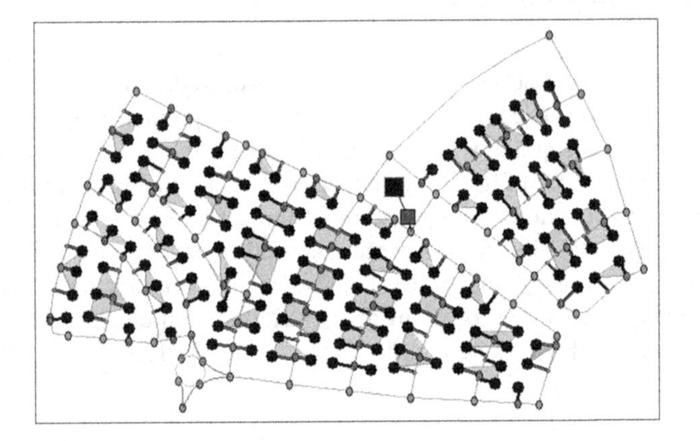

Fig. 4 Network with joint box locations

tively. The initial population was randomly generated without any prior domain knowledge. The Roulette Wheel Selection [12] was employed to choose the individuals. Finally, the crossover and mutation probability were set to 0.9 and 0.01 respectively. Those values were obtained experimentally for this problem. A large mutation probability increases the probability of searching more areas in the solution space but may prevent convergence. On the other hand, if the mutation probability was set to a very small value, premature convergence might occur at a local minima [13].

Figure 5 illustrates the duct layer after running the CST used in the GA. The lines connecting the JBs all the way to the FDH are the main ducts whilst lines connecting customer plots to JBs are considered as lead-in ducts. Compared with Fig. 4, the ducts were not laid on all of the road segments. Ducts were placed on the road segments onto which the customers are projected. They were also placed on edges that connect the anchor points to the FDH.

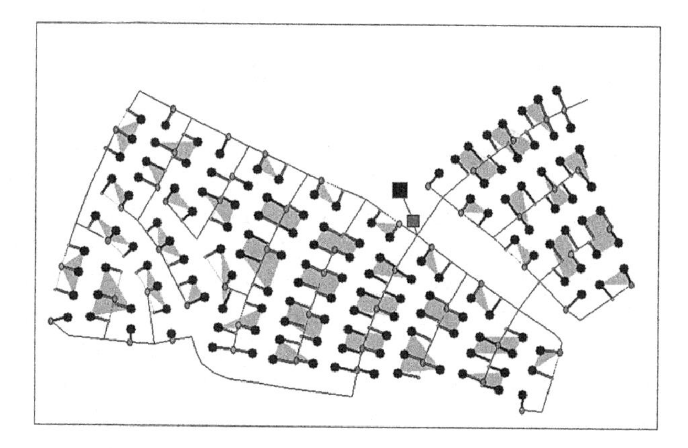

Fig. 5 Network with the ducting layer

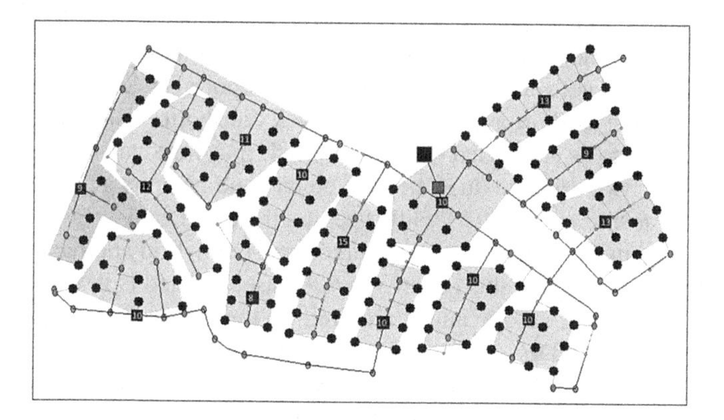

Fig. 6 Multilayer scheme design solution

5.3 Generation of Cable Layer

The cable layer model described in Sect. 4 was executed on the duct layer generated by step 2. The output is shown in Fig. 6. To adhere to a given planning rule, JBs were placed at all road junctions as a pre-process. The set of potential DC locations thus include the JBs selected by the first step and the road junctions. The DC locations selected by the model are indicated by the boxes with a value inside showing the number of connections. New polygons are created to show the cabling assignment of customers to the selected DCs. Similar to the JB clusters, DC locations are generally centric in the clusters.

It took less than 30 s in total for the three steps on a medium-sized network demonstrating that our approach is feasible for real-life network planning. In addition, the duct layer generation of step 2 shows that not all the roads are required to be trenched, which is a good indication for CAPEX reduction. As mentioned earlier, digging up the road incurs most of the capital cost.

6 Conclusion

In this paper, we have presented a hybrid approach of combining exact and AI-based algorithms to solve the multi-layer optimization problem for GPON-FTTH networks. Given a road network, the first exact method based on a Binary Integer Programming (BIP) is used to obtain the junction box (JB) locations with the consideration of customer projected points. Once the locations of JBs and the road junctions have been identified, a Genetic Algorithm is employed to create a duct layer. The advantage of using Genetic Algorithm is the simplicity of modeling the duct layer problem. We use the binary representation with a customized tree formation. Different tree structures can be evaluated in a very short period of time with an optimal or a very near optimal result. A second BIP is used to create the cabling layer with the optimal customer assignments to the selected drop closures.

There are several practical challenges to be tackled such as duct sharing among different cables, selecting optimum locations of the drop closures and identifying the path to connect the customer to the selected drop closures. As we are trying to minimize the overall network deployment cost, it is important to keep the distance of ducting and cabling as short as possible. However, the shortest cabling distance for customer A and B to be connected to drop closure C may not share the same path. If it is not the case, two separate ducts need to be laid and it incurs a higher cost. Our approach aims at sharing the ducts while keeping the cabling cost reasonably low.

From the generated results, we can conclude that our hybrid approach is suitable for the problem described in this paper. The GA-based method is flexible that it can support wide variety of modifications required by the real world application such as multiple entries of FDHs or consideration of road crossings for the duct layer. In addition, the short execution time to generate the duct and cable layers for our practical example indicates our approach is feasible for a planner to use our system. In the future, we would like to extend this work to optimize multiple FDH catchment areas and explore the possibility of combining the cable and duct generation steps into a single step.

References

1. Chardy, M., Costa, M., Faye, A., Trampont, M.: Optimizing splitter and fiber location in a multilevel optical FTTH network. Eur. J. Oper. Res. **222**, 430–440 (2012)
2. Chu, A., Poon, K., Ouali, A.: Using ant colony optimization to design GPON-FTTH networks with aggregating equipment. In: Paper presented at the IEEE Symposium on Computational Intelligence for Communication Systems and Networks, Singapore vol. 16–19 (2013)
3. CPLEX: CPLEX optimizer. http://www-01.ibm.com/software/commerce/optimization/cplex-optimizer/index.html (2011). Accessed on 12 June 2015
4. Jaumard, B., Chowdhury, R.: Location and allocation of switching equipment (Splitters/AWGs) in a WDM PON network. In: Paper presented at the 20th International Conference on Computer Communications and Networks Maui, HI, 31 July–4 Aug 2011
5. Kaminow, I., Li, T., Willner, A. (eds.): Optical Fiber Telecommunications. (Academic Press, Boston 2013)

6. Kavian, Y., Ghassemlooy, Z.: Intelligent Systems for Optical Networks Design: Advancing Techniques. IGI Global, Hershey, PA (2013)
7. Lakic, B., Hajduczenia, M.: On optimized passive optical network (PON) deployment. In: Paper presented at the Second International Conference on Access Networks and Workshops, Ottawa, Ontaria, 22–24 August 2007
8. Li, J., Shen, G.: Cost minimization planning for greenfield passive optical networks. IEEE/OSA J. Optical Commun. Network. **1**, 17–29 (2009)
9. MacQueen, J.: Some methods for classifications and analysis of multivariate observations. In: Paper presented at the 5th Berkeley Symposium on Mathematical Statistics and Probability in University of California, LA (1967)
10. Mitcsenkov, A., Paksy, G., Cinkler, T.: Geography- and infrastructure-aware topology design methodology for broadband access networks (FTTx). J. Photonic Netw. Commun. **253**, 266 (2011)
11. Ouali, A., Poon, K.: Optimal design of GPON/FTTH networks using mixed integer linear programming. In: Paper presented at the 16th European Conference on Networks and Optical Communications, Newcastle-Upon-Tyne, 20–22 July 2011
12. Phillipson, F.: Roll-out of reliable fiber to the cabinet: An interactive planning approach. IEEE/OSA J. Optical Commun. Netw. **6**, 705–717 (2014)
13. Sivanandam, S., Deepa, S.: Introduction to Genetic Algorithm. Springer, New York (2008)
14. Van Loggerenberg, S., Grobler, M., Terblanche, S.: Solving the passive optical network with fiber duct sharing planning problem using discrete techniques. Electronic Notes in Discrete Mathmatics. vol. 41, 343–350 (2013)

A Genetic Algorithm Based Approach for Workforce Upskilling

A.J. Starkey, H. Hagras, S. Shakya and G. Owusu

Abstract In large organisations with multi-skilled workforces there is a need to optimise and adapt the skill set each member of the workforce has. Some of these engineers will show a natural ability to perform well with their current skill set and will show that they are ready to progress to a more advanced skill set. However other engineers will demonstrate that they do not have the natural aptitude to sufficiently complete the tasks they are set. In the first instance it may be beneficial to upskill the engineer, in the second instance there may be a need to remove that engineer from that particular workforce. In both these instances it is necessary to evaluate the impact of these changing skill sets on the performance on the organisation as a whole. This paper presents a genetic algorithm based system for the optimal selection of engineers to be upskilled. The data presented to the system was taken from a real world mobile workforce. The results showed that using this system to select employees for training has an overall increase in employee utilisation with a smaller percentage of the workforce being trained. The results show that the first few employees to be selected for training can produce the most benefit so selecting the right people is crucial.

A.J. Starkey (✉) · H. Hagras
University of Essex, Colchester CO4 3SQ, UK
e-mail: astark@essex.ac.uk

H. Hagras
e-mail: hani@essex.ac.uk

S. Shakya · G. Owusu
British Telecom, London, UK
e-mail: sid.shakya@bt.com

G. Owusu
e-mail: gilbert.owusu@bt.com

M. Bramer and M. Petridis (eds.), *Research and Development in Intelligent Systems XXXII*, DOI 10.1007/978-3-319-25032-8_20

1 Introduction

For any company that is required to manage a multi-skilled workforce, the impact of the different skill sets can influence the utilisation of each member of the workforce. This dictates their productivity and improves the cycle time to provide services. This in turn results in greater customer satisfaction.

A common type of multi-skilled workforce is a mobile workforce in the service provisions sector. This type of workforce needs multi-skilled engineers to deal with the large variety of infrastructure job types that would be generated.

Groups of engineers form teams and each team is assigned an area to work in. these teams can be of varying size and if they are part of a mobile workforce, the team could cover geographical areas of a range of sizes.

There has been some discussion about a multi-skilled engineer workforce [1, 2]. The argument against having multi-skilled engineers is that it goes against the principles of specialisation. Where employees will perform better if given one specific task type rather than a range of task types. In the case of multi-skilling, engineers are expected to be competent in each task type they are trained for.

The argument for multi-skilled employees has been laid out in [1]. Employees with multi-skills are useful when demand get high and the company wants to maintain a high level of customer satisfaction. Another advantage put forward by [2] is that a multi-skilled workforce can help where the labor market is scarce of the types of people you want to employ. [2] also puts forward the idea that to get the most productivity out of a multi skilled engineer, the skills they should be trained in should be correlated in some way.

2 Overview of the Skill Optimisation Problem

2.1 Overview of Multi-skilled Engineers

Mobile workforce engineers could have varying numbers of skills based on the types of tasks they work on and how experienced they are. More experienced engineers are more likely to have more skills and tend to have more advanced skills.

The skill sets of the engineers will differ between the different areas groups of engineers (teams) are assigned to. So a team with a given number of engineers and an optimised set of skills for each engineer will not necessarily be the best setup for another area.

There are additional factors to consider here as well, including the number of hours an engineer works (or % of Full Time Employment, FTE). Each engineer will naturally favor certain types of tasks and perform better at those tasks compared to others they are required to complete. This is specific to each engineer and should not be generalised across the team.

The best way that any configuration of teams and skill sets can be organised is to run a simulation with the configuration that needs to be assessed. This simulation needs to, as accurately as possible, replicate how the engineers would go about completing the list of tasks they have been assigned.

Given that this simulation has to emulate a mobile workforce, it needs to have a measure to assess how long engineers spend travelling and how far they have traveled. The simulation also needs to accurately assign tasks to engineers.

[3] notes that the skill optimisation problem is a Combinatorial Optimisation (CO) problem [4]. Algorithms designed to tackle CO problems usually aim for a metaheuristic approach [5] because the optimisation has to be completed within a reasonable amount of time, as the environment changes on a frequent basis. A common approach to tackling these large scale and complex optimisation problems is Genetic Algorithms (GA) [6, 7, 8]. However [3] Attempts to solve their problem by using Ant Colony Optimisation (ACO). In this paper we will present a real value GA based approach to solving our skill optimisation problem because ACO is potentially slower in this type of problem and is potentially weaker than GA at exploring a diverse range of solutions [9].

2.2 *Objective and Constraints*

Once a suitable simulation has been constructed, there needs to be a way of evaluating the differences between different engineer setups and when running 'What-If' scenarios.

In this paper we present a number of objectives that need to be assessed when evaluating a mobile workforce.

- Maximize Coverage: Coverage is the amount of tasks that are estimated to be completed. This is measured in hours. In Eq. (1) this is represented at the sum total of all engineers completed work (E_{cw}).

$$\sum E_{CW} \tag{1}$$

- Minimize Travel: Minimizing traveling increases the amount of available time for each engineer and also decreases costs. In Eq. (2) this is represented as the sum total of all engineers travel distance (E_{td}) divided by the sum total of engineers (E) as this is represented as average travel per engineer.

$$\sum E_{td} / \sum E \tag{2}$$

- Maximize Utilisation: Utilisation is the percentage of time an engineer is completing tasks. Unutilised time is when the engineer is idol or travelling. In Eq. (3) this is the sum total of engineer completed work (E_{cw}) divided by engineer available time (E_{at}), this is then divided by the sum total number of engineers (E) to get the average utilization.

$$(\sum E_{CW}/E_{at})/\sum E \tag{3}$$

There are a number of constraints that need to be applied to the simulation. The first constraint is that engineers are only allocated tasks they are qualified to complete. However these skills can be altered to find how the addition or subtraction of skills effects the objective values.

Another constraint is the total amount of work an engineer can complete in the simulation. The simulation could emulate one day of the workforce, one week or even longer. However the amount of work each engineer has needs to be realistic for the simulation. For example if the simulation was set to look at an average day then an engineer might have a maximum of 8 h worth of work that they can do. However, those on part time contracts may only do 5 h and those on overtime may do 10. These values can be changed for capacity planning and are specific to each engineer.

3 Overview of Genetic Algorithms

Genetic Algorithms (GAs) are based on the theory of natural selection and evolution. In GAs, over time a population, or species, will adapt to its environment. This adaptation takes place through the idea of survival of the fittest.

The individuals in the population that have characteristics most suited to its environment are the individuals most likely to survive. Thus these characteristics are passed on to the next generation of the species. The individuals less suited to the environment don't survive, so these characteristics are lost in the next generation.

Each individual has a chromosome (or multiple chromosomes) where a chromosome is a collection of genes. These genes are what determine the characteristics of the individual. When the next generation of the species is created, genes from two parent individuals will be combined to determine the characteristics of the child [10]. Figure 1 shows the process of a standard GA. The first step is to initialise the population. This is done by creating N solutions and randomly assigning the genes of each solution.

These genes are then evaluated to see how suited to the environment they are. Each solution (individual) is then given a score to represent the solution fitness. Once each solution has been evaluated. The evolution process can begin. The first step in this is the selection of solutions to crossover. There are a number of selection operators out there such as tournament or roulette [11].

In tournament selection a subset of solutions from the population are chosen. Then the solution with the highest fitness will be chosen as the first parent. The process is repeated to find the second parent.

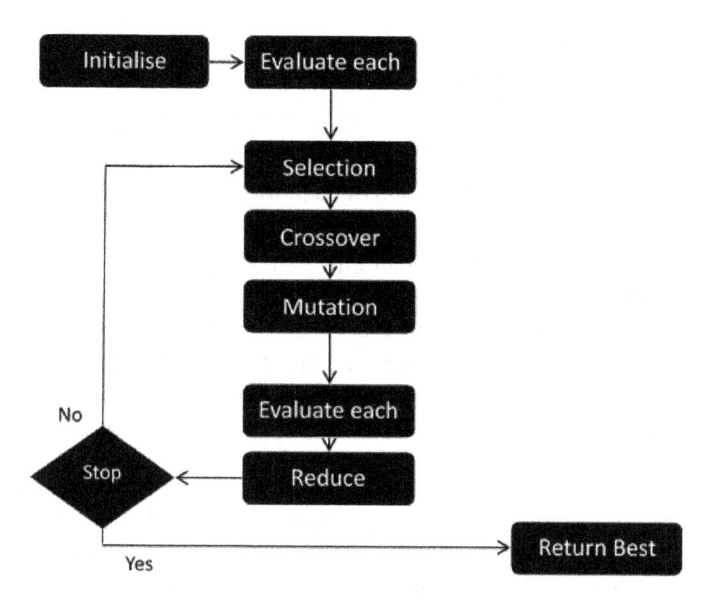

Fig. 1 GA flow chart

Roulette selection is based upon a roulette wheel where the sections of the roulette wheel each represent an individual of the population. Each section of the wheel is also weighted to the individual's fitness. Thus, the individuals in the population with a higher fitness are more likely to be selected to be one of the parents.

Once two parents have been chosen they will crossover their genes using a crossover operator (1 point, 2 point, uniform) [12]. Crossover will generate 2 child solutions that will be added to the new population set. Every so often one of the genes in a child solution will randomly change, this is known as mutation. Once enough children have been generated and the new population set is the same size as the old population set, the old population will die off and the new population will go to the fitness evaluation stage.

The fitness function applies a formula to work out how good a solution is and gives it a score. Based on the objectives we outlined in Sect. 2.2 we use the fitness function given in Eq. (4)

$$Fitness = \frac{(Coverage \times W_1) \times (Utilization \times W_2)}{(Travel \times W_3)} \tag{4}$$

W is the weighting of each objective, w_1 is the weighting of the coverage objective, w_2 is the weighting of the utilization objective and w_3 is the weighting of the travel objective. For this paper we keep all the weightings equal.

The stopping criteria decides when the GA process should stop. This can be done by setting the maximum number of generations or waiting until convergence happens. In the proposed system we are using max generation. This is due to time constraints when trying to find the best engineers to upskill.

4 The Proposed GA Based Skill Optimisation System

The system is primarily focused around a real valued genetic algorithm (RVGA). The genes in each of the solutions represents an engineer ID. The solution length (number of genes) is related to the number of upskills, where an upskill in the next logical skill set for any given engineer. Figure 2 shows an example of the real valued chromosome, where an ID of an engineer is stored within each gene. This then tells the simulation that this set of engineers needs to use their upskilled skill set.

The next logical skill set is an important aspect that should have already been decided based on the type of engineers an organisation has. These next logical skill sets are designed to build upon the skill set the engineer already has.

So for example if an engineer already has the skills of server installation the next logical skill to give this engineer might be server repair rather than air conditioning installation. The next logical skill set may also be tailored by technical managers who see engineers have an aptitude (or inaptitude) for a particular type of task.

The proposed system has one major part outside of the genetic algorithm and this is the daily simulation. This simulation is able to estimate the coverage, travel and utilisation values for any given solution, i.e. the team and their proposed skills including any new skills. This is extremely important as the values given from this simulation are fed straight into the fitness function.

The simulation starts by cycling through each team in an area then assigns tasks to each engineer from the tasks list. This task list is geographically based, so the simulation takes into account where the task are located in relation to the engineer.

As a result, the tasks assigned to the engineer will be based on distance and skill compatibility. As the simulation is route based, the closest tasks will change as the engineer moves around the area.

Once an engineer has reached their maximum number of working hours for an average day the simulation will move onto the next engineer in the list.

The crucial variable for this problem is the skill compatibility of an engineer to a task. From the GA we get the ID's of engineers to be upskilled so if the simulation comes to one of these engineers, their skill set will be different and will contain more skills than if they were not in their upskilled state.

ID: 14	ID: 65	ID: 21	ID: 78	ID: 123	ID: 105	ID: 8	ID: 53

Fig. 2 Upskilling chromosome

As a result the engineer has more tasks to choose from and their route may be different as a result. The way this simulation system is designed means that utilisation and coverage should always increase regardless of the skill configurations from solution to solution.

This is because once the N number of upskilled engineers have been chosen the order in which engineers are selected from the list to simulated will change. The list will choose the engineers with the least amount of skills first and leave the engineers with the most amount of skills until last. This means that there won't be a situation where an engineer is chosen to be upskilled and are then given some of the tasks that a lower skilled engineer could have done.

If this situation was to happen utilisation is likely to be reduced, as the low skilled engineer has less tasks to choose from, meaning either traveling more to find work (reducing utilisation) or not matching with enough compatible tasks to fill all their available hours.

However, because the upskilled engineer will be further down the list, the minimum the engineer will do is exactly the same as if he/she were not upskilled. This ultimately means that this engineer is a poor choice to spend time and money on training. This solution is then likely to be lost as the GA evolves.

5 Experiments and Results

In our data we already have a list of engineers and their current skill sets. We have the task data that would be presented to them on an average day. This enables us to simulate the overall utilisation of the engineers. This initial utilisation value gives us a base value to compare our optimisation results. This is important because this initial utilisation value has been created from the current system of choosing engineers to upskill i.e. by managers picking who they think is suitable for more training. The original results for the area being tested shown in Table 1.

In the following experiments we are trying to optimise a group of 9 teams with a total of 141 employees. There are 10 different approved skill combinations utilising 11 different skills.

Our first set of experiments aim to tune the GA to be as beneficial as possible to the outlined problem. These experiments test whether Tournament Selection or Roulette Selection is better for the problem. It also tests if a crossover value of 0.4 or 0.2 is better for this problem. Tables 2, 3, 4 and 5 outline these results. The following results are for 5 upskills. This means the system will try to pick the best 5 possible candidates to be trained to their next logical skill set.

Table 1 Original results for test area

Coverage	90.20 %
Travel per engineer	24.84 km
Average utilisation	77.03 %

Table 2 Tournament selection with crossover of 0.4

	Coverage (%)	Travel (km)	Utilisation (%)
	94.12	25.61	80.38
	94.01	26.02	80.29
	94.06	25.48	80.33
	93.94	25.23	80.23
	94.12	25.10	80.38
Average	**94.05**	**25.49**	**80.32**

Table 3 Tournament selection with crossover of 0.2

	Coverage (%)	Travel (km)	Utilisation (%)
	94.01	25.61	80.29
	94.01	25.88	80.29
	94.12	25.48	80.38
	93.94	25.23	80.23
	94.05	25.57	80.32
Average	**94.03**	**25.55**	**80.30**

Table 4 Roulette selection with crossover of 0.4

	Coverage (%)	Travel (km)	Utilisation (%)
	93.90	26.07	80.20
	93.90	25.75	80.20
	93.63	25.70	79.97
	94.01	25.63	80.29
	93.73	25.64	80.05
Average	**93.83**	**25.76**	**80.14**

Table 5 Roulette selection with crossover of 0.2

	Coverage (%)	Travel (km)	Utilisation (%)
	93.44	25.43	79.80
	93.42	25.62	79.79
	93.96	25.53	80.25
	93.90	26.16	80.20
	93.76	25.40	80.08
Average	**93.70**	**25.63**	**80.02**

From these results we can see that Tournament selection produced higher average utilisation than Roulette selection in both the 0.2 crossover probability test and the 0.4 crossover probability test. Tournament selection also produced higher average coverage in both tests too.

Perhaps more interestingly is that Tournament selection also produced a lower average travel rate. Typically we would expect travel to increase as more tasks are covered (as the engineers would have to travel to more tasks). This increase in coverage and decrease in travel explains the overall increase in utilisation.

Our second set of results focus on the number of engineers to be upskilled vs the benefit from the upskill. Tables 6 and 7 add to the results we already have from Table 2 as the following results used Tournament selection with a crossover rate of 0.4.

The maximum number of engineers who can be upskilled for the area we are experimenting with is 107 out of 141. The remaining 34 engineers already hold the maximum amount of skills available to them. The results for the maximum number of possible upskills are shown in Table 8.

Given that we now have the original results, results for 5, 10, 15 and maximum upskills we can plot them to see the level of diminishing returns for each upskill. This is important as the number of upskills directly correlates to training costs. Figures 3, 4 and 5 shows the graphs for the number of engineer upskilling versus the coverage benefit, utilisation benefit and travel cost respectively.

Figures 3 and 4 show that the most benefit per upskill is gained within the first 5. After this the benefit to both coverage and utilisation is greatly reduced. The benefit to coverage and utilisation becomes almost negligible after 15 upskills.

Figure 5 shows the average travel per engineer, which for the first 10 upskills increases linearly. However after this point travel starts to be reduced with a significant drop at maximum upskills. This is likely because engineers have many more tasks to choose from that are closer to their current location.

Table 6 Optimisation with 10 upskills

	Coverage (%)	Travel (km)	Utilisation (%)
	95.19	26.33	81.30
	95.02	25.72	81.16
	95.19	26.21	81.30
	95.00	25.62	81.13
	94.94	26.64	81.09
Average	**95.07**	**26.10**	**81.20**

Table 7 Optimisation with 15 upskills

	Coverage (%)	Travel (km)	Utilisation (%)
	95.33	25.86	81.42
	95.28	26.47	81.37
	95.09	25.52	81.21
	95.33	25.92	81.42
	95.27	26.39	81.36
Average	**95.26**	**26.03**	**81.35**

Table 8 Maximum number of upskills for test area

Coverage	95.36 %
Travel per engineer	24.75 km
Average utilisation	81.45 %

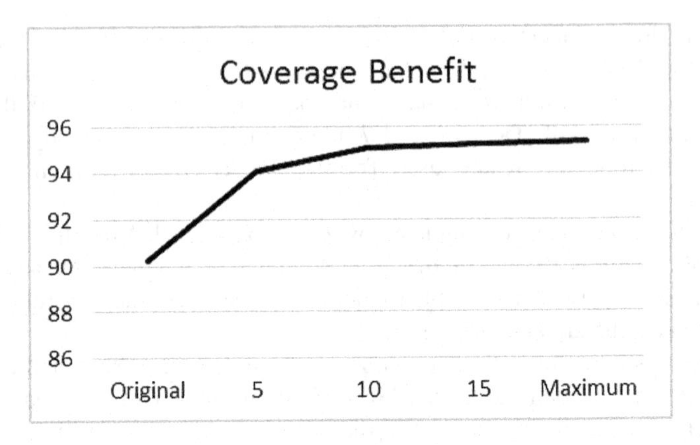

Fig. 3 Coverage benefit

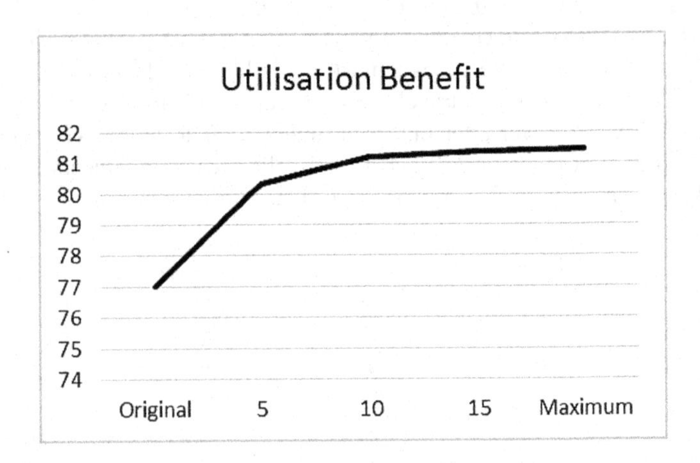

Fig. 4 Utilisaton benefit

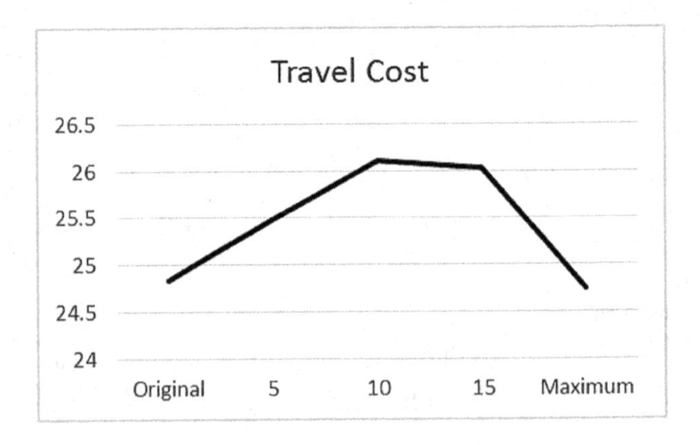

Fig. 5 Travel cost

From this we can say the only significant benefit gained after 10 upskills comes from the reduction in travel.

If we compare the genetic algorithm based system with the current manual system (where managers choose engineers they think are suitable for more training) the results show that using this system to select employees for training has a 4.27 % increase in overall employee utilisation with only 3.52 % of the workforce being trained (5 out of 141 engineers in the workforce). It also shows that there is a 5.41 % increase in overall engineer utilisation when 7.04 % is selected to be trained (10 out of 141 engineers in the workforce).

This shows that the first few employees to be selected for training can produce the most benefit so selecting the right people is crucial and hence the proposed system should be used. After this point there is a potentially an exponential level of diminishing returns on employee utilisation.

6 Conclusions and Future Work

In this paper we have presented a real value GA system for engineer upskilling recommendations. We ran a number of experiments to first find out the best settings for the RVGA to optimise with. We then we ran the optimisation on a test area to assess the potential benefit of having the system over the current manual system.

The results showed that for this particular problem Tournament selection with a crossover probability of 0.4 performed better. The parameters for the GA were tuned using empirical experiments and some of the best performing setups are reported in this paper.

Once the system was tuned to these settings the optimisation for the area found that the most benefits gained from the system were in the first 5 engineers which were upskilled. However it also found that upskilling all possible engineers is the best way to reduce travel costs.

Our future work will focus on having the system suggest the best number of upskills, rather than have this as one of the systems initial parameters.

We also plan to look at a multi-objective solution to this problem that could potentially find solutions that both increase coverage and utilisation whilst also reducing travel at the same time.

References

1. Koole, G., Pot, A., Talim, J.: Routing heuristics for multi-skill call centers. In: Proceeding of the 2003 Simulation Conference, vol. 2, pp. 1813–1816 (2003)
2. Lin, A., Ahmad, A.: SilTerra's experience in developing multi-skills technician. In: IEEE International Conference on Semiconductor Electronics, pp. 508–511 (2004)

3. Hu, Z., Mohd, R., Shboul, A.: The application of ant colony optimization technique (acot) for employees selection and training. In: First International Workshop on Database Technology and Applications, pp. 487–502 (2009)
4. Turchyn, O.: Comparative analysis of metaheuristics solving combinatorial optimization problems. In: 9th International Conference on the Experience of Designing and Applications of CAD Systems in Microelectronics, pp. 276–277 (2007)
5. Fanm, W., Gurmu, Z., Haile, E.: A bi-level metaheuristic approach to designing optimal bus transit route network. In: 3rd Annual International Conference on Cyber Technology in Automation, Control and Intelligent Systems, pp. 308–313 (2013)
6. Domberger, R., Frey, L., Hanne, T.: Single and multiobjective optimization of the train staff planning problem using genetic algorithms. In: IEEE Congress on Evolutionary Computation, pp. 970–977 (2008)
7. Liu, Y., Zhao, S., Du, X., Li, S.: Optimization of resource allocation in construction using genetic algorithms. In: Proceedings of the 2005 International Conference on Machine Learning, pp. 18–21 (2005)
8. Tanomaru, J.: Staff Scheduling by a Genetic Algorithm with Heuristic Operators. In: International Conference on Evolutionary Computation, pp. 456–461 (1995)
9. Alhanjouri, M., Alfarra, B.: Ant colony versus genetic algorithm based on travelling salesman problem. Int. J. Comput. Tech. Appl. **2**(3), 570–578 (2013)
10. Hossain, K., El-Saleh, A., Ismail, M.: A comparison between binary and continuous genetic algorithm for collaborative spectrum optimization in cognitive radio network. In: IEEE Student Conference on Research and Development, pp. 259–264 (2011)
11. Zhong, J., Hu, X., Gu, M., Zhang, J.: Comparison of performance between different selection strategies on simple genetic algorithms. In: Proceeding of the 2005 International Conference on Computational Intelligence for Modelling, Control and Automation, and International Conference on Intelligent Agents, Web Technologies and Internet Commerce, pp. 1115–1121 (2005)
12. Murata, T., Ishibuchi, H.: Positive and negative combination effects if crossover and mutation operators in sequencing problems. In: Proceeding of the IEEE International Congress on Evolutionary Computation, pp. 170–175 (1996)

Applications of Intelligent Agents and Evolutionary Techniques

Multi Agent Based Simulation Using Movement Patterns Mined from Video Data

Muhammad Tufail, Frans Coenen, Jane Hurst and Tintin Mu

Abstract A mechanism for extracting movement patterns from video data with which to drive Multi Agent Based Simulations (MABS) is described. Two types of movement pattern are considered: absolute and relative. The proposed mechanism is fully described in the context of a rodent behaviour MABS. To evaluate the resulting MABS a process is adopted whereby the simulation is "videoed" and the movement pattern generation process repeated (thus completing the cycle). The nature of the simulated movement patterns is then compared with the video data movement patterns. The advantage of relative movement patterns over absolute movement patterns is that they are more generic and this is illustrated in the paper using a case study.

1 Introduction

Computer simulation is an important tool used with respect to many application domains such as industrial engineering, management science and operations research [4, 5, 11, 14]. It has many advantages. Firstly it allows users to conduct what-if style experiments without the need for the resource required for real life experimentation. Secondly it allows users to investigate phenomena of interest multiple times with full control of parameters. And thirdly they are non-intrusive.

M. Tufail (✉) · F. Coenen
Department of Computer Science, The University of Liverpool, Liverpool L69 3BX, UK
e-mail: M.Tufail@liv.ac.uk

F. Coenen
e-mail: coenen@liv.ac.uk

J. Hurst
Institute of Integrative Biology University of Liverpool, Leahurst Campus, Chester High Road, Neston CH64 7TE, UK
e-mail: jane.hurst@liv.ac.uk

T. Mu
Department of Electrical Engineering and Electronics, The University of Liverpool, Liverpool L69 3BX, UK
e-mail: T.Mu@liv.ac.uk

M. Bramer and M. Petridis (eds.), *Research and Development in Intelligent Systems XXXII*, DOI 10.1007/978-3-319-25032-8_21

Multi Agent Based Simulation (MABS) is a type of computer simulation where the simulation is realised using agent based technology. Using MABS the individual "characters" that make up a simulation are represented as interacting agents. MABS is a good option with respect to real world scenarios that involve entities that behave in an autonomous manner.

The challenge of computer simulation, and by extension MABS, is how to build the model in a manner that produces the most realistic operation possible. In the case of MABS the traditional approach is to "hand craft" agent behaviour [1, 2], however this approach is both time consuming and error prone. An alternative approach, and that considered in this paper, is to learn the desired agent behaviour direct from video data. This avoids the disadvantages associated with hand crafting agent behaviour; and, it is suggested here, results in more effective behaviour capture (behaviour that leads to more realistic simulations than that which could otherwise be achieved).

The work presented in this paper proposes a process for mining video data to extract what we have called *movement patterns* that can then be used to effectively drive a MABS platform. Two types of movement patterns are considered: absolute and relative. The proposed processes for mining and utilising such patterns in a MABS context is fully described. The effectiveness of the resulting MABS are analysed by "closing the loop"; videoing the simulation and using this video to mine a second set of movement patterns the nature of which can be compared with the original set of patterns extracted from the input video data. To act as a focus for the work we consider rodent behaviour MABS, more specifically mouse behaviour simulation in the context of "mouse in a box" scenarios. The motivation for this application was that behaviourolgists are interested in analysing mouse behaviour from a pest control perspective.

The rest of this paper is organised as follow. An overview of some previous work is presented in Sect. 2. In Sect. 3 the nature of the video data used for illustrative purposes with respect to this paper is discussed. The pattern mining framework is presented in Sect. 4, followed in Sect. 5 by the MABS framework in which the mined movement patterns are utilised. Section 6 presents an evaluation of the operation of the MABS in terms of the movement patterns used, while Sect. 7 presents a case study. A summary and some conclusions are presented at the end of this paper, together with some suggestions for future work, in Sect. 8.

2 Previous Work

The proposed movement pattern mining process is founded on the concept of video data mining. Video data mining deals with the extraction of implicit knowledge from video data [13] using some sort of object tracking system. For example in [7] a video object tracking system is described that tracks mice with respect to mouse in a box scenarios, similar to those considered in this paper, where a video camera is suspended over the box. Further examples of tracking mice in video data can be found in [3, 6, 8, 9]. The system described in [9] tracks mouse movements by following a pattern "painted" on the back of each animal using hair bleach. In [3] a system is described that uses a combination of video and radio frequency tracking to obtain behavioural profiles. A criticism of this latter approach is that the nature of the behaviour may be affected by the presence of the radio frequencies. In [8] a computer vision process is used to analyse AVI (Audio Video Interleave) files to capture the behaviour of mice with respect to what are known as Morris Water Maze Tests.[1] In [6] a mechanism called Mice Profiler was proposed which allows the capture of information about the relative position, orientation and distance between pairs of rodents.

The distinction between the above mentioned tracking systems and that presented in this paper is that in the case of the work presented in this paper the video object tracking is the first stage of a comprehensive learning process. In this paper we are interested in discovering movement patterns to support the operation of (rodent behaviour) MABS. In the context of MABS for animal behaviour simulation the work described in [1, 2] is of some interest in that it is directed at the creation of a Mammalian Behaviour MABS (MBMABS) framework. However, in [1, 2] the operation of the MABS is hand crafted in the traditional manner. More specifically the concept of a manually constructed behaviour graph is used where nodes represent states and edges state changes.

3 Video Data Set

In this section the nature of the raw video data used with respect to the work presented in this paper is described. As noted above the focus for the work was mouse in a box scenarios as used by mouse behaviourolgists. The data was obtained by suspending a video camera over a 1.2 m^2 box into which one or more rodents and objects were introduced. A still from one of these videos is given in Fig. 1. In the case off the depicted scenario the box contained: (i) three equal size smaller boxes (areas) connected by two tunnels, (ii) three enclosed "nests" (the circular objects with air holes featured in the figure) and (iii) an open nest in the middle box (at the top of the middle box in Fig. 1). In the figure, although difficult to see, the mouse is located in the upper half of the right-hand area. A similar set up was used with respect to earlier work conducted by the authors and presented in [12]. To move from one area to another a rodent (a mouse in this case) must use one of the tunnels.

For video processing a software system was design and developed that featured a "Blob tracking" mechanism as described in [10]. This developed software process the video data "frame by frame". The software is semi-automated in that it requires user intervention because on some occasions the mouse location is lost, for example when it disappears into a nest, and cannot be automatically rediscovered. The reason for this is that the quality of the video is not particularly good and because the light intensity and colour of the moving object does not always remain constant.

[1]This is a recognised task for studying rodent learning where a rodent is required to find a submerged platform.

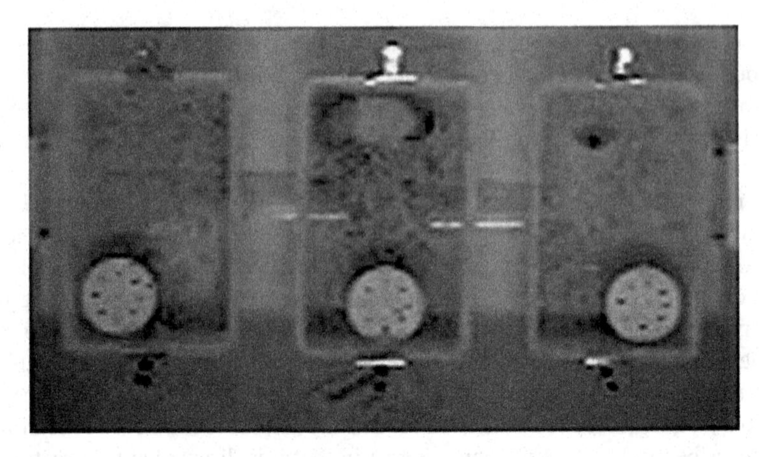

Fig. 1 Still from mouse video data featuring a scenario comprising three interconnected boxes

Using the video tracking software the location of the rodents that featured in the videos was recorded using a sample interval s measured in terms of a number of video frames. For the purpose of the evaluation and case study considered later in this paper $s = 5$ was used (note that 5 frames equates to 20 millisecond of video time). The video data used was sufficiently extensive so that all locations were covered (this was important with respect to the intended MABS).

4 Pattern Mining Framework

In this section we describe the proposed pattern mining framework. As already noted the focus for the work is mouse-in-a-box scenarios, where the box measures $1.2\,\text{m}^2$. We refer to the "floor space" of these boxes as the environment (the environment in which the envisaged agents will move around). The nature of these environments was captured using a grid representation. In other words the environments were conceptualised in terms of a tile world. An example grid representation, with respect to the video still given in Fig. 1, is given in Fig. 2. In this case the grid measures 19×8 (thus 152 grid cells). The grid representation is discussed in further detail in Sect. 4.1 below. Using the proposed video tracking system mouse agent locations were recorded according to the relevant grid number. Once the tracking was complete the recorded data was processed so that a set of movement patterns was obtained. Two types of movement pattern were considered: (i) absolute patterns and (ii) relative patterns. Both types of pattern are considered in Sect. 4.2 below.

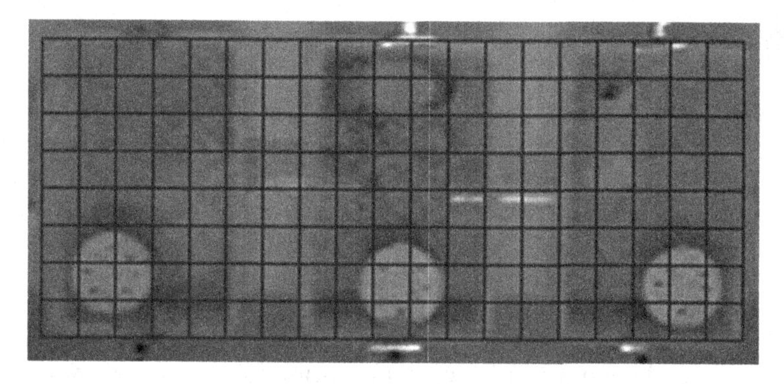

Fig. 2 Grid for environment with respect to video still presented in Fig. 1

4.1 Grid Representation

Using the proposed grid representation each cell is assigned a sequential identification number between 1 and n. We say that each grid cell has an *address* and use the variable A to indicate the complete set of addresses. The effect of the numbering is to linearised the space as shown in Fig. 3. The advantage of this enumeration is that movement is always described by a numeric constant k. For example to move one cell to the "north", in the case of the grid numbering presented in Fig. 3, $k = -19$; and to move one cell to the south east $k = 20$. Note that the value of k captures both distance and direction, hence we refer to such values as *movement vectors*. Note that $k = 0$ indicates no movement.

As noted above rodent locations were recorded using a sample interval of s. For each sample point, and each rodent agent, its location in terms of the cell number of the cell in which it was located was recorded. Pairs of samples separated by s thus represented a movement pattern as discussed in further detail in the following subsection. We also consider the idea of areas. These are collections of grid cells that may be considered to form a group. For example in Fig. 1 we can identify three areas each separated by a tunnel. The significance of areas is that special consideration is required when a rodent agent moves from one area to another. This will be explained in further detail in Sect. 5.

1	2	3	4	5	6	7	8	9	10	11	12	13	14	15	16	17	18	19
20	21	22	23	24	25	26	27	28	29	30	31	32	33	34	35	36	37	38
39	40	41	42	43	44	45	46	47	48	49	50	51	52	53	54	55	56	57
58	59	60	61	62	63	64	65	66	67	68	69	70	71	72	73	74	75	76
77	78	79	80	81	82	83	84	85	86	87	88	89	90	91	92	93	94	95
96	97	98	99	100	101	102	103	104	105	106	107	108	109	110	111	112	113	114
115	116	117	118	119	120	121	122	123	124	125	126	127	128	129	130	131	132	133
134	135	136	137	138	139	140	141	142	143	144	145	146	147	148	149	150	151	152

Fig. 3 Grid numbering for grid presented in Fig. 2 illustrating concept of movement vectors

4.2 Movement Patterns

Movement pattens comprise a tuple of the form $\langle a, V, p\rangle$, where: (i) a is a location (the "from" location), (ii) V is a set of movement vectors of the form described above and (iii) p is the probability of the movement vector occurring (a real number between 0.0 and 1.0). The size of V depends on how far we wish to look ahead. With respect to the evaluation and case study presented later in this paper $|V| = 5$ was used. Using $|V| > 1$ means that our rodent agents have a "memory", they have a planned route they wish to follow. Thus where $|V| > 1$ we have a sequence of locations $\{a_1, a_2, \ldots, a_{|V|}\}$ were $a_{|V|}$ is the end location and the remaining locations are intermediate locations which we refer to as *waypoints*. Previous work by the authors reported in [12] used $|V| = 1$.

Whatever the case, and as noted above, two types of movement pattern were considered with respect to the work presented in this paper: (i) absolute patterns and (ii) relative patterns. The distinction between the two, as the terminology suggests, is that in the first case locations are recorded relative to the origin of the environment while in the second the location is recorded relative to the local surroundings. Absolute locations are therefore expressed in terms of a specific address, thus $a \in A$. While when using relative patterns locations are represented using descriptors. The significance is that absolute patterns can only be used with respect to simulations that feature the same environment, while relative patterns are more versatile and can be used for a variety of simulations. However, relative descriptors are more complex. (A further advantage is that for some scenarios the number of relative patterns identified may be far fewer than the number of absolute patterns identified.)

Location descriptors comprised a tuple of the form $\langle D, a\rangle$ where: (i) D is a set of nine location type labels for the 3×3 sub-grid centred on the location in question linearised from top-left to bottom-right, and (ii) a is an area label. The ground type labels are taken from the set $L = \{b, g, i, n, o, w, t, -\}$, where: (i) b is a blocked location (an illegal location for a rodent location because, for example, it forms part of an obstruction), (ii) g is a gate location (the entrance/exit to a tunnel the significance of which will become clear later in this paper), (iii) i is an uncovered nest site, (iv) n is a covered nest site, (v) o is open space (effectively a location that does not belong to any of the other ground types), (vi) w is a wall location (a location next to the perimeter of the environment or next to an obstruction), (vii) t is a tunnel location, and (viii) $-$ a location outside of the environment (this is relevant with respect to cells located next to the environment boundary). The set of areas depends on the nature of the environment. In the case of the environment shown in Fig. 3 we can identify three areas. For convenience, and with respect the rest of this paper, we will labels thes as: L (left), M (Middle) and R (Right). Figure 4 shows the environment given in Fig. 3 with the grid cells labeled with their ground type. Table 1 gives some example location patterns for the environment shown in Fig. 4.

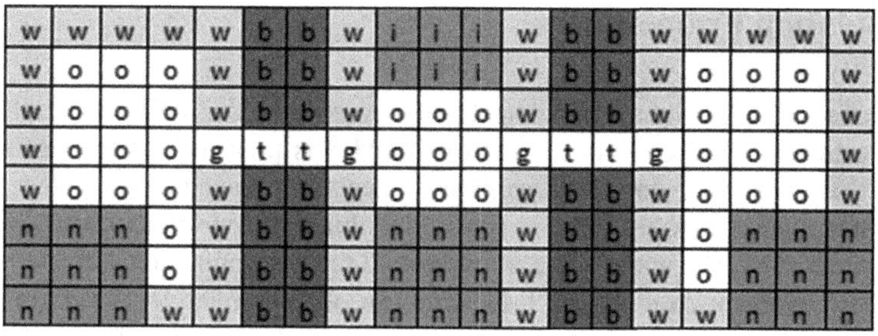

w	w	w	w	w	b	b	w	i	i	i	w	b	b	w	w	w	w	w
w	o	o	o	w	b	b	w	i	i	i	w	b	b	w	o	o	o	w
w	o	o	o	w	b	b	w	o	o	o	w	b	b	w	o	o	o	w
w	o	o	o	g	t	t	g	o	o	o	g	t	t	g	o	o	o	w
w	o	o	o	w	b	b	w	o	o	o	w	b	b	w	o	o	o	w
n	n	n	o	w	b	b	w	n	n	n	w	b	b	w	o	n	n	n
n	n	n	o	w	b	b	w	n	n	n	w	b	b	w	o	n	n	n
n	n	n	w	w	b	b	w	n	n	n	w	b	b	w	w	n	n	n

Fig. 4 Environment grid given in Fig. 3 with the grid cells annotated with ground type codes taken from the set $L = \{b, g, i, n, o, t, w, -\}$

Table 1 Example location descriptors for the environment considered in Figs. 2 and 3

Cell num	Descriptor	Cell num	Descriptor	Cell num	Descriptor
1	*− − − − ww − woL*	24	*wwbowbowbL*	47	*wiiwoogooM*
2	*− − −wwwwooL*	25	*wbbwbbwbbB*	48	*iiiooooooM*
3	*− − −wwwoooL*	26	*bbwbbwbbwB*	49	*iiwoowoogM*
4	*− − wwwoowL*	27	*bwibwibwoM*	50	*iwbowbogtM*
5	*− − −wwbowbL*	28	*wiiwiiwooM*	51	*wbbwbbgttB*
6	*− − −wbbwbbB*	29	*iiiiiiioooM*	52	*bbwbbwttgB*
7	*− − −bbwbbwB*	30	*iiwiiwoowM*	53	*bwobwotgoR*
8	*− − −bwibwiM*	31	*iwbiwbowbM*	54	*woowoogooR*
9	*− − −wiiwiiM*	32	*wbbwbbwbbB*	55	*oooooooooR*
10	*− − −iiiiiiM*	33	*bbwbbwbbwB*	56	*oowoowoowR*
11	*− − −iiwiiwM*	34	*bwwbwobwoR*	57	*ow − ow − ow − R*
12	*− − −iwbiwbM*	35	*wwwwoowooR*	58	*−wo − wo − woL*
13	*− − −wbbwbbB*	36	*wwwooooooR*	59	*woowoowooL*
14	*− − −bbwbbwB*	37	*wwwoowoowR*	60	*ooooooooooL*
15	*− − −bwwbwoR*	38	*ww − ow − ow − R*	61	*oowoogoowL*
16	*− − −wwwwooR*	39	*−wo − wo − woL*	62	*owbogtowbL*
17	*− − −wwwoooR*	40	*woowoowooL*	63	*wbbgttwbbT*
18	*− − −wwwoowR*	41	*oooooooooL*	64	*bbwttgbbwT*
19	*− − −ww − ow − R*	42	*oowoowoogL*	65	*bwotgobwoM*
20	*−ww − wo − woL*	43	*owbowbogtL*	66	*woogoowooM*
21	*wwwwoowooL*	44	*wbbwbbgttB*	67	*oooooooooM*
22	*wwwooooooL*	45	*bbwbbwttgB*	68	*oowoogoowM*
23	*wwwoowoowL*	46	*bwibwotgoM*	69	*owbogtowbM*

5 Simulation Framework

The mechanism whereby the movement patterns, generated as described in the foregoing section, were used with respect to a MABS is presented in this section. At the start of simulation one or more mouse agents are placed at some *legal location*. A legal location is defined as a grid cell within the environment whose ground type is not b or $-$. The agents then move around the environment as directed by the extracted movement patterns. Each location (absolute or relative) will have one or more movement patterns associated with it.[2] Some of these may be illegal in the sense that if adopted they would result in the rodent agent either moving outside off the environment or moving to a "blocked" location (a cell with a group type of—or b respectively).

Note that the relative descriptors are rotation variant, thus the number of descriptors could be decreased further if rotation invariant descriptors were used. Tables 2 and 3 give some example absolute and relative movement patterns respectively using $|V| = 1$. The probability value p in each case is calculated according to Eq. 1 where ϕ is the total number of occurrences for the location in question. Note that the set of p values associated with the set of movement patterns $MP_i = \{mp_{i1}, mp_{i2}, \dots\}$ for a specific location a_i will sum to 1. Recall also that $k = 0$ indicates no movement (the rodent agent stays where it is).

$$p = \frac{v}{\phi} \tag{1}$$

The simulation should operate so that the *sample* time (interval at which locations were extracted from the video data) is maintained. However, for visualisation purposes the simulation time should be less (as otherwise the rodent agents appear to jump from location to location rather than move from location to location). Simulation time was calculated using Eq. 2 where q is some constant. Empirical results indicated that $q = 5$ produced a good result. Thus a set of "way points" needs to be calculated for each path in V. The total number of way points is thus $|V| \times q$, 25 with respect to the case study presented later in this paper. Once the way points have been calculated we next have to check that none of the way points represent illegal locations.

An additional complication, that requires special consideration, is where the relevant set of movement patterns includes locations in more than one area. In this case it will be highly likely that the "line of sight" travel line will pass through blocked areas. Where this happens the movement pattern will not be deemed to be illegal, but instead when implemented the line of travel should be via the relevant gate locations so that the rodent agent passes through the relevant tunnel (or tunnels).

Out of the legal set of movement patterns one will be elected in a probability driven random number using the p value associated with each movement pattern. Note that if there are illegal movement patterns than the values for p will need to

[2]The situation where we have an incomplete set of movement patterns is a subject for future work, currently we extract large numbers of movement patterns so as to avoid this situation.

Table 2 Example absolute movement patterns

Absolute location	Movement vector V	Probability p
93	−23	0.094
93	−22	0.031
94	−22	0.059
94	−3	0.059
94	0	0.059
97	−33	0.333
97	−32	0.333
97	−31	0.333
98	−29	0.033
98	0	0.024
99	−38	0.007
99	−31	0.129
99	−30	0.072
99	−29	0.065
99	0	0.007
99	1	0.014
99	6	0.014
100	−65	0.010
100	−37	0.005
100	−14	0.021
100	−13	0.047
100	0	0.073

be temporarily recalculated. It is possible that there is no legal movement that can be adopted (more so in the case of absolute movement patterns than in the case of relative movement pattens because the first are more specific). To date, with respect to the experiments that have been conducted, this has not happened because of the large number of movement patterns available, however for future work consideration clearly needs to be given to this issue.

$$simulation\ time = \frac{sample\ time}{q} \tag{2}$$

6 Evaluation

It is difficult to evaluate the operation of simulations (MABS or otherwise) with respect to any "Gold standards". The novel mechanism adopted with respect to the work presented in this paper is to "complete the loop". The operation of individual

Table 3 Example relative movement patterns

Relative location	Movement vector V	Probability p
bbwttgbbw	−17	0.010
bbwttgbbw	−16	0.003
bbwttgbbw	−3	0.132
bbwttgbbw	−2	0.295
bbwttgbbw	−1	0.014
bbwttgbbw	1	0.051
bbwttgbbw	2	0.231
bbwttgbbw	3	0.156
bwnbwnbwn	−18	0.750
bwobwnbwn	−18	1.000
bwobwobwo	−38	0.125
bwobwobwo	−19	0.875
bwobwotgo	0	0.019
bwobwotgo	1	0.009
bwobwotgo	18	0.074
bwobwotgo	19	0.898
bwotgobwo	−37	0.004
bwotgobwo	−19	0.006
bwotgobwo	−18	0.063
bwotgobwo	−17	0.061
bwwbwobwo	0	0.019

MABS runs was evaluated by videoing the simulation and repeating the process of mining movement patterns. If the patterns extracted from the video data were similar to the patterns extracted from the simulation video data it could be argued that the simulation was realistic. Of course the simulation run time and the video run time have to be the same for the comparison to be meaningful.

In the context of this paper the operation of the proposed rodent behaviour MABS using movement patterns was evaluated using the scenario presented earlier in Figs. 1, 2, 3 and 4. The results are presented in Fig. 5. The figure shows the number of absolute and relative movement patterns recorded using the simulation data compared with the number of movement patterns obtained using the original video data. The identified patterns are grouped according to the nine different area combinations featured in the evaluation scenario (*LtoL*, *LtoR*, *LtoM*, *MtoL*, *MtoR*, *MtoM*, *RtoL*, *RtoR* and *RtoM*). The Y-axis represents the number of extracted movement patterns (of course in many cases the extracted movement patterns will be duplicates with existing patterns). From the figures it can firstly be observed that there is a good correspondence between the simulation data and the video data indicating that the simulation is realistic. It should also be noted that a large number of patterns are obtained in both cases, hence the chance of there being no movement patterns associated with a particular location is minimised.

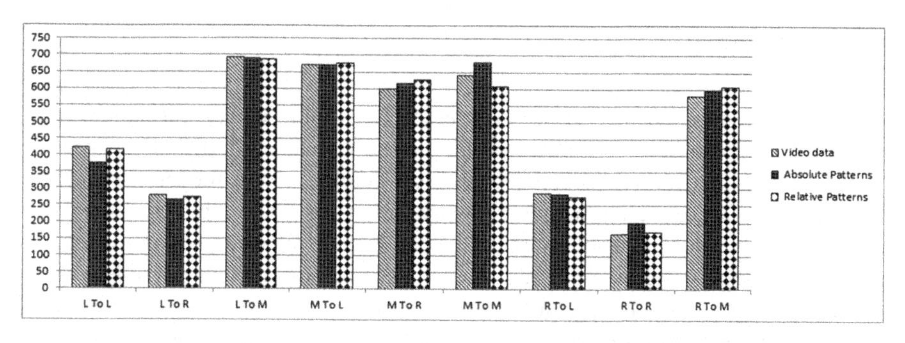

Fig. 5 Comparison of Simulated (Absolute and Relative) versus video data relative movement patterns

w	w	w	w	w	b	b	w	i	i	i	w	b	b	w	i	i	i	w	b	b	w	i	i	i	w	b	b	w	i	i	i	w	b	b	w	w	w	w	w
w	o	o	o	w	b	b	w	i	i	i	w	b	b	w	i	i	i	w	b	b	w	i	i	i	w	b	b	w	i	i	i	w	b	b	w	o	o	o	w
w	o	o	o	w	b	b	w	o	o	o	w	b	b	w	o	o	o	w	b	b	w	o	o	o	w	b	b	w	o	o	o	w	b	b	w	o	o	o	w
w	o	o	o	g	t	t	g	o	o	o	g	t	t	g	o	o	o	g	t	t	g	o	o	o	g	t	t	g	o	o	o	g	t	t	g	o	o	o	w
w	o	o	o	w	b	b	w	o	o	o	w	b	b	w	o	o	o	w	b	b	w	o	o	o	w	b	b	w	o	o	o	w	b	b	w	o	o	o	w
n	n	n	o	w	b	b	w	n	n	n	w	b	b	w	n	n	n	w	b	b	w	n	n	n	w	b	b	w	n	n	n	w	b	b	w	o	n	n	n
n	n	n	o	w	b	b	w	n	n	n	w	b	b	w	n	n	n	w	b	b	w	n	n	n	w	b	b	w	n	n	n	w	b	b	w	o	n	n	n
n	n	n	w	w	b	b	w	n	n	n	w	b	b	w	n	n	n	w	b	b	w	n	n	n	w	b	b	w	n	n	n	w	b	b	w	w	n	n	n

Fig. 6 Case study environment (*colour coding* same as that used in Fig. 4)

7 Case Study

One of the advantages claimed for relative patterns in Sect. 4.2 above is that they are more versatile than absolute patterns in that they can be used with respect to environments not identical to those from which they were generated (unlike in the case of absolute movement pattens). In this section a case study is presented where the relative movement patterns extracted from the scenario presented earlier in Figs. 1, 2, 3 and 4 (examples of which are given in Table 3) were used with respect to an alternative environment of the form shown in Fig. 6 (the colour coding is the same as that used in Fig. 4). From the figure it can be seen that the "playing area" is larger (40×11) and features six areas whereas the original environment featured three.[3] Simulations run using this environment demonstrated that the previously generated movement patterns were entirely suited to generating realistic simulations using this environment and similar alternative environments (of course absolute movement patterns could not be used for this purpose).

[3]Note that the original area lebelling, $\{L, M, R\}$, had to be reinterpreted with respect to this alternative scenario.

8 Conclusion

In this paper a mechanism has been discussed for mining movement patters from video data that can be incorporated into a rodent behaviour MABS. Two types of movement patterns were considered, absolute and relative. The movement patterns have probabilities associated with them which were used to select patterns in a probability driven random manner so as to drive a MABS. When selecting movement patterns only legal patterns could be chosen, those that do not result in a rodent agent passing through or ending up at a location outside of the environment or a blocked location (a location with a ground type of—or *b*). An added complication was where a rodent agent moves from one area to another as this had to be realised using the tunnels connecting areas (at least with respect to the scenario used as a focus with respect to this paper). The operation of the MABS was conducted by "completing the loop". The simulations were videoed and these videos were processed in the same manner as the original input data. The nature of the identified movement patterns form the simulated data were then compared with the movement patterns generated from the video data. Good levels of comparison were obtained suggesting that realistic simulations were produced using the proposed mechanism. The added claimed advantage of relative movement patterns is that they can be used with respect to alternative environments than those from which they were originally extracted. This was illustrated using a case study. Overall the authors have been very encouraged by the results produced. For further work the intention is to consider more complex scenarios featuring additional obstructions and types of area which will necessitate the use of more versatile forms of relative location descriptions.

References

1. Agiriga, E., Coenen, F., Hurst, J., Beynon, R., Kowalski, D.: Towards large-scale multi-agent based rodent simulation: the mice in a box scenario. In: Max Bramer, Miltos Petridis, and Lars Nolle, editors, Research and Development in Intelligent Systems XXVIII, pp. 369-382. Springer, London (2011)
2. Agiriga, E., Coenen, F., Hurst, J., Kowalski, D.: A multiagent based framework for the simulation of mammalian behaviour. In: Max Bramer and Miltos Petridis, editors, Research and Development in Intelligent Systems XXX, pp. 435-441. Springer International Publishing (2013)
3. Aharon, W., Alexander, S., Genadiy, V., Liat, E., Molly, D., Assif, Y., Libi, H., Ofer, F., Tali, K.: Automated long-term tracking and social behavioural phenotyping of animal colonies within a semi-natural environment. Nat. Commun. **4** (2013)
4. Bunn, D.W., Oliveira, F.S.: Agent-based simulation-an application to the new electricity trading arrangements of england and wales. **5**, 493–503 (2001)
5. Davidsson, P., Henesey, L., Ramstedt, L., Rnquist, J.T., Wernstedt, F.: An analysis of agent based approaches to transport logistics. Trans. Res. Part C: Emerg. Technol. **13**(4), 255–271 (2005)
6. Giancardo, L., Sona, D., Huang, H., Sannino, S., Manag, F., Scheggia, D., Papaleo, F.: Automatic visual tracking and social behaviour analysis with multiple mice, vol. 08, September 2013

7. Hueihan, J., Estibaliz, G., Yu, X., Vinita, K., Tomaso, P., Steele, A.D.: Automated home-cage behavioural phenotyping of mice. Nat. Commun. **1**, 68 (2010)
8. Leon, G.M., Moreno-Baez, A., Sifuentes-Gallardo, C., Garcia-Dominguez, E., Valencia, M.G.: Analysis of avi files for mice behavior experiments in the morris water maze. In: Electronics, Robotics and Automotive Mechanics Conference (CERMA), 2011 IEEE, 131-136, November 2011
9. Ohayon, S., Avni, O., Taylor, A.L., Perona, P., Roian Egnor, S.E.: Automated multi-day tracking of marked mice for the analysis of social behaviour. J. Neurosci. Methods **219**(1), 10–19 (2013)
10. Opencv-2.3 blob-tracking module. http://www.enl.usc.edu/enl/trunk/aqua/OpenCV-2.3.../Blob_Tracking_Modules.doc. Accessed 01 Dec 2013
11. Shen, W., Norrie, D.H.: Agent-based systems for intelligent manufacturing: a state-of-the-art survey. Knowl. Inf. Syst. **1**(2), 129–156 (1999)
12. Tufail, M., Coenen, F., Mu, T.: Mining movement patterns from video data to inform multi agent based simulation. In: Proceedings 10th International Workshop on Agents and Data Mining Interaction (ADMI-14) hosted as part of the International Conference on Autonomous Agents and Multi-Agent Systems (AAMAS-2014) (2014)
13. Vijayakumar, V., Nedunchezhian, R.: A study on video data mining. Int. J. Multimed. Inf. Retr. **1**(3), 153–172 (2012)
14. Witten, I.H., Frank, E., Hall, M.A.: Data Mining: Practical Machine Learning Tools and Techniques: Practical Machine Learning Tools and Techniques. The Morgan Kaufmann Series in Data Management Systems. Elsevier Science, Amsterdam (2011)

Assembly of Neural Networks Within a Federation of Rational Agents for the Diagnosis of Acute Coronary Syndromes

J.J. Sprockel and E. Gonzalez

Abstract The process of clinical diagnosis poses a complex problem, it is the first step of any process in health care. Acute coronary syndromes are one of the main causes of death worldwide; its most frequent presentation, acute chest pain, requires for diagnosis, take into consideration aspects of the clinical record, the electrocardiogram, and markers of heart disease. Intelligent systems tools have been successfully tested for the diagnosis. Multi-agent systems are a promising way for the computer modeling of this process. In theory, relevant data can be gathered and adaptability and learning capabilities can be added. The present work presents a federation of agents which are the product of an analysis made through the AOPOA methodology, which integrates the diagnosis of chest pain focused on acute coronary syndromes by means of a neural network assembly system (some of the neural networks are specialized in special populations), exhibiting a high level of diagnostic accuracy.

1 Introduction

Clinical diagnosis is a fundamental competence for medical practice in all of its branches, it is the first step to determine any process in health care; a prompt and accurate diagnosis is mandatory in areas such as the emergency service. Chest pain represents a particularly important entity, although it is caused mainly by benignant entities, fast identification of life-threatening situations has a direct impact on a successful treatment [1]. The top life-threatening situation is acute myocardial infarction (AMI), which is part of the acute coronary syndromes (ACS), it is one of

J.J. Sprockel (✉)
Fundación Universitaria de Ciencias de La Salud—Hospital San José, Bogotá, Colombía
e-mail: jjsprockel@fucsalud.edu.co

E. Gonzalez
Pontificia Universidad Javeriana, Bogotá, Colombía
e-mail: eguerrero@javeriana.edu.co

M. Bramer and M. Petridis (eds.), *Research and Development in Intelligent Systems XXXII*, DOI 10.1007/978-3-319-25032-8_22

the main causes of death worldwide [2]. Proper characterization of ACS requires the combination of a complete medical record, electrocardiograms (ECG) follow up, and myocardial damage markers.

Big efforts have been made in the last decades to develop tools such as clinical algorithms and regression models [3], computational algorithms [4] or intelligent systems to support the diagnosis of ACS. Intelligent systems are those attempts to emulate human intelligence, they favor reasoning and learning in uncertain or inaccurate environments; multiple strategies of this type have been applied exhibiting good rates of success; among them are decision trees [5], Bayesian networks [6], k nearest neighbors [7], artificial neural networks (ANN) [8], fuzzy logic [9], evolutionary computation [10], and support vector machines [11].

Multi-agent systems (or collaborative rational agents) are a collection of independent agents that communicate with one another in order to cooperate in the joint solution of a complex task [12]. Multi-agent systems are becoming a promising way to model the diagnosis process by means of a computer system. In theory, relevant data can be gathered and add adaptability and learning capabilities. There are very few works approaching this area and is an opportunity to exploit the strengths of this strategy in benefit of the patients. Multi-agent systems (MAS) have the possibility to cooperate, encapsulating the tasks for the diagnosis and favor the re-utilization of the code already created, allowing the incorporation of agents that bring a variety of contributions to the whole process.

The objective of the present work is to present the design and evaluation of a system of collaborative rational agents to support the diagnosis of acute chest pain showing the results of the experiments in the first phase focused on the diagnosis of acute coronary syndromes in patients consulting the emergency service based on the decision assembly made by a set of artificial neural networks. Section 2 describes the methodology used; Sect. 3 presents the results of the experiments showing the performance of each of the individual neural networks, the assembly by different systems, and the addition of specialized neural networks. Section 4 ends the present work with a discussion that compares the results with the literature revised and proposes possible future works.

2 Methodology

The methodology used in this work includes three components: analysis and design of the multi-agent system, the description of the cooperation system and the experiments carried out to select the assembly system for the neural networks.

2.1 Analysis and Design of the Multi-agent System

The analysis and design of the multi-agent system followed the steps of the AOPOA methodology (Agent-Oriented Programming Organizational Approach) for the construction of multi-agent systems through the decomposition of a complex problem in an iterative, recursive way, based on the concept of goals and identification of tasks in a structured way with an organizational approach [13].

After analyzing of requirements, it was determined that the problem of chest pain must be approached by consecutive phases on which the diagnosis by clinical entities could be performed. Thus, the system consists on a federation of collaborative agents which has a *general leading agent* that sets the pathway throughout the whole diagnosis process, and a set of *phase managing agents* which are activated in a sequential manner for the diagnosis of each group of diseases (Fig. 1).

The *general leading agent* begins the first *phase managing agent* for the ACS diagnosis; this agent requests the diagnosis judgment to a set of *clinical agents.* Some of the consulted agents are specialized in certain special groups for diagnosis; when they respond, an assembly method performs the final diagnostic decision. When a positive result case appears, the diagnosis is assigned to an ACS by the *phase managing agent* and *general leading agent* gives such a report and terminates the process; in a negative case, a start signal is sent to the *second phase managing agent*, as shown in Fig. 1, leading to the diagnosis of pulmonary embolism which in turn utilizes the concept of a group of specialized agents, in this case, it will be composed of a new set of *clinical agents*. Thus the process progresses increasingly including diseases that manifest with chest pain covered by the plan that is controlled by the general leading agent.

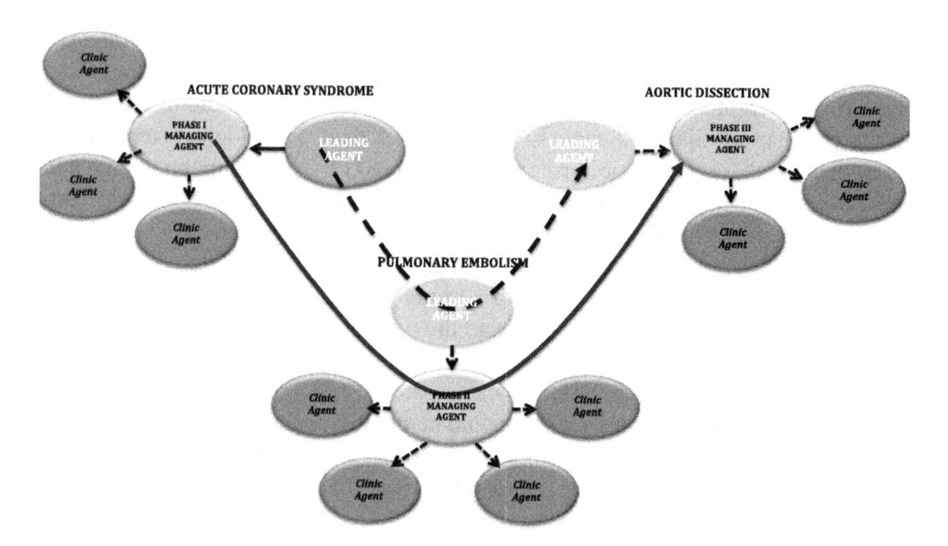

Fig. 1 Model of the federation of rational agents

2.2 Description of the Cooperation Model

In the late 90s Ferber proposed a model to describe the cooperation into a multi-agent system [14]. Following such 5Cs model the following components are described: task assignment (collaboration) is carried out in a centralized manner by direct assignment through the general leading agent and then by the managing agents of each phase, the main resource is the consultation report which is operated by the general leading agent. Another resource, the patient´s clinical record, will be accessed freely by each of the clinical agents in an independent manner.

With respect to coordination, this process follow a set of established basic sequential steps, the point for synchronization is that of the decision within each of the phases which be worked out on an individual basis for each phase, the rest of the tasks are sequential. No conflict resolution issues are expected as there are compatible objectives and the clinical record has open access. Communication between agents mainly takes place by means of a request-reply model.

2.3 Selection of Assembly Model

Different systems have been developed in order to combine multiple pattern recognition techniques. These are called assembly systems, among which are: simple voting or weighted, the Bayesian average model, bagging, boosting, tree-based models, conditioned mixed models (mix of linear regression, logistic, and expert models); other mechanisms described are the adaptative linear neural network model and conditional random fields.

There are several reasons to use assembly-based systems [15]: statistics, attempting to improve the capability of generalization; when there is a large amount of data or scarce data; and also the divide-and-rule principle and the need for fusing data from different sources. It can be seen how the aforementioned items are related to the problem of clinical diagnosis, which makes their use a logical step. In this case, the database population is extremely varied, having subpopulations that exhibit special behavior, which makes that their individualized management contributes to improve the diagnostic performance. Upon having multiple associations of variables that have been recognized in the medical field, structured in several scales that could identify ACS cases, the training of ANNs from these scales and their assembly could lead to gain data and the possibility to improve the capability of generalization.

In order to take advantage of the characteristics of MAS, it was decided that this initial experiment would integrate a variable number of neural networks including a group specialized in the diagnosis of subgroups that have special characteristics for diagnosis purposes. A number of options were taken into consideration to accomplish the integration, including argumentation; given the scope of the present work an assembly system by weighted or simple voting was selected. Then, a diagnostic tests study was performed which was divided into two phases. In the first phase a set of ANN were trained and tested for the diagnosis of ACS. In the second phase, this set of ANN was integrated under different forms of assembly for the purpose of improving the performance in the diagnosis.

2.3.1 Description of the Database

The different neural networks were trained using the data of 307 patients coming from two databases (Table 1): the first database has 159 patients aged 18 + who attended consultation at the emergency service due to chest pain between February 20 and October 30 2012; this cohort is the result of an institutional study that assessed the implementation of a critical pathway for ACS at the Hospital San José of Bogotá [16]. The second database has 148 patients, selected in a prospective manner, aged 18 + hospitalized with clinical suspicion of ACS between July 25 2013 and August 1 2014; this database was designed and collected specifically for the present work.

The diagnosis of the infarctions was based on the third redefinition of infarction [17] and the anginas were classified in accordance with the compatible clinical data (mainly chest pain, recent-onset, *in crescendo* or rest) together with a positive result in any of the invasive or non-invasive coronary stratification strategies.

Table 1 Database description

	First database N = 159	Second database N = 148
Collection time	Feb–Oct 2012	Jul 2013–Aug 2014
Diagnosis, n (%)		
Angina	41 (25, 8 %)	28 (18, 9 %)
Infarction	37 (23, 3 %)	68 (46, 0 %)
Non coronary	81 (50, 9 %)	52 (35, 1 %)
Age (mean)	61,4 y	64,1 y
Women, n (%)	72 (45,3 %)	61 (41, 2 %)
Diabetes, n (%)	33 (20, 8 %)	29 (19, 6 %)
Renal failure, n (%)	–	26 (17, 6 %)
Aged 75 +, n (%)	32 (20, 1 %)	42 (28, 4 %)
Prior coronary dissease	49 (30, 8 %)	39 (26, 4 %)
Coronary stratification, n (%)	100 (62, 9 %)	126 (85, 1 %)
Arteriography, n (%)	50 (31, 4 %)	99 (66, 9 %)

2.3.2 Training and Neural Network Tests

The neural networks were trained and tested in the general population and also in the different particular subgroups by applying a genetic algorithm for the selection of the weights of the networks. During the training process, the population was divided into 70 % for training and 30 % for test purposes. The evolutionary strategy was developed as follows:

- Representation of the individuals: two vectors made up by the weights of the first and second layers of the neural network, divided into groups according to the number of neurons in the hidden layer (4, 6, 8, and 10) and the input data coming from the different scales.
- Population size: 200 individuals, 50 per each group of neurons in the hidden layer (4, 6, 8, and 10).
- Initial population: the value of each of the weights was assigned at random.
- Mechanism of evolution: mutation (10 %) and recombination (90 %).
- Selection: a tournament approach was used to select the individuals.
- Fitness function: consisted in assessing the result for each network in each of the training examples, then the performance was calculated by assessing the diagnostic accuracy of each network (ratio of number of correct diagnoses to total number of cases assessed).
- Termination criterion: 80 iterations.

2.3.3 Tests with the Assembly Models

In order to detect the best ways of assembly for the neural networks within a federation of collaborative rational agents for the diagnosis of acute coronary syndromes, a preliminary factorial 3^3 experiment was designed within a fixed effects model, not replicated, with a 5 % significance level. All the combinations of the different levels of each of the factors were assessed. Diagnostic accuracy of each of the assembly architectures was selected as the response variable, tested in 30 % of the total set of data (92 examples).

The following factors and their different levels were studied:

(1) The number of data items that were taken into consideration for the training and test of the networks; this factor is divided into three levels: low (close to 50); *medium* (around 90), and *high* (above 140).
(2) The number of neural networks selected from each of the training data sets (scales of Braunwald, Florencia, and Sanchis); the levels were set as: *high* (eight), *medium* (five), and *low* (two).
(3) The heuristics employed within the decision assembly system:
    a. *Simple voting:* consists on the count of those agents having a positive result, in case of being more than half of the total number of agents the final result was positive.

b. *Weighted voting by predictive values:* the respective predictive value (positive or negative) was added to the count of positives or negatives; the predictive value was already known at the test phase of the neural network upon its training; if the count of positives was greater than the count of negatives, the result was considered positive.
c. *Weighted voting by likelihood ratio* (likelihood ratios, LR): it was obtained by getting a value that equally weighted both positive or negative results, obtained from LR+/LR−, positive and negative results were summed separately and then the diagnosis was assigned to the higher value.

This cycle described above was executed, including also the assembly of the results produced by the voting of the specialized agents. In particular, the specialized groups of diagnosis include: diabetics, women, chronic renal patients, and/or senior patients (aged 75+).

3 Results

Upon running the genetic algorithm, 84 networks were selected which had a accuracy level above 70 %, divided as follows:

- 72 for the diagnosis of acute coronary syndromes based on the whole population (8 networks per each of the three levels of number of training data in the 3 input data coming from each scale),
- 12 in specific subpopulations (women, aged 75+, diabetics, and those with an antecedent of coronary disease).

For these individual neural networks, an average 76.7 % accuracy was found in the training phase. The average accuracy during test phase was 72 %, ranging between 60 and 81 %. The lowest average was found in the case of the neural network trained with the lowest number of examples developed from the scale of Sanchis. The highest accuracy was found for the cases based on the scale of Braunwald. Figure 2 compiles the obtained results.

Significant variability was found in the global performance results for the different combinations examined, obtaining in most cases high levels of sensitivity (91–100 %) and lower levels of specificity (36–67 %). If the analysis focuses on the results of the global accuracy of the test, as shown in Fig. 2 and Table 2, it can be observed that in most cases the performance of the different treatments is above that of the individual agents (range between 72 and 85 %) and the weighted systems have better performance than simple voting.

The ANOVA encountered that the factors assessed are significant; on average the weighted system by predictive values was better than the other two systems assessed. Upon performing the tests including the neural networks specialized in subgroups, a significant increase in diagnostic accuracy was found. The treatments

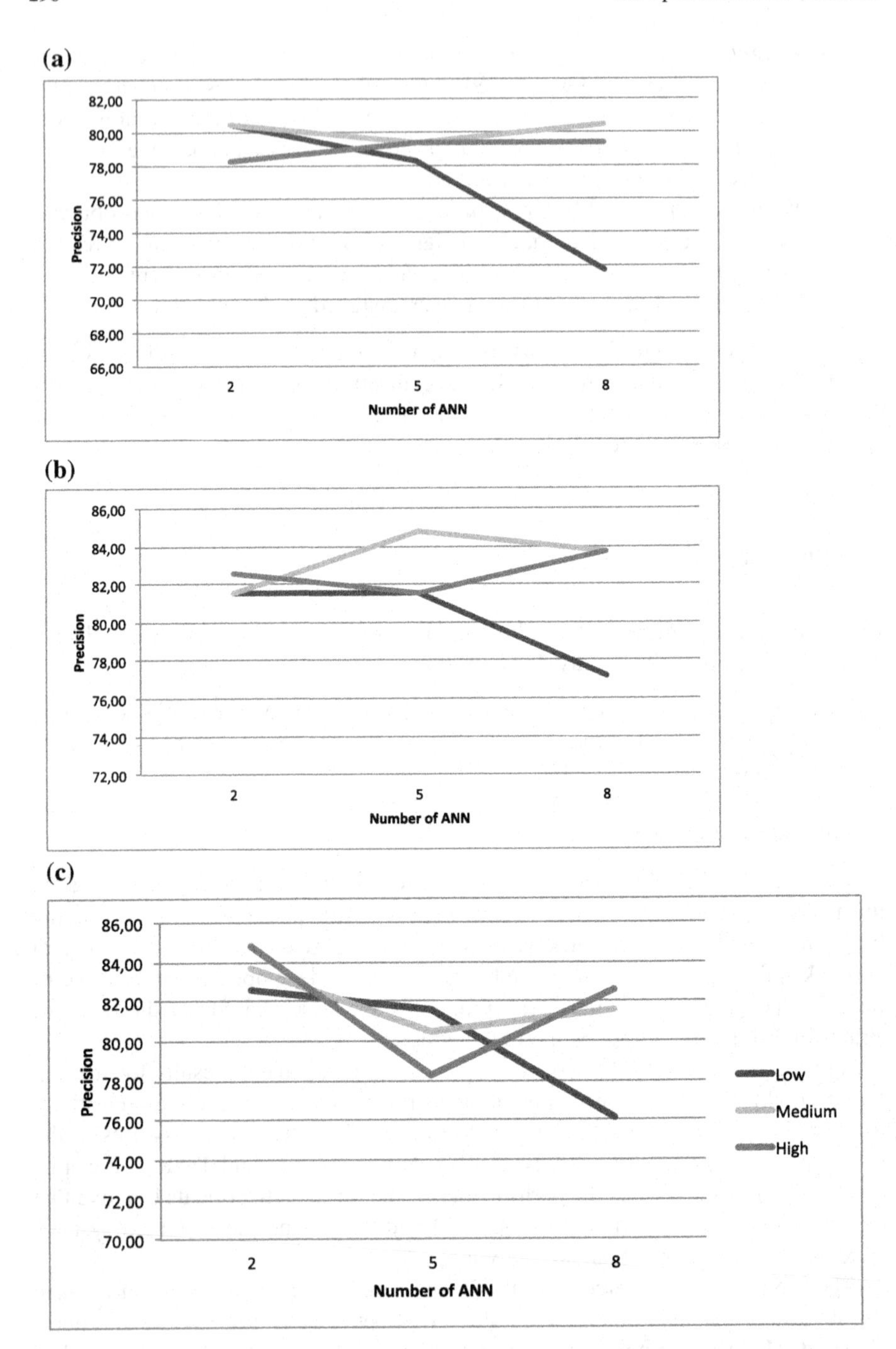

Fig. 2 Results of the experiments according to each factor assessed. LR: Likelihood ratios, ANN: Artificial Neural networks. **a**. Simple voting. **b**. Weighted voting by predictive values. **c**. Weighted voting by LR

Table 2 Results of the assembly experiments

Ann number	Level of training data	Type of assembly	Acuracy	Subgrups accuracy
2	High	Simple	80	80
5	High	Simple	80	83
8	High	Simple	78	80
2	Medium	Simple	78	80
5	Medium	Simple	79	80
8	Medium	Simple	79	81
2	Low	Simple	**72**	**76**
5	Low	Simple	80	79
8	Low	Simple	79	82
2	High	Weighted by PVs	82	86
5	High	Weighted by PVs	82	**88**
8	High	Weighted by PVs	83	**88**
2	Medium	Weighted by PVs	82	83
5	Medium	Weighted by PVs	85	85
8	Medium	Weighted by PVs	82	86
2	Low	Weighted by PVs	77	83
5	Low	Weighted by PVs	**84**	84
8	Low	Weighted by PVs	**84**	84
2	High	Weighted by LRs	83	85
5	High	Weighted by LRs	**84**	**89**
8	High	Weighted by LRs	**85**	**88**
2	Medium	Weighted by LRs	82	80
5	Medium	Weighted by LRs	80	82
8	Medium	Weighted by LRs	78	80
2	Low	Weighted by LRs	76	78
5	Low	Weighted by LRs	82	82
8	Low	Weighted by LRs	83	84

*PVs: Predictive values, LR: Likelidhood ratios, ANN: Artificial neural networks

exhibiting the best performance were those with networks trained with a high level of examples, with 5 and 8 agents for each scale used, and weighted both by LR and predictive values (maximum accuracy of 89 and 88 %).

4 Discussion

The results for the different types of assembly show an improvement in diagnostic accuracy as compared with each of the individual neural networks. The inclusion of diagnostic decisions by an additional group of specialized agents applied on subpopulations of interest due to their particular behavior at the moment of their presentation: women, diabetics, coronary patients, and those aged 75+, translated into an increase in accuracy up to 89 %.

Two previous works were found which used systems based on decision assembly–neural networks for the diagnosis of ACS:

- In the first [21], Bagging and cross-splitting were used for the analysis of the data coming from 634 electrocardiograms (130 diagnosed with ACS); the results resembled those of the present work, exhibiting lower performance though; area under the curve ROC 0,81; the performance was above that of each of the individual neural networks that were part of the system.
- On the other hand, a second study developed and tested a multi-agent system in which intelligent systems tools were assembled for the diagnosis of coronary disease, through several boosting techniques: AdaBoost, MultiBoost, and LogitBoost [22]; the database had 270 patients with 13 variables and was performed at Cleveland Clinic Foundation, with 120 patients exhibiting coronary disease. High levels of performance were achieved, having AdaBoost the highest (accuracy 97,6 %, sensitivity 97,6 %, specificity 97,6 % with AUC 0,989). This very good results are related to the fact that the individual neural networks already have a high performance.

The process described shows a structured and systematic analysis applying a methodology developed for this purpose. The use of multi-agent systems allows dividing tasks, with a potential for learning and adaptability within the context of a complex problem such as the comprehensive diagnosis of chest pain. Its application allows the exchange of agents facilitating the evolution towards better performance or the use in specific population groups from a geographic or any other perspective, the consideration of other diagnoses having the same approach and the potential support through suggestions for studies and management, thus considering the temporal context, represents unique contributions from the perspective of multi-agent systems.

The present study is limited by the fact that it has not yet undergone a process of cross-validation, thus being a prototype. At this moment, a study of this type has started in the same institutions. On the other hand, it is unknown if the benefit of the assembly remains when using neural networks having high initial performance (above 90 %), such as those found in the revision of the literature.

It can be considered as a strength the fact of using aspects of medical knowledge in the problem of diagnosing chest pain, taking advantage of the scales applied at hospitals for the classification of acute coronary syndromes through the data provided by the scales of Braunwald, Sanchis, and Florencia. In this work, a large number of neural networks were developed coming from diverse data and training populations, the database included data from two hospitals, and a robust experimental framework was structured analyzing several ways to carry out the assembly.

5 Conclusions

It is possible to model the process to diagnose chest pain through a federation of rational agents, as a result of its analysis using the AOPOA methodology, ending up dividing the process into different phases for the diagnosis of clinical entities. The present work describes the results of the first phase which focuses on acute coronary syndromes, upon the assembly of a set neural networks trained by means of a genetic algorithm that included a number of networks specialized in special subgroups exhibiting particular behavior in their clinical presentation (diabetics, women, senior patients, and renal patients). A maximum diagnostic accuracy of 89 % was achieved by means of an assembly based on voting weighted by the operational characteristics of the test (likelihood ratio or predictive values); thus, it is possible to use this assembly within a system that supports clinical decisions.

There are a number of possibilities for future works, such as the use of other forms of integration of the decisions like argumentation or any of the more sophisticated assembly systems such as Bagging, Boosting, linear model of adaptive neural network or conditional random fields. It would also be possible to continue the work on other phases for the diagnosis of other diseases, further tools of intelligent systems other than neural networks could also be included such as vector support machines, decision trees, or Bayesian networks within the clinical agents, even the concepts of specialist physicians and laboratories could be included upon exposing the case.

Acknowledgments The authors are indebted to the members of the Department of Research and Statistics and to the medical staff and residents of the internal medicine service of the Fundación Universitaria Ciencias de la Salud—Hospital San José of Bogotá.

References

1. Erhardt, L., et al.: Task force on the management of chest pain. Eur. Heart J. **23**, 1153–1176 (2002)
2. Ministerio de Salud y Protección Social.: Prevención enfermedades NO transmisibles: Enfermedades cardiovasculares. Enfermedades cardiovasculares. http://www.minsalud.gov.co/salud/publica/PENT/Paginas/enfermedades-cardiovasculares.aspx (2013)
3. Hess, E.P., et al.: Diagnostic accuracy of clinical prediction rules to exclude acute coronary syndrome in the emergency department setting: a systematic review. Can. J. Emerg. Med. **10**, 373–382 (2008)
4. Steurer, J., Held, U., Schmid, D., Ruckstuhl, J., Bachmann, L.M.: Clinical value of diagnostic instruments for ruling out acute coronary syndrome in patients with chest pain: a systematic review. Emerg. Med. J. **27**, 896–902 (2010)
5. Coskun, O., Eren, A., Eren, M.: A computer based telemedicine protocol to predict acute coronary syndrome in patients with chest pain at home. Int. Heart J. **47**, 491–500 (2006)
6. Vila-Francés, J., et al.: Expert system for predicting unstable angina based on Bayesian networks. Expert Syst. Appl. **40**, 5004–5010 (2013)

7. Salari, N., Shohaimi, S., Najafi, F., Nallappan, M., Karishnarajah, I.: Application of pattern recognition tools for classifying acute coronary syndrome: an integrated medical modeling. Theor. Biol. Med. Model. **10**, 57 (2013)
8. Sprockel, J., Diaztagle, J.J., Alzate, W., Gonzalez, E.: Redes neuronales en el diagnóstico del infarto agudo de miocardio. Revista Colombiana de Cardiología **21**, 215–223 (2014)
9. Muthukaruppan, S., Er, M.J.: A hybrid particle swarm optimization based fuzzy expert system for the diagnosis of coronary artery disease. Expert Syst. Appl. **39**, 11657–11665 (2012)
10. Sprockel, J., Alzate, W.: Aplicación de la computación evolutiva en el diagnóstico del infarto agudo del miocardio. Repertorio de Medicina y Cirugía **23**, 199–203 (2014)
11. Conforti, D., Guido, R.: Kernel-based Support Vector Machine classifiers for early detection of myocardial infarction. Optim. Methods Softw. **20**, 401–413 (2005)
12. Moreno, A., Garbay, C.: Software agents in health care. Artif. Intell. Med. **27**, 229–232 (2003)
13. González, E., Torres, M.: AOPOA: Organizational approach for agent oriented programming. In: ICEIS 2006—8th International Conference on Enterprise Information Systems, Proceedings **SAIC,** pp. 75–80 (2006)
14. Ferber, J.: Multi-agent systems: an introduction to distributed artificial intelligence. (Addison-Wesley Professional, Reading 1999)
15. Polikar, R.: Ensemble based systems in decision making. IEEE Circ. Syst. Mag. **6**, 21–45 (2006)
16. Sprockel, J.J., Diaztagle, J.J., Chaves, W.G., Simon, C., Hernandez, J.I.: Estructuración e implementación de una ruta crítica para el manejo de pacientes con síndrome coronario agudo en el Servicio de Urgencias del Hospital San José de Bogotá. Revista Colombiana de Cardiología **20**, 262–274 (2013)
17. Thygesen, K., et al.: Third universal definition of myocardial infarction. Circulation **126**, 2020–2035 (2012)
18. Braunwald, E., et al.: Diagnosing and managing unstable angina: agency for health care policy and research. Circulation **90**, 613–622 (1994)
19. Conti, A., et al.: A new simple risk score in patients with acute chest pain without existing known coronary disease. Am. J. Emerg. Med. **28**, 135–142 (2010)
20. Sanchis, J., et al.: New risk score for patients with acute chest pain, non-ST-segment deviation, and normal troponin concentrations: a comparison with the TIMI risk score. J. Am. Coll. Cardiol. **46**, 443–449 (2005)
21. Green, M., et al.: Comparison between neural networks and multiple logistic regression to predict acute coronary syndrome in the emergency room. Artif. Intell. Med. **38**, 305–318 (2006)
22. Mandal, I., Sairam, N.: Accurate prediction of coronary artery disease using reliable diagnosis system. J. Med. Syst. **36**, 3353–3373 (2012)

Cluster Analysis of Face Images and Literature Data by Evolutionary Distance Metric Learning

Wasin Kalintha, Taishi Megano, Satoshi Ono, Kenichi Fukui and Masayuki Numao

Abstract Evolutionary distance metric learning (EDML) is an efficient technique for solving clustering problems with some background knowledge. However, EDML has never been applied to real world applications. Thus, we demonstrate EDML for cluster analysis and visualization of two applications, i.e., a face recognition image dataset and a literature dataset. In the facial image clustering, we demonstrate improvement of the cluster validity index and also analyze the distributions of classes (ages) visualized by a self-organizing map and a K-means clustering with K-nearest neighbor centroids graph. For the literature dataset, we have analyzed the topics (i.e., a cluster of articles) that are the most likely to win the best paper award. Application of EDML to these datasets yielded qualitatively promising visualization results that demonstrate the practicability and effectiveness of EDML.

W. Kalintha (✉)
Graduate School of Information Science and Technology, Osaka University, 8-1 Mihogaoka Ibaraki, Osaka 567-0047, Japan
e-mail: wasin@ai.sanken.osaka-u.ac.jp

T. Megano · S. Ono
Graduate School of Science and Engineering, Kagoshima University, 1-21-40 Kohrimoto, Kagoshima 890-0065, Japan
e-mail: sc108052@ibe.kagoshima-u.ac.jp

S. Ono
e-mail: ono@ibe.kagoshima-u.ac.jp

K. Fukui · M. Numao
The Institute of Scientific and Industrial Research (ISIR), Osaka University, 8-1 Mihogaoka Ibaraki, Osaka 567-0047, Japan
e-mail: fukui@ai.sanken.osaka-u.ac.jp

M. Numao
e-mail: numao@ai.sanken.osaka-u.ac.jp

M. Bramer and M. Petridis (eds.), *Research and Development in Intelligent Systems XXXII*, DOI 10.1007/978-3-319-25032-8_23

1 Introduction

Distance metric learning (DML) [15] attempts to optimize a metric to improve clustering or classification. Example approaches include nearest neighbor classification [13], clustering [7, 14], and data visualization [6, 8].

Figure 1a shows data points with three classes, i.e., circles, squares, and stars, and three initial partitions (or clusters) in Euclidean space. Note that one of the clusters has data points in all three classes. To cluster all data points correctly, the data space transformation stretches the partitions as shown in Fig. 1b.

In contrast to unsupervised clustering, e.g., K-means clustering, which only desires to put similar data points into the same cluster, and vice versa, semi-supervised clustering methods [1, 7, 11, 14], including our evolutionary distance metric learning (EDML) [5], make use of available class label to group instance of the same class label into the same cluster.

EDML optimizes any cluster validity index, such as purity, F-measure, or entropy, depending on the clustering purpose. Since EDML is a semi-supervised clustering, in each individual cluster, it contains instances which are identical class and similar features. Furthermore, by using smoothed cluster validity indices [4] to consider neighbor relations in the data space, EDML can capture the neighbor relations of the clusters. Because an objective function based on cluster validity is massively multimodal in changing the distance metric, EDML uses a parallel differential evolution with self-adaptive control parameters and generalized opposition-based learning (GOjDE) algorithm [12], which can deal with multimodality without manual adjustment of its control parameters and improve candidate solutions quality.

In this paper, we are interested in cluster analysis using EDML, because until now no research work about cluster analysis using EDML has been reported. Thus, we selected the data set that is easy to vistualize and analyze the cluster structure, whether good or not, particularly a face recognition image dataset and a literature dataset. In the face image dataset, we first demonstrate that EDML can improve a cluster validity index and discuss the distributions of classes (ages) visualized by two variations of EDML to obtain clusters with neighbor relations by self-organizing map (SOM) [9] and by K-means clustering with K-nearest neighbor centroids graph. Then with the literature dataset, we analyze which topic, i.e., a cluster of articles, tends to win the best paper award.

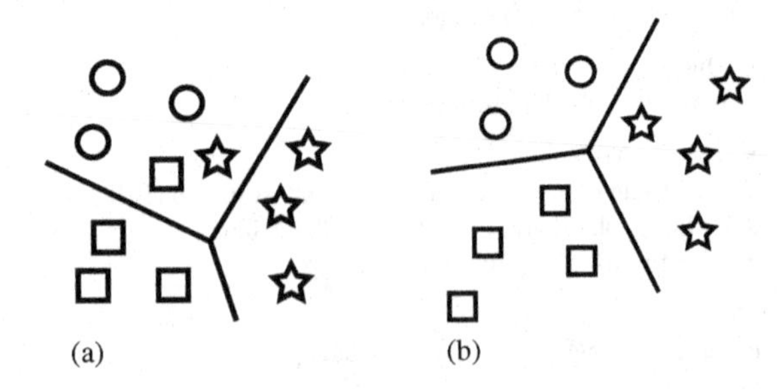

Fig. 1 Conceptual diagram of distance metric transformation. **a** Euclidean space. **b** After transformation

2 Evolutionary Distance Metric Learning

2.1 Global Distance Metric Learning

In this work, a Mahalanobis-based distance is used, similar to many global DML methods. Given a dataset $\mathscr{D} = \{\mathbf{x}_i = (x_{i,1}, \dots, x_{i,v})^t \in \mathbb{R}^v\}_{i=1}^N$, the Mahalanobis-based distance can be defined as follows:

$$d_{i,j}^2 = (\mathbf{x}_i - \mathbf{x}_j)^t \mathbf{M} (\mathbf{x}_i - \mathbf{x}_j), \quad (1)$$

where $\mathbf{M} = (m_{k,l})$ is a $v \times v$ matrix. In the original Mahalanobis distance, $\mathbf{M}$ is given by the inverse of the variance-covariance matrix of the input data, i.e., $\mathbf{M} = \Sigma^{-1}$. In DML, the elements of $\mathbf{M}$ are variables to be learned that represent a transformation of the input data. In this case, $\mathbf{M}$ must be a symmetric positive semi-definite matrix to satisfy the distance propositions.

The proposed metric learning approach optimizes a clustering index $Eval$ (introduced in the next section) as follows:

$$\begin{aligned} &\text{Maximize } Eval(Clustering(d_{i,j}^2)), \\ &\text{s.t. } |m_{k,k}| > \sum_{l(k \neq l)} |m_{k,l}|, \\ &0 < m_{k,k} \leq 1, \ -1 \leq m_{k,l} \leq 1 \ (k \neq l), \end{aligned} \quad (2)$$

where $Clustering(d_{i,j}^2)$ denotes a clustering result obtained using a distance metric $d_{i,j}^2$—i.e., $Clustering() : \mathbf{x} \mapsto c \in \mathbf{C}$, where $\mathbf{C}$ is a set of cluster identifiers. Moreover, matrix $\mathbf{M}$ must be a *diagonally dominant matrix*—i.e., $|m_{i,i}| > \sum_{j(i \neq j)} |m_{i,j}|$— and the diagonal elements must be positive to ensure that $\mathbf{M}$ is a positive semi-definite matrix.

2.2 Neighborhood Smoothing of Clustering Index

This paper focuses on using extended F-measure or $F_1 score$ by using the class label as external criteria for evaluation of a clustering result. Generally, an external clustering index evaluates individual cluster quality. Fukui and Numao [4] introduced neighborhood smoothing in the set-based and the likelihood in pairwise-based measures by adding a weighting function $h_{i,j}$. Then the F-measure can be extended.

- Weighted F-measure: wFme
  wFme is an extension of set-based cluster validity indices [5]. Given dataset $\mathscr{D}$ with cluster set $\mathbf{C}$ and class set $\mathbf{T}$, let $N_{s,i}$ is the number of data points with class $s \in \mathbf{T}$ and the ith cluster $C_i \in \mathbf{C}$, N_s denotes the number of data points in class s,

N_i denotes the number of data points in ith cluster, and N is the total numbers of data points in $\mathscr{D}$. These values are smoothed by $h_{i,j}$, as follows:

$$N'_{s,i} = \sum_{C_j \in \mathbf{C}} h_{i,j} N_{s,j}, \tag{3}$$

$$N'_i = \sum_{s \in \mathbf{T}} N'_{s,i} = \sum_{s \in \mathbf{T}} \sum_{C_j \in \mathbf{C}} h_{i,j} N_{s,j}, \tag{4}$$

By using Eqs. (3) and (4) Precision ($\boldsymbol{Prec}(s, C_i)$) and Recall ($\boldsymbol{Rec}(s, C_i)$) can be extended. Then, wFme can be obtained from the original formula as follows.

$$\text{wFme}(\mathbf{C}) = \sum_{s \in \mathbf{T}} \frac{N_s}{N} \max_{C_i \in C} F(s, C_i), \tag{5}$$

where $F(s, C_i)$ is a harmonic average of $\boldsymbol{Prec}(s, C_i)$ and $\boldsymbol{Rec}(s, C_i)$.
And $\boldsymbol{Prec}(s, C_i) = N'_{s,i}/N'_i$, $\boldsymbol{Rec}(s, C_i) = N'_{s,i}/N_s$.

- Weighted Pairwise F-measure: wPF
  wPF, an extension of pairwise-based cluster validity indices, is based on the degree by which the data pairs belong to the same cluster by a smoothing class and a cluster confusion matrix (Table 1) of data pairs using $h_{i,j}$ [4].
  Given $c(k)$ and $t(k)$ denoting the cluster/class assignment for the instance x_k. $likelihood(c(i) = c(j))$ or $h_{i,j}$ has been proposed to indicate a degree that data pair x_i, x_j belongs to the same class instead of the actual number of data pairs. Therefore, each value in class and cluster confusion matrix of data pairs show in Table 1 is replaced by summation of likelihoods as follows:

$$a' = \sum_{\{i,j|t(i)=t(j)\}} h_{i,j}, \tag{6}$$

$$b' = \sum_{\{i,j|t(i)\neq t(j)\}} h_{i,j}, \tag{7}$$

$$c' = \sum_{\{i,j|t(i)=t(j)\}} (1 - h_{i,j}) = a + c - a'. \tag{8}$$

  Finally, wPF is a $F_1 score$ which calculated using these extended a', b', and c'.

In this study, $h_{i,j}$ is a Gaussian function. Therefore, $h_{i,j} = \exp(-r_{i,j}/\sigma)$, where $r_{i,j}$ denotes the inter-cluster distance between C_i and C_j, and $\sigma(> 0)$ is a smoothing (neighborhood) radius.

Both extended F-measures is used as *Eval*() in Eq. (2).

Table 1 Class and cluster confusion matrix of data pairs

	t(i) = t(j)	t(i) ≠ t(j)
c(i) = c(j)	a	b
c(i) ≠ c(j)	c	–

2.3 Self-organizing Map

SOM [9] has been used as a base clustering method as well as for visualizing DML results. With SOM, high-dimensional data can be visualized by a two-dimensional map. In addition, SOMs can be used to visualize clustering results that facilitate intuitive analysis.

2.4 K-Means Clustering with K-Nearest Neighbor Centroids Graph

For another option for obtaining clusters with neighbor graph, K-means clustering with K-nearest neighbor centroids graph (KMN-KNN) [4] has been used in order to easily analyze obtained cluster structures especially in case of micro-clusters or cluster number is larger than class number. With this, we can study entire data structures via the inter-cluster connectivity and also visualize neighbor relations among cluster centroids.

2.5 Evolutionary Distance Metric Learning Framework

The EDML framework is summarized in Fig. 2 [5]. First, candidates of metric transform matrix **M** are generated by GOjDE. Next, the obtained **M** is used in Eq. (1) to acquire clusters with neighbor structures. This cluster structure can be obtained using

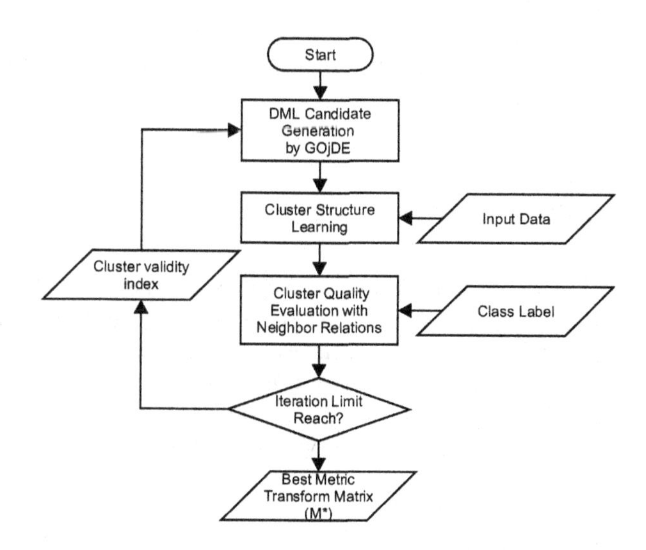

Fig. 2 EDML framework

any clustering algorithm. In this study, vector quantization with topology preservation by a SOM or KMN-KNN is used. Then, the quality of the cluster structure can be evaluated with class labels through the smoothed clustering index. Here wFme and wPF are used with the face image and literature dataset, respectively. This is followed by feeding the evaluating value back into GOjDE as the fitness for candidate **M**. GOjDE selects candidates based on fitness to evolve and generate the next candidates by mutation and crossover with certain probabilities. These steps are repeated until the termination condition, i.e., the iteration limit, is satisfied. Finally, the best metric transform matrix $\mathbf{M}^*$ is obtained in terms of the most smoothed clustering index among the overall generations of candidates.

2.6 DE with Self-Adapting Control Parameters and Generalized Opposition-Based Learning: GOjDE

We focus on applying EDML to real world problems, which typically involve very high-dimensional data. High-dimensional global optimization is one such high-complexity problem. Therefore, GOjDE [12], which is a parallel DE based on graphics processing units which improve the quality of candidate solutions, is used rather than jDE [3].

3 Application to Face Recognition

To demonstrate the usefulness of EDML, we first applied EDML to face image data obtained from a face recognition project.[1] In this application, wFme is used as objective function for EDML. Clustering of face images was improved by utilizing labels from the face data. We then visualized the data distribution using a SOM and KMN-KNN.

3.1 Experimental Settings

The face data contains monochrome facial images with 99 features. We first applied principal component analysis (PCA) for dimensional reduction. Each face image has age, sex, skin color, and facial expression. In our experiment, only the age category was used as the class label for DML, namely Child, Teen, Adult and Senior, and for each class its corresponding class number is 1, 2, 3 and 4 respectively. We selected 50 images from each category (200 images in total).

[1]http://courses.media.mit.edu/2004fall/mas622j/04.projects/faces/.

Table 2 Dataset information

Dataset	#Instances	#Features	#Classes	DML representation
faceR-D10C4	200	10	4	Full
faceR-D55C4	200	55	4	Diagonal

Note that the number of dimensions D depends on the representation of the matrix **M** in Eq. (1), i.e., a diagonal (dig) or a full matrix. Consequently, $D_{dig} = N_A$, and $D_{full} = N_A\ (N_A + 1)/2$, where N_A is the number of features. We set both D_{dig} and D_{full} to 55, where 55 features from PCA are used for D_{dig}, and 10 features are used for D_{full} (Table 2). In addition, population size in the diagonal case was set to five times the number of dimensions and twice the number of dimensions for the full matrix case. We conducted five trials with 10,000 iterations (generations) per trial.

3.2 Results

3.2.1 Effect of DML

The transition of the average fitness for each iteration is shown in Fig. 3. First, by comparing DML to the Euclidean distance method (e.g., a normal SOM) as the baseline, a diagonal and a full matrix DML (even with random search) can improve clustering performance. Technically, a full matrix has richer expressive power than a diagonal matrix. However, when the number of features is equal, a diagonal representation empirically outperforms a full matrix DML. In addition, by comparing different evolutionary algorithms, it was observed that GOjDE can slightly improve clustering performance compared to jDE.

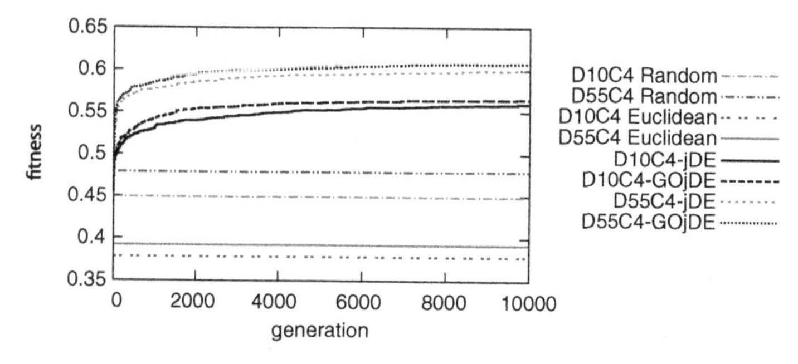

Fig. 3 Trend of the fitness

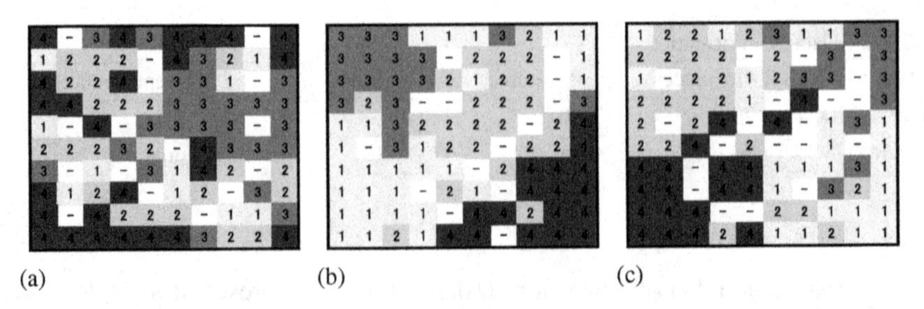

Fig. 4 SOM visualization. **a** No DML (faceR-D55C4). **b** Full matrix DML (faceR-D10C4). **c** Diagonal DML (faceR-D55C4)

3.2.2 SOM Visualization

SOM is employed to visualize the cluster structure because SOM is often useful for instinctively understanding learning results. We also used SOM to obtain cluster with neighbor relations.

Figure 4 shows the SOM visualization results. The number of nodes was set to 10×10. Each cell represents a SOM node, which is a micro-cluster. In Fig. 4, a representative class label obtained by majority voting is also indicated, and "-" denotes that there is no corresponding sample for the node, which indicates a gap between adjacent nodes. The shade of the color labels represents the age of the majority class of the cluster. Younger class is shown with brighter color. For both EDML with a full matrix (Fig. 4b) or a diagonal matrix (Fig. 4c), the same class is much denser than the SOM without DML (Fig. 4a). In addition, Fig. 5 shows how each class in Fig. 4c is distributed. Fig. 5a–d show that the data of the same class are allocated closely and different classes are allocated in different places in the map. These show how

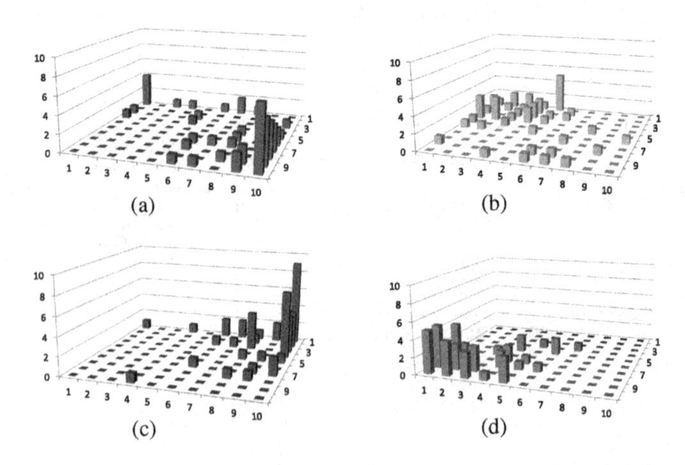

Fig. 5 Distribution of each class by SOM (faceR-D55C4). **a** class 1. **b** class 2. **c** class 3. **d** class 4

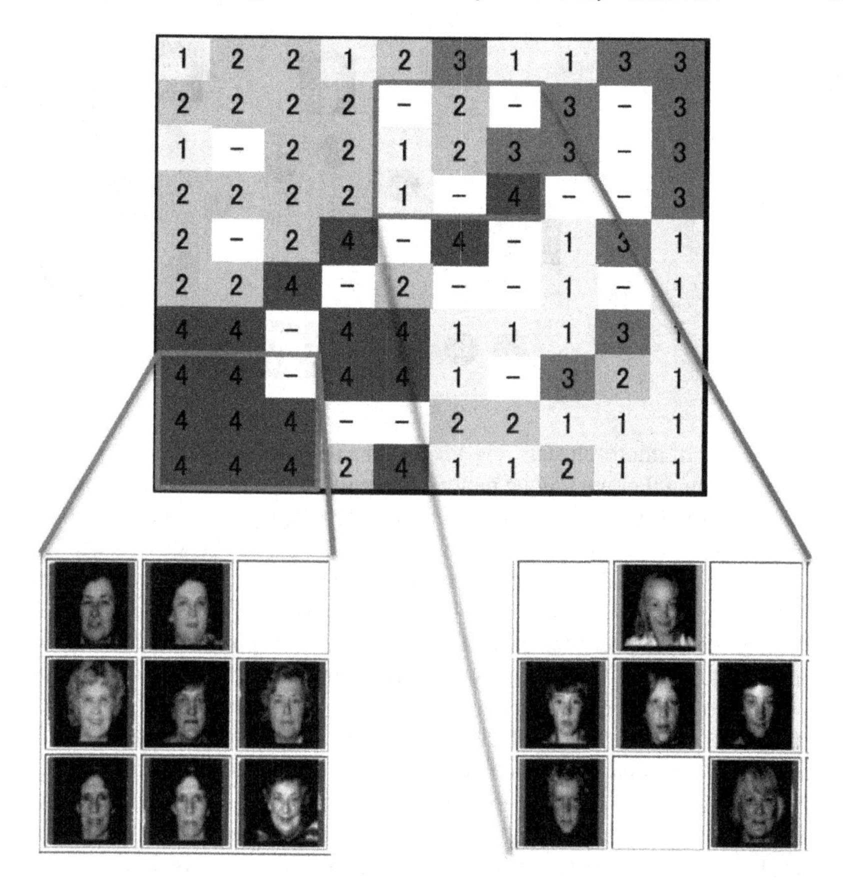

Fig. 6 Face image example (faceR-D55C4)

the same class is well clustered. Finally, Fig. 6 shows an example of a representative face image in each node according to class label in 3×3 nodes. Although bottom left clusters indicated in Fig. 6 are almost all seniors (class 4), upper clusters suggest that adults or seniors (class 3 and 4) in these clusters look young since the image features are closer to child and teen (class 1 and 2).

3.2.3 KMN-KNN Visualization

We also applied EDML with KMN-KNN for further analysis on neighborhood relation of clusters. In this visualization, KMN-KNN, with 20 clusters and 5 mutual neighborhoods, is employed to obtain cluster with neighborhood cluster structure. Figure 7 shows the KMN-KNN structure. Let C_i denotes the ith cluster. Each node represents C_i, where an edge between two clusters is indicated a neighborhood relation. Similar to SOM visualization, the shade of the color nodes represents the age of the majority class of the cluster. EDML is capable of investigating neighbor relations

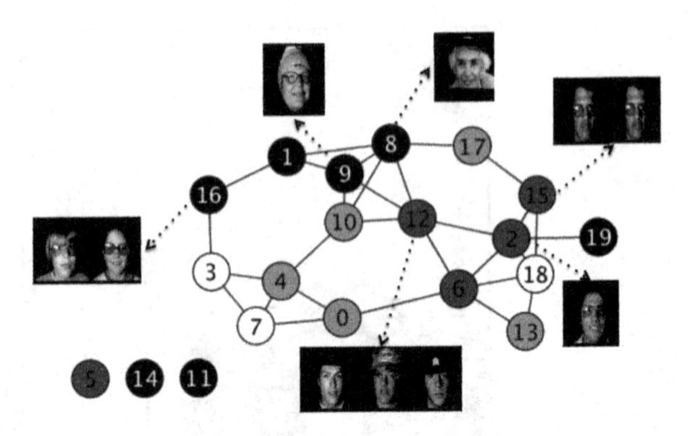

Fig. 7 KMN-KNN visualization (faceR10C4)

of large number of micro-clusters, even lager than the number of classes. The micro-cluster lead to an unknown unique pattern recognition by categorizing it together in one micro-cluster. Figure 7 also shows an example image of these unique pattern in micro-clusters, e.g., person who wears glasses (C_2, C_4 and C_{15}) or hat (C_8, C_9 and C_{12}). Then we analyze KMN-KNN structure, neighboring clusters with the same majority class, e.g., C_2 and C_{15} (Adult), and C_3 and C_7 (Child) and so on. They usually have a relation between each other. Moreover, it is possible to have a relation between the clusters whose majority class is different (C_6 and C_{13}). In order to investigate that, we examine inside each micro-cluster, i.e., how many instances are contained, according to class label (Table 3). Majority class in each cluster is indicated by a **bold face**. The cluster in which the class of the samples is identical, i.e., C_5, C_{11} and C_{14} (Indicated by a ***boldface-italic***) are isolated in the graph (Fig. 7), no neighborhood relationship. On the other hand, the micro-cluster contains unidentical class neighborhood relation are falled into two categories: first, a neighborhood among the clusters with the same majority class, e.g., C_1, C_8 and C_9 has class "Senior" as a majority class. Sometimes it forms a complete graph. This also occurs with C_2, C_6, C_{12}; second, micro-cluster, which produces a comparable number of majority class and minority class (C_6 and C_{12}) tends to extend their neighborhood to the clusters with the same minority class. To summarize the results, EDML with KMN-KNN can preserve both class labels and similarity of image features in the neighborhood graph of clusters.

4 Application to Literature Data

The **I-Scover dataset** IEICE Knowledge Discovery (I-Scover)[2] is a literature dataset that contains approximately 160,000 documents. From the 160,000 documents, we extracted Japanese documents according to title, abstract, field, and best paper award results. In total, approximately 20,000 documents were extracted.

[2]http://www.ieice.org/jpn/nyukai/pdf/senmon.pdf.

Table 3 The number of samples of each class in each cluster, where # indicates cluster ID

	#0	#1	#2	#3	#4	#5	#6	#7	#8	#9	#10	#11	#12	#13	#14	#15	#16	#17	#18	#19
Child	3	0	5	**6**	7	0	2	**9**	0	0	3	0	1	2	0	4	0	3	**4**	1
Teen	**5**	0	2	3	**9**	0	0	2	1	1	**6**	0	3	**5**	0	3	1	**7**	1	1
Adult	0	1	**15**	0	2	***1***	**8**	1	0	0	2	0	**10**	0	0	**7**	0	0	2	1
Senior	0	**7**	1	0	1	0	0	1	**13**	**7**	2	***2***	1	0	***1***	2	**9**	1	0	**2**

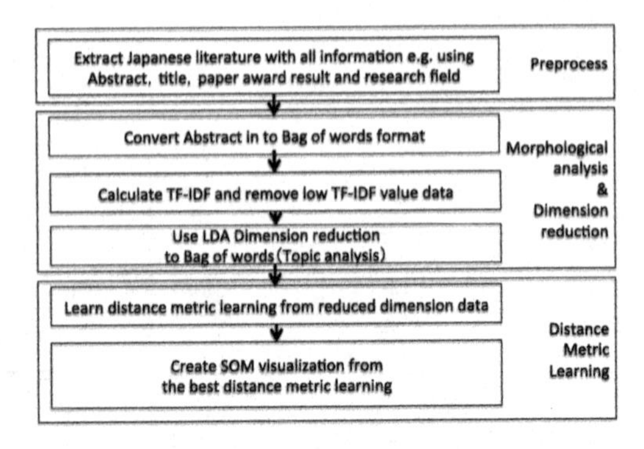

Fig. 8 I-Scover process flow

4.1 Preprocessing and Settings

A flowchart that describes the application of EDML to the I-Scover dataset is shown in Fig. 8. We used abstracts to extract term frequency-inverse document frequency (TF-IDF) features and latent Dirichlet allocation (LDA) [2] is used for topic extraction and dimensional reduction. The number of topics estimated by LDA was set to 15; therefore the input to EDML was 15 dimensions. In addition, a weighted pairwise F-measure was used in EDML. We applied EDML with two types of class categories, i.e., best paper awards and field of research (Table 4).

4.2 Results

Visualization results with the representative class of SOM nodes are shown in Fig. 9. The size of SOM map was set to 10×10. As seen in the figure, both best paper award or field results gave higher fitness (wPF) compared to Euclidean distance. However, SOM representation of category 1 (best paper awards) did not demonstrate confirmable improvement because of a high imbalance between papers that did and did not receive awards. Conversely, with category 2 (fields), SOM representation is organized to some extent in relation to class 1 (fundamentals and boundary) and class 4 (information and system).

Table 4 I-Scover training dataset

Class No.	Category 1	#	Category 2	#
1	Best paper award	14	Fundamentals and boundary	377
2	Information systems society award	15	Communications	173
3	Paper award	103	Electronics	135
4	No award	1022	Information and system	469
			Total	1154

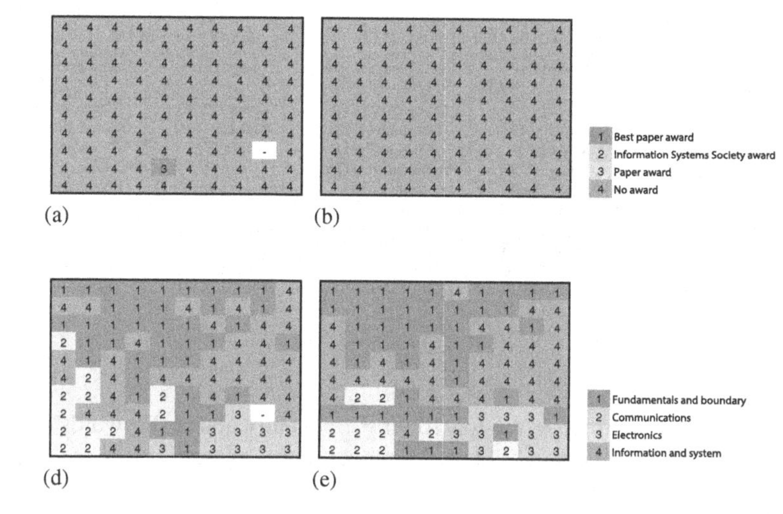

Fig. 9 SOM visualization (I-Scover data). **a** Cat.1: Euclid Fitness: 0.149. **b** Cat.1: EDML Fitness: 0.174. **c** Cat.2: Euclid Fitness: 0.105. **d** Cat.2: EDML Fitness: 0.114

A prototype Web Viewer[3] is a tool that can be used to check clustering results in a Web browser. Each circle represents a SOM micro-cluster, where the normalized number of articles in the micro-cluster is indicated. They are normalized by the greatest number of articles in all micro-clusters. A user can choose a category, i.e., paper awards or fields, and after selecting a category to analyze micro-clusters, the corresponding bibliography is shown to the right of the micro-cluster. Furthermore, there is a link to access the detail pages about the article. The Web Viewer is expected to obtain new knowledge about the relationship between the articles that cannot be found in the original keywords.

4.3 Discussion

Analysis of each SOM node, i.e., a micro-cluster, shows that we can cluster the topic that has a higher chance to win best paper awards, i.e., an attractive topic, has large number of awards, and an upcoming topic, the small number of papers but higher ratio of award. Currently, papers about sound, electronics, coding and decoding, and image-vision get a lot of researcher attraction since they have a large number of awards 15, 10, 6, and 5, respectively. Despite, character recognition has only 10 papers in total, approximately 30 % of these paper received awards. Thus character recognition is going to be an upcoming topic.

[3]http://mediaeng.ibe.kagoshima-u.ac.jp/iscover/microcluster.html (in Japanese).

Table 5 Top 10 most frequent words that appear in the winning papers in each field

Rank	Sound (音声系)	Electronics (エレクトロニクス系)	Coding and decoding (符号・復号系)	Image-vision (画像・ビジョン系)	Character recognition (文字認識系)
1	Translation (翻訳)	Beam (ビーム)	Experiment (試験)	Image (映像)	Character (文字)
2	Phoneme (音素)	Radiation (放射)	Detection (検出)	Edge (エッジ)	Recognition (認識)
3	Denominator (分母)	Light (光)	Coding (符号)	Face (顔)	Cluster (クラスタ)
4	Character (文字)	Loading (装荷)	Break-down (故障)	Color (色)	Matching (マッチング)
5	Cluster (クラスタ)	Arrival (到来)	Scale (尺度)	Point of view (視点)	Search (検索)
6	Recognition (認識)	Human body (人体)	Approach (アプローチ)	Texture (テクスチャ)	Similarity (類似)
7	Kanji (漢字)	Polarized (偏)	Principle (原理)	Shooting (撮影)	Scale (尺度)
8	Matching (マッチング)	Refraction (屈折)	Inquiry (問合せ)	Movement (動き)	Section (区間)
9	Document (文書)	Resonance (共振)	Convolution (畳)	Vehicle (車両)	Translation (翻訳)
10	Search (検索)	Sheet (板)	VTR	Graphic (図形)	Document (文書)

In addition, we summarize the top 10 most frequently appear keywords in each field of the winning papers in Table 5. Moreover, these micro-clusters are empirically located near the corners and borders of the SOM map: coding and decoding (upper left corner), image-vision (upper right corner), sound, character recognition (rightmost border; sixth and seventh rows), and electronics (bottom right). Thus, it is possible to analyze the micro-clusters to examine which documents are likely to win a paper award or which field has attracted researchers' attention.

5 Conclusion and Future Work

This study has demonstrated the application of EDML functionality to facial image and literature data. EDML increases the cluster validity index and simplifies the distribution of classes visualized by SOM or KMN-KNN. As a result, we can easily analyze micro-clusters and provide beneficial information, i.e., group a same feature e.g., wearing hat and glasses images together in micro-cluster, maintain both class labels and similarity of features in the neighborhood graph of clusters, and determine which article topics have a higher chance of winning awards.

Given the promising results, in the future we wish to apply evolutionary multi-objective distance metric learning (MOO-EDML) [10], which is helpful in finding the attributes that affect the trade-off relationship among objective functions, in order to simultaneously maximize every objective function. MOO-EDML can yield a better transform matrices than single-objective optimization. This will help us to obtain admirable cluster structure.

Acknowledgments This work was partially supported by the Kayamori Foundation of Informational Science Advancement, and by the cooperative research program of "Network Joint Research Center for Materials and Devices".

References

1. Bilenko, M., Basu, S., Mooney, R.J.: Integrating constraints and metric learning in semisupervised clustering. In: Proceedings of the 21st International Conference on Machine Learning, pp. 81–88. ACM (2004)
2. Blei, D.M., Ng, A.Y., Jordan, M.I.: Latent dirichlet allocation. J. Mach. Learn. Res. **3**, 993–1022 (2003)
3. Brest, J., Greinero, S., Boskovic, B., Mernik, M., Zumer, V.: Self-adapting control parameters in differential evolution: a comparative study on numerical benchmark problems. IEEE Trans. Evol. Comput. **10**(6), 646–657 (2006)
4. Fukui, K., Numao, M.: Neighborhood-based smoothing of external cluster validity measures. In: Proceedings of the 16th Pacific-Asia Conference on Knowledge Discovery and Data Mining (PAKDD-12), pp. 354–365 (2012)
5. Fukui, K., Ono, S., Megano, T., Numao, M.: Evolutionary distance metric learning approach to semi-supervised clustering with neighbor relations. In: Proceedings of 2013 IEEE 25th International Conference on Tools with Artificial Intelligence (ICTAI), pp. 398–403 (2013)
6. Goldberger, J., Roweis, S., Hinton, G., Salakhutdinov, R.: Neighbourhood components analysis. In: Advances in Neural Information Processing Systems, pp. 513–520 (2004)
7. Hertz, T., Bar-Hillel, A., Weinshall, D.: Boosting margin based distance functions for clustering. In: Proceedings of the 21st International Conference on Machine Learning (ICML-04), pp. 393–400 (2004)
8. Kaski, S., Sinkkonen, J.: Principle of learning metrics for exploratory data analysis. J VLSI Signal Process. Syst. Signal Image Video Technol. **37**, 177–188 (2004)
9. Kohonen, T.: Self-Organizing Maps. Springer, New York (1995)
10. Megano, T., Fukui, K., Numao, M., Ono, S.: Evolutionary multi-objective distance metric learning for multi-label clustering. In: Proceedings of 2015 IEEE Congress on Evolutionary Computation (CEC-15), pp. 2945–2952 (2015)
11. Wagstaff, K., Cardie, C., Rogers, S., Schrödl, S.: Constrained k-means clustering with background knowledge. In: Proceedings of the International Conference on Machine Learning (ICML-01), pp. 577–584 (2001)
12. Wang, H., Rahnamayan, S., Wu, Z.: Parallel differential evolution with self-adapting control parameters and generalized opposition-based learning for solving high-dimensional optimization problems. J. Parallel Distrib. Comput. **73**, 62–73 (2013)
13. Weinberger, K.Q., Blitzer, J., Saul, L.K.: Distance metric learning for large margin nearest neighbor classification. J. Mach. Learn. Res. (JMLR) **10**, 207–244 (2009)
14. Xing, E.P., Ng, A.Y., Jordan, M.I., Russell, S.J.: Distance metric learning with application to clustering with side-information. In: Advances in Neural Information Processing Systems (NIPS), pp. 505–512 (2002)
15. Yang, L.: Distance metric learning: A comprehensive survey. Tech. Rep. 16, Michigan State Universiy (2006)

AI Applications

A Novel K-Means Voice Activity Detection Algorithm Using Linear Cross Correlation on the Standard Deviation of Linear Predictive Coding

M.K. Mustafa, Tony Allen and Kofi Appiah

Abstract This paper presents a novel Voice Activity Detection (VAD) technique that can be easily applied to on–device isolated word recognition on a mobile device. The main speech features used are the Linear Predictive Coding (LPC) speech features which were correlated using the standard deviation of the signal. The output was further clustered using a modified K-means algorithm. The results presented show a significant improvement to a previous algorithm which was based on the LPC residual signal with an 86.6 % recognition rate as compared to this new technique with a **90 %** recognition rate on the same data. This technique was able to achieve up to **97.7 %** recognition for female users in some of the experiments. The fast processing time makes it viable for mobile devices.

1 Introduction

Speech recognition is the phenomenon by which a device translates human utterances into a machine understandable format for further processing. One very important stage of this recognition is the pre-processing stage, where the prospective audio signal that will be used for recognition is processed for use by the recognizer but there is less emphasis being placed on it in publication. There are different ranges of pre–processing which include but are not limited to voice activity detection, filtering, speech encoding and feature enhancement schemes. These have laid the foundation on which different speech recognition systems process speech signals. These different pre–processing techniques are achieved

M.K. Mustafa (✉) · T. Allen · K. Appiah
Nottingham Trent University, NG1 4BU Nottingham, UK
e-mail: mohammed.mustafa2009@my.ntu.ac.uk

T. Allen
e-mail: tony.allen@ntu.ac.uk

K. Appiah
e-mail: kofi.appiah@ntu.ac.uk

M. Bramer and M. Petridis (eds.), *Research and Development in Intelligent Systems XXXII*, DOI 10.1007/978-3-319-25032-8_24

either by the use of different filters or even much more sophisticated algorithms that use a lot of statistical analysis.

This paper is about the pre–processing applied before speech recognition to detect the part of the speech signal with voice. This is to enable the speech recognizer to only process the information in the voiced speech segments as opposed to the unvoiced. Voice activity pre–processing has been applied to speech analysis for various applications. This is used in GSM devices to discontinue transmission of speech to conserve battery [1]. It has also been applied for channel bandwidth conservation.

The development of an automatic speech recognition system for mobile devices involves taking into consideration a technique or scheme to stop the listening or rather recording process of the microphone in order for the recognition part of the process to begin. The predominant application of speech recognition is done with an on–line approach in order to get around the challenges of mobile device power, computation time and memory. The schemes used to discontinue the transmission of speech in order for the recognition to begin is hidden from the general public and literature due to the market competition involved between the predominant speech recognition systems for mobile phones which include Google's Talk to Google, Apple's SIRI and the most recent one to join the group being The Cortana from Microsoft.

The rest of this paper is structured in sections describing the background literature as well as related work and subsequently the proposed system. Section 2 gives a brief introduction about Voice Activity Detection and the different works in the literature related to this work. Section 3 is based on the Linear Predictive Coding (LPC) and gives the background information about the speech feature used. The rest of the paper starting from Sect. 4 is based on the proposed algorithm and the different stages of experiments. Section 5 is the methodology used while Sect. 6 is the audio data used for experimentation. Section 7 provides the results of the experiments and Sect. 8 is the conclusion and discussion. Section 9 is the future work.

2 Voice Activity Detection (VAD)

Voice Activity Detection is regarded as one of the pre–processing techniques that is utilised in some speech recognition systems. In VAD the principal objective is to find the respective regions of speech within an audio signal while ignoring the regions of silence for use in further processing. Not only does this help in reducing the processing cost but the processing time as well because it is a well-known fact that speech recognition is a computationally exhaustive task that imposes heavy demands on the processing power and storage capacity of any target device [2] and there are no extra merits for passing a huge amount of speech data to such a device [3]. This is why an efficient pre—processing algorithm can help reduce this work load.

In [4] LPC was used but instead of the standard deviation, the mean of the noise was used and then the speech is binarized with a certain threshold. This highlights the importance of noise in any VAD algorithm. In [5], auto correlation was used in conjunction with the Zero Crossing rate. However, this poses a dual calculation situation which is not desirable for the ultimate goal of mobile based VAD application. The consideration of the long term speech information is important to a VAD algorithm as argued by [6, 7]. In [6] the Long term speech divergence between the speech and non- speech frames of the entire signal is considered. This is then used to come up with thresholds by comparing the long term spectral envelope to the average noise spectrum. However, in [7] it is argued that the adoption of the average noise spectrum magnitude information is not attainable because in practice there is no stability or rather stationary noise. In [7] Long term signal variability is considered. This is a measure computed using the last frames of the observed signal with respect to the current frame of interest. The only problem with a consideration of the last few frames of the speech signal in [7] or even the first few ones as presented in [6] is that the availability of the desired noise spectrum cannot be guaranteed to lie in those regions.

We put emphasis on [8] where it introduced the concept of linear cross correlation (LCC). This technique was very effective for VAD purposes in a speech verification system. This paper gets it antecedents from the same paper [8] and tends to use the same concept of linear cross correlation for VAD in speech recognition. The proposed algorithm improves the processing time for the linear cross correlation algorithm for use on a mobile device. We also adopt the concept of the long term spectral information. However, the proposed algorithm does not use a predetermined region of the speech signal to find the noise spectrum as argued by [6, 7]. We adopt a dynamic approach to find the region with a higher noise spectrum.

3 Linear Predictive Coding (LPC)

The choice of speech features for this paper is the Linear Predictive Coding technique. The main idea behind this method is that a speech sample is approximated as a linear combination of past speech samples [9]. Linear prediction is mostly used in low bit rate transmission of speech or storage [9]. This is achieved by minimizing the sum of squared differences between the actual speech samples and the linearly predicted ones. LPC has also become the predominant technique for estimating the basic speech parameters such as formants, spectra, vocal tract area functions [9]. The Levinson—Durbin approach is used to compute the LPC features in this paper. The prediction of a Speech sample **x[n]** is given in Eq. 1;

$$\tilde{x}(n) = \sum_{i=1}^{p} \alpha_i x(n-i) \tag{1}$$

Where $\tilde{x}(n)$ is the sample being predicted, **x(n−i)** is the previous sample, P is the order and **αi** is the prediction coefficient.

4 Proposed System

The proposed system is based on the desire to improve mobile device speech recognition using an on-device implementation. In a previous work [10], the computation times for a few number of algorithms on a mobile device with the cross correlation being one of them is presented.

In this paper we present a two stage/process based VAD algorithm. This algorithm is broken into two stages so they can be independently implemented. This can provide a multipurpose outlook algorithm, where the user or researcher can use the first stage for further processing with their own algorithm whilst ignoring the second stage which is strictly our choice of implementation in this paper. The two stages involved are;

- Linear cross correlation
- K-means clustering (Modified)

4.1 Linear Cross Correlation (LCC) Stage

The Linear cross correlation between two variables X and Y is given in Eq. 2. The correlation value is within the range of 0 and 1. If there is correlation between the two variables, the correlation value is high (closer to 1) and low (closer to 0) for variables without any correlation.

$$LCC(x,y) = \frac{N\left(\sum_{i=1}^{N} x_i y_i\right) - \left(\sum_{i=1}^{N} x_i\right)\left(\sum_{i=1}^{N} y_i\right)}{\sqrt{N(\sum_{i=1}^{N} x^{2}{}_{i}) - \left(\sum_{i=1}^{N} x_i\right)^2}\sqrt{N(\sum_{i=1}^{N} y^{2}{}_{i}) - \left(\sum_{i=1}^{N} y_i\right)^2}} \tag{2}$$

where x_i and y_i are two vectors of N samples.

This technique was applied to Discrete Fourier Transform (DFT) speech frames where it correlates the DFT frames against each other, thereby resulting in a higher correlation value for frames containing speech and low value for frames without [8]. This concept was directly applied to the LPC features but the results were not as distinctive as the DFT results. Further work applied the same principle to the LPC residual signal (prediction error) which was even more inconclusive.

In this paper LCC was applied differently to the LPC. After the computation of the LPC features using the Levinson—Durbin method. The frames containing LPC features of the entire signal are taken and Eq. 3 is used to compute the standard deviation (SD) of the respective frames. SD is a measure of noise and interference within a signal [11].

$$\sqrt{\sigma^2} = \frac{1}{N-1} \sum_{i=0}^{N-1} (x_i - \mu)^2 \tag{3}$$

where σ^2 is the variance, x_i is the signal and μ is the mean of the signal and N total number of signals.

The frame with the minimum standard deviation (MSD) is chosen as the primary frame. This is then used as the Frame for the LCC algorithm, it is used as the candidate frame to be correlated with every other frame including itself. This gives a correlation value for every frame within the audio signal. The frames with the lower correlation value are the frames with speech and vice versa. A pseudo code implementation is given in Table 1.

After computing the algorithm we are left with each frame having its corresponding correlation value with regards to the frame with the minimum SD. Figure 1 gives the initial output of the correlation algorithm, where the frames with speech are at the lower regions of the plot.

Table 1 Pseudo implementation of LCC code

A pseudo code Implementation Of the LCC Algorithm
Break A into Frames of 128 samples
for every F
Compute LPC with Order 12 using Eqn. 1
End
for every LF
Compute the SD using Eqn. 3
End
Using MSD as x in Eqn. 2
for every LF as y in Eqn. 2
Compute LCC using Eqn. 2
End

A = Audio Signal
F = Frame
LF = Frame containing LPC
SD = Standard Deviation
MSD = Minimum SD

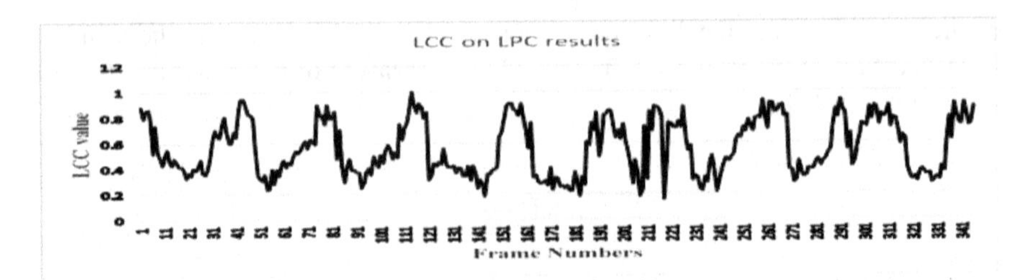

Fig. 1 LCC on LPC results

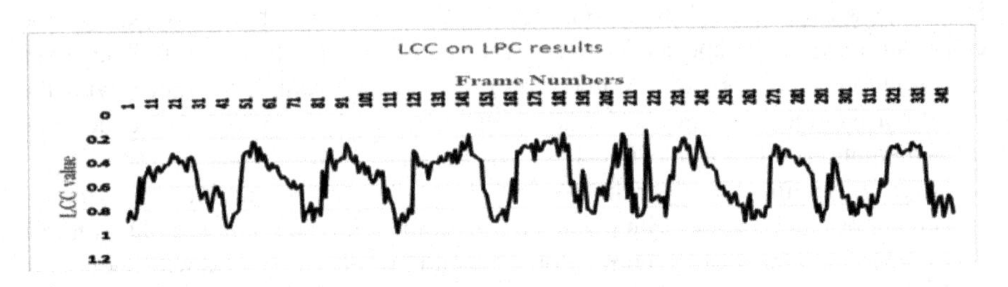

Fig. 2 LCC on LPC results flipped

Reversing the vertical axis because the regions with speech are in the lower side of the plot will give Fig. 2.

The reversed plot shows the position of the respective digits in the plot. This result can then be used for the stage 2 of our proposed system. However, this result can be used for a different algorithm to pick out the respective digits.

4.2 K-Means Clustering Stage

The K-means algorithm was applied to the correlation results in Fig. 2. The K-means algorithm was used to cluster the correlation values into two clusters of speech and silence. K-means assigns feature vectors to clusters by the minimum distance assignment principle [12, 13].

$$J = \sum_{j-1}^{k} \sum_{i=1}^{n} \left\| x_i^{(j)} - c_j \right\|^2 \tag{4}$$

where $\left\| \mathbf{x}_i^{(j)} - \mathbf{c}_j \right\|^2$ is a chosen distance measure between a data point $\mathbf{x}_\mathbf{i}^{(\mathbf{j})}$ and the cluster centre $\mathbf{c}_\mathbf{j}$.

4.2.1 Modifications to K–Means

The standard K-means algorithm was modified with data specific choices for each of the correlated results. A bias was introduced to help lower the threshold in order to do a more effective separation of the correlation results. This bias pushes the threshold for the algorithm above the median of the data. This allows enough of the speech data in the lower region to be captured. The equations for the bias calculation are given in Eqs. 5–8.

$$C_A = \left(\sum_{i=1}^{N} X_i \right) \div N \tag{5}$$

$$C_B = \left(\sum_{i=1}^{N} X_i \right) \div N \tag{6}$$

$$\mathbf{bias} = (\mathbf{C_A} - \mathbf{C_B}) * \mathbf{0.33} \tag{7}$$

$$\mathbf{C_A} = \mathbf{C_B} + \mathbf{bias} \tag{8}$$

where X_i is the mean of the ith frame within the cluster, C_A is the Centroid of the first cluster while C_B is the centroid of the second cluster and N the total number of frames within the cluster.

The pseudo code implementation of the Modified K-means is given in Table 2 for a clear understanding of how the decisions are made.

After the implementation of the K-means clustering algorithm, the result is two clusters, one with speech frames and the second with silence. The cluster with speech is used for further processing or verification.

5 Methodology

Despite the proposed work stated above, the experiments conducted did include some more features of the audio signal. Experiments were conducted using the mean of the audio signal, this was used to choose the winner frame to be correlated with the rest of the frames. This also yielded interesting results and it is a worthy venture to investigate further about the feasibility of the application of this particular set of results into any algorithm or for the overall purpose of speech recognition. However, the reason behind not choosing the mean for this experiments and why we do recommend the use of the standard deviation is because the resolution provided after the correlation of the frame with the minimum mean is not as rich and distinctive as that of the minimum standard deviation.

The maximum standard deviation or rather the frame with the highest standard deviation was also used as a candidate for the correlation process, this also yielded interesting results but as stated this was also not as distinctive as that of the

Table 2 K-means implementation

A Pseudo code Implementation Of the K-means Algorithm
for Iteration = 1
Compute Frame with Min and Max Correlation Value of A
Assign Min as centroid of C_A & Max as centroid of C_B
for every CF applying bias in Eqn. 8
if ED of (C_A - CF) < (C_B - CF)
Assign to C_A
else if ED of (C_B - CF) < (C_A - CF)
Assign to C_B
End
Calculate C_A using Eqn. 5 and C_B using Eqn. 6
If(new C_A and C_B = Previous C_A and C_B) or (Iteration = 100) terminate
End

A = Audio Signal
C_A = Cluster A
C_B = Cluster B
CF = Correlated Frame
ED = Euclidean Distance

minimum standard deviation. In our experiments and the results we present, the minimum standard deviation was used but we also extended this to use the frame with the next minimum standard deviation. This is also presented in the results section together with that of the minimum standard deviation. The next minimum even proved better in some cases as compared to the minimum.

6 Experimental Data

A total of 20 samples from 20 different speakers with the digits 1–9 (total of 180 digits) were used for these algorithms. These samples are from the Centre for Spoken Language Understanding (CSLU2002) database [14]. This is a commercially available database from the Oregon Health & Science University. These audio samples have a sampling rate of 8 kHz and encoded in 16 bits. The first 10 were used as a guide set to modify and apply the different variable adjustments to come up with an initial starting point with regards to the adjustable variables of the experiment explained in the next paragraph and the results are presented in Table 3.

Table 3 10 Initial samples

Standard deviation	Frame distance	Inter frame distance	Recognized male (%)	Recognized female (%)
Minimum	>=2	<=6	77.70	91.10
Next minimum	>=2	<=6	71.10	95.50
Minimum	>=2	<=7	77.70	93.30
Next minimum	>=2	<=7	66.60	93.30
Minimum	>=2	<= 8	75.50	91.10
Next minimum	>=2	<= 8	71.10	93.30
Minimum	>=2	<=9	73.30	93.30
Next minimum	>=2	<=9	68.80	95.50
Minimum	>=2	<=10	75.50	93.30
Next minimum	>=2	<=10	66.60	97.70
Minimum	>=3	<=6	75.50	86.60
Next minimum	>=3	<=6	71.10	95.50
Minimum	>=3	<=7	75.50	86.60
Next minimum	>=3	<=7	68.80	95.50
Minimum	>=3	<=8	80	88.80
Next minimum	>=3	<=8	71.10	93.30
Minimum	>=3	<=9	82.20	88.80
Next minimum	>=3	<=9	71.10	95.50
Minimum	>=3	<=10	77.70	88.80
Next minimum	>=3	<=10	68.80	97.70
Minimum	>=4	<=6	82.20	84.40
Next minimum	>=4	<=6	62.20	97.70
Minimum	>=4	<=7	80	82.20
Next minimum	>=4	<=7	57.70	97.70
Minimum	>=4	<=8	84.40	84.40
Next minimum	>=4	<=8	62.20	97.70
Minimum	>=4	<=9	80.00	82.20
Next minimum	>=4	<=9	62.20	93.30
minimum	>=4	<=10	75.50	82.20
Next minimum	>=4	<=10	53.30	93.30

The second set of 10 audio samples was used to test the algorithm against a previous work [10], where the same samples were used. The results are given in Table 4.

A third algorithm was developed to test the respective digits. These digits were grouped using 2 variables for the algorithm to make its decision. This is also an improvement to the test algorithm. Only the cluster with speech is passed on to this algorithm to try and find out where the respective digits are. These two variables are;

Table 4 10 samples from the previous paper [10]

Standard deviation	Frame distance	Inter frame distance	Recognized (%)
Minimum	>=2	<=6	76.60
Next minimum	>=2	<=6	80
Minimum	**>=2**	**<=10**	**90**
Next minimum	>=2	<=10	86.60
Minimum	>=3	<=10	85.50
Next minimum	>=3	<=10	85.50
Minimum	>=4	<=6	87.70
Next minimum	>=4	<=6	83.30
Minimum	>=4	<=8	87.70
Next minimum	>=4	<=8	83.30

- Frame Distance: This variable is used to check for the pauses between the respective digits. There are 9 digits with pauses in between them. This variable is the distance between the frames containing the respective digits. This can be adjusted and there is no definite size to this particular variable as the speed of speech cannot be controlled because it depends completely on the individual speaking. However we made attempts to try to generalize this as this will be shown in the results.
- Inter Cluster Distance: This particular variable is used to complement the effort of the first in the grouping of the digits. There are cases where there is a little silence in the utterance of a digit. An example of this is the digit 6. Looking at Fig. 2 it can be seen that for the digit 6 there is a slight break between the "SI" part of it and the "X" part of 6. This break using the first variable could group them into two different groups as separate digits. The inter frame distance checks every respective group to see if it meets a set minimum. If it does not attain the set minimum, this then tries to assign it to the group closest to it using the Euclidean Distance measure.

Different adjustments to these variables were used to come up with an acceptable optimal number. A number of experiments were conducted in this regard and then the same 90 from our previous work [10] is used to compare and see if there was any improvement. However, it is worthy of note that as a mobile device intended system it is important that we have variables that can be adjusted. This can be adopted to the system as measures of customization.

After the adjustments of the respective variables, the algorithm developed is in the form of a mobile application. This is deployed on the mobile phone and this is then used to verify the respective digits. The application plays the respective digits and this is audibly verified. Only the set of digits that were fully recognised were recorded as recognised. The digits that were still broken or split by the algorithm were ignored in the computation of the final result numbers.

Table 5 Linear cross correlation computation times

Algorithm computation time in seconds	
DFT LCC	LPC LCC
49.7546	2.76

7 Results

See Tables 3, 4 and 5.

8 Conclusion

It can be seen from the results that this algorithm and detection outperforms the algorithm and detection algorithm in the last paper. The optimal performance of the algorithm in that paper was **86.6 %** and that result compared to the result in this experiment on the same data with **90 %** shows a significant improvement with regards to the algorithm detection technique.

The performance of the female samples over the male samples in Table 3 is due to the smoothness and pitch of the female voice. However, due to the average sound of male speakers, there is a lot of vibration in the male speech. This can be seen by the performance of this algorithm by simply comparing the female speakers which is **97.7 %** at best with the male speakers which is **84.4 %**. It is also worthy of note that the only comparable leap achieved was with the minimum SD, where the performance for both was **84.4 %**.

In Table 5 the computation time for the respective algorithms on the mobile device is given. The LCC stage of this experiment takes **2.76 s** as compared to the original DFT in [10] which took **49.75 s**. The proposed technique and new method of extracting the LCC value of the LPC frames shows a faster processing time on the same mobile device and as such makes this technique applicable to mobile devices. Time constraint can be regarded as one of the primary focus in designing on-device mobile based speech recognition systems.

The task of assigning the frame distance and inter frame distance is very tricky. However, for an on–device mobile device implementation of speech recognition these variables can prove very important as they allow a degree of customization to the particular user of that mobile phone. This will improve the problem of generalization with regards to speech recognition and as such helps to customise a system to a particular user.

It is worthy of note that both the minimum and the next minimum standard deviation cannot be ignored for implementation purposes. The results in Table 3 shows that in the case of female speakers, the next minimum outperforms the minimum. This can also be an advantage to the implementation of a user specific system for this detection. These variables can be implemented in a dynamic approach to try and adjust them to a user.

9 Future Work

This algorithm will be used as a pre—processor for mobile device speech recognition to properly evaluate how this can affect the choices of the recognizer to be used.

Acknowledgements The authors wish to thank the Petroleum Technology Development Fund (PTDF) for their continued support and sponsorship of this research. Dr. S Mustafa, Aishatu Mustafa and colleagues who helped in conducting experiments.

References

1. Wu, B., Wang, K.: Voice activity detection based on auto-correlation function using wavelet transform and teager energy operator. Comput. Linguist. Chin. Lang. Process. **11**, 87–100 (2006)
2. Waheed, K., Weaver, K., Salam, F.M.: A robust algorithm for detecting speech segments using an entropic contrast: circuits and systems. In: 2002. MWSCAS-2002. The 2002 45th Midwest Symposium on IEEE, vol. 3, pp. III-328–III-331 (2002)
3. Alarifi, A., Alkurtass, I., Al-Salman, A.: Arabic text-dependent speaker verification for mobile devices using artificial neural networks. In: Machine Learning and Applications and Workshops (ICMLA), 2011 10th International Conference on, vol. 2, pp. 350–353, IEEE (2011)
4. Huang, H., Lin, F.: A speech feature extraction method using complexity measure for voice activity detection in WGN. Speech Commun. **51**, 714–723 (2009)
5. Ghaemmaghami, H., Baker, B.J., Vogt, R.J., Sridharan, S.: Noise robust voice activity detection using features extracted from the time-domain autocorrelation function. In: Proceedings of Interspeech (2010)
6. Ramırez, J., et al.: Efficient voice activity detection algorithms using long-term speech information. Speech commun. **42.3,** 271–287 (2004)
7. Prasanta Kumar, G., Tsiartas, A., Narayanan, S.: Robust voice activity detection using long-term signal variability. IEEE Trans. Audio Speech Lang. Process. **19.3,** 600–613 (2011)
8. Tashan, T., Allen, T., Nolle, L.: Speaker verification using heterogeneous neural network architecture with linear correlation speech activity detection. Expert Syst. (2013). doi:10.1111/exsy.12030
9. Rabiner, L.R., Schafer, R.W.: Digital Processing of Speech Signals. (IET, Stevenage 1979)
10. Mustafa, M.K., Allen, T., Evett, L.: A review of voice activity detection techniques for on-device isolated digit recognition on mobile devices. In: Bramer, M., Petridis, M. (eds.) Research and Development in Intelligent Systems XXXI, (Springer International Publishing, Switzerland 2014)
11. Smith, S.W.: The Scientist and Engineer's Guide to Digital Signal Processing. (FreeTech Books, San Diego 2003)
12. Looney, C.G.: A fuzzy clustering and fuzzy merging algorithm, CS791q Class notes (1999)
13. Žalik, K.R.: An efficient k′-means clustering algorithm. Pattern Recogn. Lett. **29**, 1385–1391 (2008)
14. CSLU Database.: http://www.cslu.ogi.edu/corpora/isolet/

Opinionated Explanations for Recommendation Systems

Aonghus Lawlor, Khalil Muhammad, Rachael Rafter and Barry Smyth

Abstract This paper describes a novel approach for generating explanations for recommender systems based on opinions in user-generated reviews. We show how these opinions can be used to construct helpful and compelling explanations at recommendation time. The explanation highlights how the pros and cons of a recommended item compares to alternative items. We propose a way to score these explanations based on their content. The scores help to identify compelling explanations, providing a strong reason why the item being explained is better or worse than the alternatives. We describe the results of offline experiments and a live-user study based on TripAdvisor data to demonstrate the usefulness of this approach.

1 Introduction

Recommender systems learn about our likes and dislikes to make suggestions to help us decide what to watch, read, and buy. But generating a list of suggestions is just part of the recommendation process. Explaining recommendations can make it easier for users to make good decisions, increasing conversion rates for businesses, and leading to more satisfied users; see [1, 2, 5, 6, 12, 14, 16]. For example, early work explored the utility of explanations in collaborative filtering with [6] reviewing various ways to explain movie suggestions using ratings, meta-data, neighbours and testing different presentation styles (histograms, confidence intervals, text).

A. Lawlor (✉) · K. Muhammad · R. Rafter · B. Smyth
Insight Centre for Data Analytics, University College Dublin, Belfield, Dublin 4, Ireland
e-mail: aonghus.lawlor@insight-centre.org

K. Muhammad
e-mail: khalil.muhammad@insight-centre.org

R. Rafter
e-mail: rachael.rafter@insight-centre.org

B. Smyth
e-mail: barry.smyth@insight-centre.org

M. Bramer and M. Petridis (eds.), *Research and Development in Intelligent Systems XXXII*, DOI 10.1007/978-3-319-25032-8_25

Bilgic and Mooney [1] used keywords to justify items arguing that the goal of an explanation should not be to *"sell"* the user on the item but to help the user to make an informed judgment. They found users overestimated item quality when presented with similar-user style explanations. Keyword approaches were also developed by [14] to generate explanations in the style of *"Item A is suggested because it contains feature X & Y that are also included in items B, C, & D, which you also like."*; see [16] for related ideas based on user tags. Explanations like this relate one item to others. Pu and Chen [12] build explanations that emphasise the tradeoffs between items, such as *"Here are laptops that are cheaper and lighter but with a slower processor"*; see also [13]. In Zhang et al. [17] a hybrid matrix factorisation framework for personalised recommendations was developed based on user-feature and item-feature relationships, and feature-level explanations are designed which highlight the features pushing the item into the top-K list.

In this work we generate *feature-based, personalised, opinionated* explanations (see also [15]). Like the work of [12, 13], our explanations relate items to alternative recommendations, to help the user to better understand the trade-offs and compromises that exist within a product-space; see also [9]. We also leverage the opinions in user-generated reviews as our primary source of item and recommendation knowledge. This paper pays particular attention to the clarity and helpfulness of these opinionated explanations, thereby complementing related work by [11], which focused on the role of opinionated explanations in ranking.

2 Opinionated Recommendation

This paper builds on recent work about mining opinions from user reviews to generate item descriptions for recommender systems. The work of [4] describes how shallow NLP, opinion mining, and sentiment analysis can be used to extract rich feature-based product cases. It is not possible to fully cover these techniques here and the interested reader is referred to [3, 4]. However, we will provide a brief summary based on Fig. 1. We will rely on a TripAdvisor dataset of hotels and reviews and so from time to time we will refer to this data without loss of generality.

2.1 Review Feature Extraction

As in [3, 4], for review r_i we mine *bi-gram* features and *single-noun* features; see [7, 8]; e.g. bi-grams which conform to a noun followed by a noun (e.g. *bath tub*) or an adjective followed by a noun (e.g. *double room*) are considered, excluding bi-grams whose adjective is a sentiment word (e.g. *excellent, terrible* etc.). Separately, single-noun features are validated by eliminating nouns that are rarely associated with sentiment words in reviews as per [7], since such nouns are unlikely to refer to item features. We refer to each of these extracted features, f_j's as *review features*.

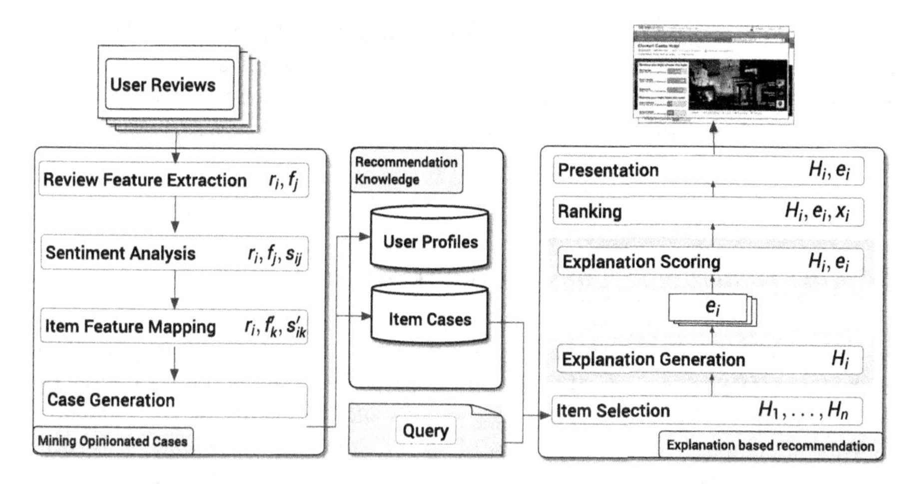

Fig. 1 An overview of the explanation-based recommendation architecture

2.2 *Sentiment Analysis*

For a review feature f_k we determine if there are any sentiment words in the sentence containing f_k. If not, f_k is marked *neutral*, otherwise we identify the sentiment word w_{min} with the minimum word-distance to f_k. Next we determine the part-of-speech (POS) tags for w_{min}, f_k and any words that occur between w_{min} and f_k. This POS sequence is an *opinion pattern*. We compute the frequency of all opinion patterns in all reviews; a pattern is *valid* if it occurs more than average. For valid patterns, we assign sentiment to f_k based on the sentiment of w_{min} and subject to whether the corresponding sentence contains any negation terms within 4 words of w_{min}. If there are no negation terms, then the sentiment assigned to f_k is that of the sentiment word in the sentiment lexicon; otherwise this sentiment is reversed. If an opinion pattern is not valid then we assign a *neutral* sentiment to each of its occurrences within the review set; see [10] for a fuller description. The end result of sentiment analysis is that we determine a sentiment label s_{ik} for each f_k in review r_i.

2.3 *Item Feature Mapping*

These review features often refer to very specific hotel details (e.g. the *orange juice* at breakfast) which are often too fine-grained for explanation purposes; although [3, 4] have shown their utility in recommendation. Therefore we map these low-level features to higher-level *item features* which correspond to features such as *bar/restaurant*, *room quality*, *breakfast*, etc. To automate this mapping, we apply a k-means clustering to the set of review sentences to find words that tend to co-occur frequently in review sentences. After some manual adjustment the resulting clusters can be labelled with a known set of high-level features; we use TripAdvisor's *amenities*. Thus, each cluster contains a set of low-level *review features* and is mapped to a high-level *item feature*. Using this information we can automatically map each (r_i, f_k, s_{ik}) review feature tuple to a corresponding (r_i, f'_j, s'_{ij}) item feature tuple.

2.4 Case Generation: Constructing Item Cases

For each item/hotel H we have review features $\{f_1, \ldots, f_m\}$ mined from *reviews*(H). Each review feature is mapped to an item feature f'_j and we aggregate the review feature's mentions and sentiment scores to associate them with the corresponding f'_j. We can compute various properties of each f'_j: the fraction of times it is mentioned in reviews (its *importance* and the degree to which it is mentioned positively or negatively (its *sentiment* as in Eqs. 1 and 2; note, $pos(f'_j, H)$ and $neg(f'_j, H)$ denote the number of times that feature f'_j has positive or negative sentiment in reviews for H, respectively. Thus, each hotel can be represented as a *case*, *item*(H), which aggregates item features, importance and sentiment data as in Eq. 3.

$$imp(f'_j, H) = \frac{count(f'_j, H)}{\sum_{\forall f' \in H} count(f', H)} \tag{1}$$

$$sent(f'_j, H) = \frac{pos(f'_j, H)}{pos(f'_j, H) + neg(f'_j, H)} \tag{2}$$

$$item(H) = \{(f'_j, sent(f'_j, H), imp(f'_j, H)) : f'_j \in features(H)\} \tag{3}$$

2.5 Case Generation: Constructing User Profiles

User profiles are produced in a similar way: for user U review features are mined from U's reviews; each review feature is mapped to an item feature; and we aggregate these item features and their popularity scores as a profile. Currently, we don't store sentiment user profiles as it is not required at the present time. However, in the future we intend to consider this option in more detail as such information may prove valuable when it comes to understanding a user's rating tendency.

$$user(U) = \{(f'_j, imp(f'_j, U)) : f'_j \in reviews(U)\} \tag{4}$$

3 From Opinions to Compelling Explanations

In what follows we assume the target user U_T is presented with a set of hotel recommendations $\{H_1 \ldots H_n\}$. Our task is to generate an explanation for each of these recommendations in turn and we will refer to the current one as the *target* hotel or H_T and the other items as the *alternatives* or H'.

3.1 Generating a Basic Explanation Structure

We will describe the construction of a *basic* explanation structure, which begins with the data structure shown in Fig. 2.

Explanations come in two parts. The *pro* part is a set of positive features; reasons you might choose the hotel. The *con* part comprises the negative features; reasons to reject the hotel. These features are selected on the basis of three key components:

1. Sentiment Score—item feature $f'_j \in H_T$ is a *pro* if it has a majority of positive sentiments ($sent(f'_j, H_T) > 0.7$ in our TripAdvisor data) otherwise it is a con.
2. Relationship to Alternatives—To be a pro (or con) item feature f'_j must have a sentiment score that is *better than* (or *worse than*) at least one of the alternatives.
3. Importance to User—item feature f'_j must be contained within the user profile to ensure it has been mentioned by the user in his or her past reviews.

We generate explanations with these pro and con features using Eqs. 5–10. In Fig. 2 we can see an example of how this selects pros such as *Bar/Lounge* and *Free Breakfast*, which are important to the user, positive in sentiment, and better than some of the alternative recommendations. Likewise, we see cons such as *Leisure Centre*, which are also relevant to the user, but this time less favourably reviewed and worse than some of the alternatives.

Hotel: Clontarf Castle
User: Peter Parker

	Feature	Importance	Sentiment	BetterThan/WorseThan
PROS	Bar/Lounge *	0.25	0.60	75%
	Free Breakfast	0.22	0.79	10%
	Room Quality *	0.18	0.98	90%
	Restaurant *	0.15	0.80	80%
	Shuttle Bus	0.06	0.75	10%
CONS	Airport Shuttle *	0.21	0.20	90%
	Leisure Centre *	0.11	0.32	80%
	Swimming Pool	0.10	0.45	33%
	Room Service	0.5	0.46	20%
	⋮	⋮	⋮	⋮

Fig. 2 An example of an explanation structure showing pros and cons that matter to the user along with associated importance, sentiment, and better/worse than scores

$$betterThan(f_j', H_T, H') = \frac{\sum_{H_c \in H'} 1[sent(f_j', H_T) > sent(f_j', H_c)]}{|H'|} \tag{5}$$

$$worseThan(f_j', H_T, H') = \frac{\sum_{H_c \in H'} 1[sent(f_j', H_T) < sent(f_j', H_c)]}{|H'|} \tag{6}$$

$$pro(f_j', U_T, H_T, H') \leftrightarrow \\ sent(f_j', H_T) > 0.7 \wedge betterThan(f_j', H_T, H') > 0 \wedge imp(f_j', U_T) > 0 \tag{7}$$

$$con(f_j', U_T, H_T, H') \leftrightarrow \\ sent(f_j', H_T) < 0.7 \wedge worseThan(f_j', H_T, H') > 0 \wedge imp(f_j', U_T) > 0 \tag{8}$$

$$Pros(U_T, H_T, H') = \\ \{(f_j', v, m) : pro(f_j', H_T, H') \wedge v = betterThan(f_j', H_T, H') \wedge m = imp(f_j', U_T)\} \tag{9}$$

$$Cons(U_T, H_T, H') = \\ \{(f_j', v, m) : con(f_j', H_T, H') \wedge v = worseThan(f_j', H_T, H') \wedge m = imp(f_j', U_T)\} \tag{10}$$

3.2 Filtering Compelling Explanations

This basic explanation structure can be made up of many features, which may complicate the decision making if presented to the end user in this way. Moreover, many of the pros might be better than only a small fraction of the alternative recommendations, and conversely for cons, thereby limiting their usefulness as compelling reasons to choose or avoid the hotel in question. However, we can filter features based on how strong a reason they represent for choosing or rejecting the target hotel. To do this we define a *compelling* feature to be one that has a *betterThan* (pro) or *worseThan* (con) score of >50 % instead of just >0. Thus, a compelling pro is *better* than a *majority* of alternative recommendations and a compelling con is *worse* than a *majority* of alternatives. A compelling pro may be a strong reason to choose the target hotel while a compelling con may be a strong reason to avoid it.

We define a *compelling explanation* as a non-empty explanation which contains only compelling pros and/or compelling cons. For instance, referring back to Fig. 2, we have marked compelling features with an asterisk after their name; so, the compelling explanation derived from this basic explanation structure includes *Bar/Lounge, Room Quality, Restaurant*, as pros, and *Airport Transport* and *Leisure Centre* as cons. These are all features that matter to the user and they distinguish the hotel as either better or worse than a majority of alternatives.

3.3 From Explanations to Ranking

As an aside it is worth highlighting another aspect of this work: the idea that explanations might also be used for the *ranking* of recommendations. We can estimate the quality of an explanation numerically and use this for ranking purposes. To do this we use a straightforward scoring function to measure the strength of an explanation as the weighted difference of its pros and cons as shown in Eq. 11; this can be applied to either basic or compelling explanation structures.

$$
\begin{aligned}
& strength(U_T, H_T, H') = \\
& \quad \sum_{f \in Pros(U_T, H_T, H')} betterThan(f, H_T, H') \times imp(f, U_T) - \\
& \quad \sum_{f \in Cons(U_T, H_T, H')} worseThan(f, H_T, H') \times imp(f, U_T)
\end{aligned} \tag{11}
$$

A further discussion of the role of explanations in recommendation ranking is beyond the scope of this paper. However, the interested reader is referred to the work of [11] for a more in-depth treatment of this idea.

4 The Explanation Interface

How can explanation information be presented in a helpful way to users? Fig. 3 shows three example treatments that could be presented alongside a given hotel description. In each, features that matter to the user are separated into pros and cons; we also only present compelling features. The features are ranked based on their importance to the user and each is associated with a sentiment bar to indicate the percentage of positive sentiments expressed by reviewers. Treatments Fig. 3b, c further enrich the explanation by relating each feature to the other recommendation alternatives at different levels of precision. Figure 4 shows an example of one of these explanation types in context in TripAdvisor.

5 Evaluation

Next, we describe a pair of evaluations designed to explore the form and function of our explanations in the context of a TripAdvisor dataset and user judgements.

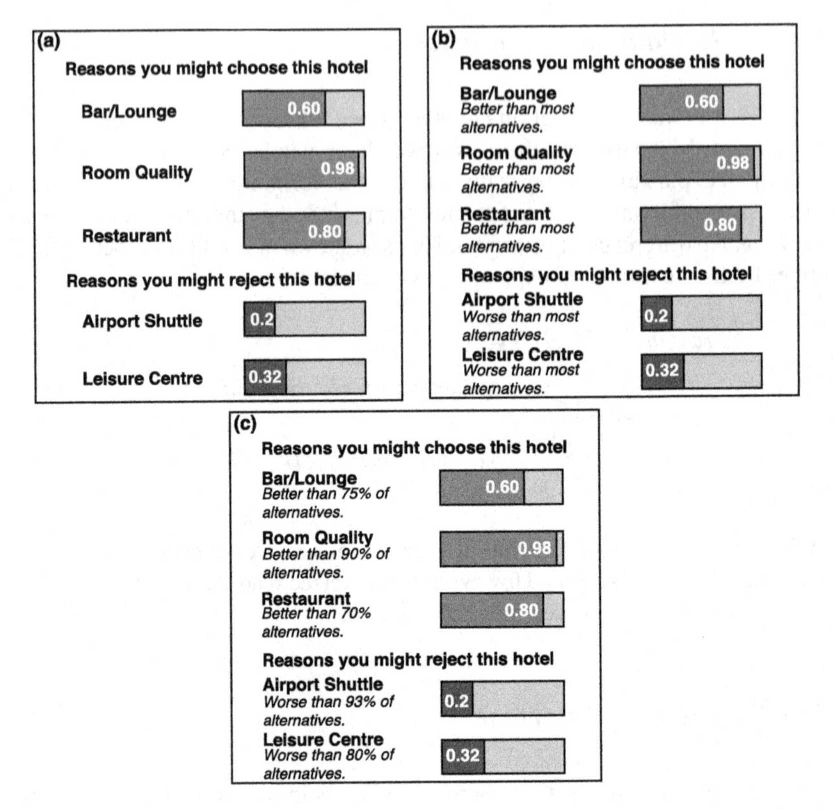

Fig. 3 Explanation styles: **a** sentiment only; **b** sentiment plus alternatives; **c** sentiment plus alternatives plus percentages

5.1 Offline Evaluation

For the first part of our evaluation, we use a large TripAdvisor dataset as a source of user profiles, reviews, and hotel cases. It contains 10,000 users who have each written at least 10 hotel reviews for $2,062$ hotels. In addition, we had more than 220,000 reviews by almost 150,000 reviewers available for the hotel cases. For each target user U_T we know the hotel they booked, H_B, and the related hotels recommended by TripAdvisor; we understand that TripAdvisor generates these using a combination of location, similar users, and meta-data. Thus, we can generate approximately 100,000 *user sessions*, one for each user booking and containing the booked hotel and the related TripAdvisor suggestions. Next we generate basic and compelling explanations for each of the hotels in a user session—that's approximately 1,000,000 explanations of each type—and analyse their form, focusing on the number and type of features that are commonplace in the resulting explanations.

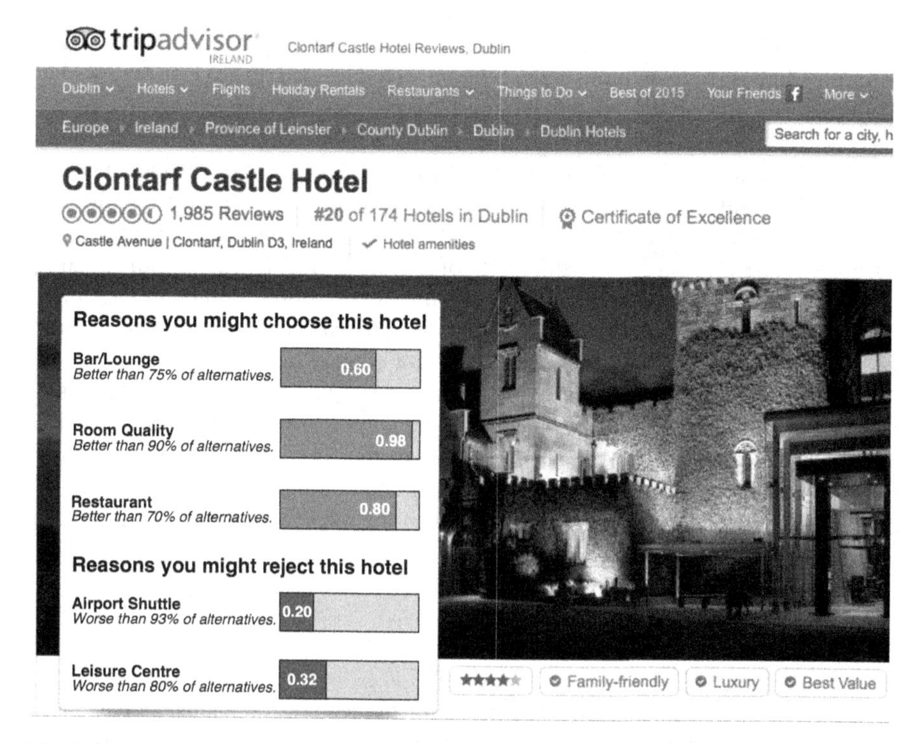

Fig. 4 An example explanation (sentiment plus alternatives plus percentages) in context. By mousing over the sentiment bars, the user sees a preview of relevant review fragments. It is also feasible to use the explanation as a navigation aid so that by clicking on the sentiment bars or explanation text the user can navigate to corresponding reviews or alternative candidates

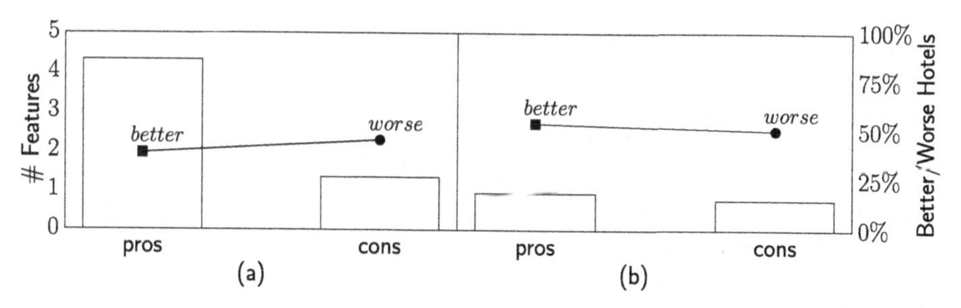

Fig. 5 The average number of pros and cons and the average *betterThan* and *worseThan* per explanation for basic and compelling explanations. **a** Basic. **b** Compelling

5.1.1 Pros versus Cons, Better versus Worse

Figure 5a, b shows the average number of pros and cons (left y-axis), and the average *betterThan*/*worseThan* scores (right y-axis), in basic and compelling explanations. We see that on average we are recommending about 4 pros versus only 2 cons in basic explanations compared to 2 pros and 2 cons in compelling explanations. The extra pros in basic explanations reflect the positive bias in TripAdvisor reviews but it is interesting that approximately half of these pros are not compelling.

This bias is also suggested by the difference between the average *betterThan* score for pros (49 %) and the *worseThan* score for cons (69 %). For a typical hotel, its basic pros will typically be better than about 49 % of the alternatives in the recommendation session. In contrast, when it comes to the basic cons, it is usually the case that the hotel in question does worse than most of the alternatives in the recommendation session. A similar pattern is seen for compelling explanations, although the difference now is less pronounced; 70 % average *betterThan* scores for compelling pros versus approximately 75 % *worseThan* scores for compelling cons.

Overall we see that compelling explanations are simpler than basic explanations—they contain fewer pros and cons—and they are more compelling because their features are better or worse than a large majority of the alternative recommendations. Intuitively this combination of simplicity and compellingness should make compelling explanations particularly effective when it comes to helping users to decide whether to accept or reject a given recommendation.

5.1.2 On the Frequency of Explanation Features

Figure 6 shows the frequency distributions for the features contained in basic and compelling explanations. In each histogram, the individual bars refer to a specific item feature, and each bar shows the number of times that the feature occurs as a pro and as a con. The histograms also show the average *betterThan* and *worseThan* scores for these features, based on their pro and con occurrences, respectively.

We see that a handful of item features tend to dominate in explanations. Features like *free breakfast* and *bus service* appear very frequently compared to others such as *fitness centre*, *high-speed wifi*, and *kids activities*. We also see the strong positive review bias in TripAdvisor as a majority of features present mostly as pros. For example, in Fig. 6a we can see that *free breakfast* appears as a con in 47,263 explanations but as a pro in 238,577 explanations; it is worth noting the unusually high negative sentiment associated with the *fitness centre* and *high speed wifi* features, both of which appear more frequently as cons than pros. This also explains the relatively high value for the *worseThan* scores compared to *betterThan* scores mentioned previously. It is relatively unusual for a feature to be listed as a con (<20 % of the time in most cases) and so if a feature is a con it is likely that it is a pro in the alternative recommendation candidates and so it is likely to have a sentiment score that is worse than a majority of alternatives. In contrast if a feature is a pro in an explanation it is also likely to be a pro in the explanations of the alternatives and so it is less likely to have a sentiment score that is better than most alternatives.

This data tells us about the features that matter the most to users (based on their reviews) but it also indicates whether a particular feature is likely to appear as a compelling pro or a compelling con in an explanation. For example, *free breakfast* is the most common feature to appear in explanations and it appears as a pro over 85 % of

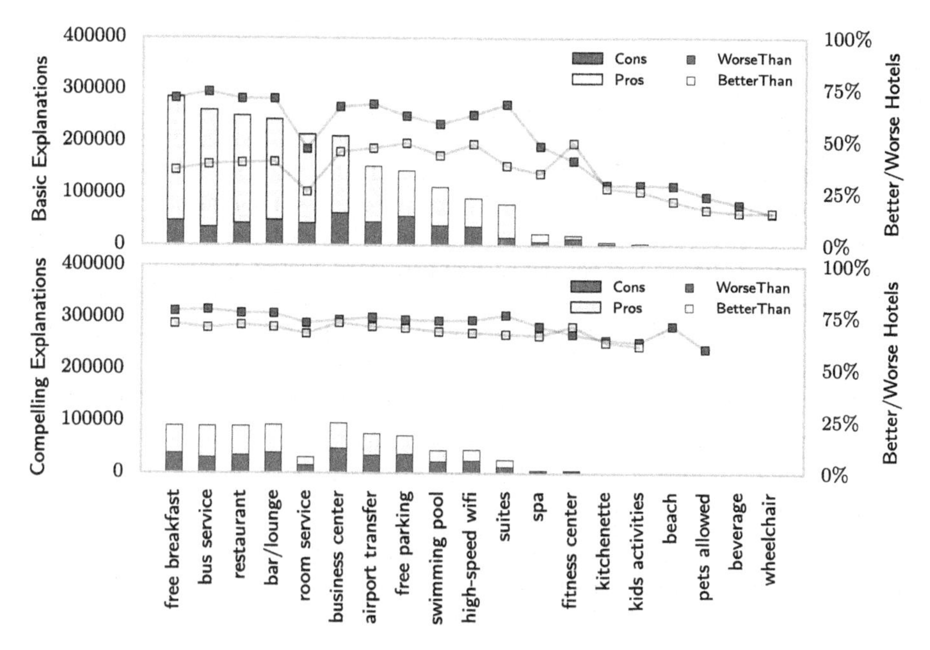

Fig. 6 An analysis of the relative frequency of features in the pros and cons of explanations and their corresponding *betterThan* and *worseThan* scores

the time and a con just under 15 % of the time. However, as a pro it has an average *betterThan* score of only about 30 %, whereas as a con it has a *worseThan* score of almost 80 %. Therefore, this feature is less likely to appear as a pro in a compelling explanation whereas it is very likely to appear as a con in a compelling explanation. This is evident in Fig. 6b which shows the corresponding data for compelling explanations. This time *free breakfast* appears as a con in 38, 271 compelling explanations and as a pro in 52, 763 explanations. As a hotel owner, if your hotel's *free breakfast* is being negatively reviewed then there is a strong likelihood that this feature will be exposed as a compelling con in any explanation generated for your hotel. As a user who has a preference for *free breakfast*, you will likely be influenced by this feature as a con in compelling explanations.

5.2 Live-User Study

The true test of this approach will depend on the opinions of users in a live setting and whether or not the explanations help users make better decisions in the long-term. This is a challenging evaluation setting and it is beyond the scope of the present work to fully explore this broader issue. That being said, we have completed an initial user study to gather initial impressions of different explanation styles and types of information and we will summarise the results of this study in what follows.

5.2.1 Setup

Our user study took the form of an online questionnaire, which placed participants in a simple hotel booking setting, asking them to evaluate the 3 styles of explanation interface presented earlier (Fig. 3) in the context of TripAdvisor as per Fig. 4. In what follows we will refer to these 3 styles as S (pros and cons with sentiment only), $S + A$ (pros and cons with sentiment and comparison to alternatives), and $S + A + \%$ (pros and cons with sentiment and percentage comparison to alternatives).

48 people participated in the user study, mostly Ph.D. students and researchers in our research centre. They were presented with each interface in turn—varying the presentation order—and they were asked to express their agreement on a scale of 1 (strongly disagree) to 10 (strongly agree) with each of the following two statements:

1. *Clarity:* The explanation is clear and easy to understand.
2. *Helpfulness:* The explanation will help me to make a choice about whether or not to choose or reject this hotel.

Finally, each participant was asked to rate the usefulness of the various explanation components used in these interfaces on a scale of 1 (not useful) to 10 (very useful) by responding to the following questions:

1. How useful was it to separate the amenities into groups of pros (positive sentiment) and cons (negative sentiment)?
2. How useful did you find the sentiment bars?
3. How useful did you find the explanations that compared the hotel to alternative recommendations?
4. When comparing the hotel to alternatives how useful did you find the precise percentage information?

5.2.2 Results

The results are presented in Fig. 7a for each of the 3 interfaces. We can see that overall participants found the interfaces clear and helpful with a preference for interfaces 2 and 3, which included extra information about how the hotel compared with alternative recommendations in addition to simple sentiment information.

Figure 7b shows the average utility ratings for each of the various explanation components. In general these ratings are high across all of the different explanation components with an average overall rating that is greater than 7. We can see that participants found the separation of amenities into pros and cons particularly useful (an average rating of 8.48) followed by the use of sentiment information in the explanations (7.60). There is little difference expressed between the purely text-based comparison to alternatives (e.g. "... better than most alternatives") and a more precise comparison (e.g. "... better than 93 % of alternatives") with both components scoring above 7 on average.

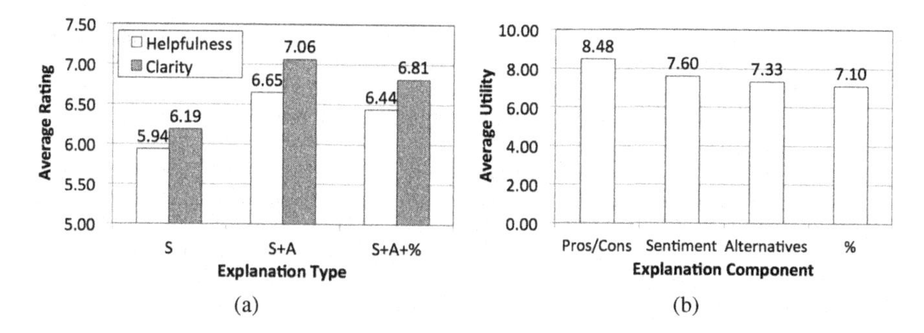

Fig. 7 In **a** we show the ratings for different explanation types. In **b** we show the average utility scores for each component in the explanations

The results, preliminary as they may be, do suggest that users are perceiving value in the type of explanations that we are generating. The combination of sentiment and a comparison to alternatives presents as the preferred interface type with users reporting high levels of clarity and helpfulness.

6 Conclusions

This work brings together ideas from case-based reasoning, opinion mining, and recommender systems [3, 4]. We have described an approach to generating explanations for recommender systems from user reviews. We have evaluated these explanations using a combination of offline and online evaluations using large-scale TripAdvisor data and live-users. As part of our future work, we plan to make progress on a more extensive live-user evaluation involving real-time recommendation sessions. It will also be interesting to incorporate additional information as part of our explanations. For example, in the work presented we compare recommendation candidates to alternative recommendations but we could also consider the relationship to a user's previous bookings. In this way, our explanations could help the user to understand how a particular hotel/item relates to alternative recommendations but also to hotels they have booked in the past.

Acknowledgments This work is supported by Science Foundation Ireland through through the Insight Centre for Data Analytics under grant number SFI/12/RC/2289.

References

1. Bilgic, M., Mooney, R.J.: Explaining recommendations: satisfaction versus promotion. In: Proceedings of Beyond Personalization Workshop at the 2005 International Conference on Intelligent User Interfaces, vol. 5, pp. 13–18 (2005)

2. Coyle, M., Smyth, B.: Explaining search results. In: Proceedings of the 19th International Joint Conference on Artificial Intelligence, pp. 1553–1555 (2005)
3. Dong, R., O'Mahony, M.P., Smyth, B.: Further experiments in opinionated product recommendation. In: Proceedings of the 22nd International Conference on Case-Based Reasoning, pp. 110–124 (2014)
4. Dong, R., Schaal, M., O'Mahony, M.P., Smyth, B.: Topic extraction from online reviews for classification and recommendation. In: Proceedings of the 23rd International Joint Conference on Artificial Intelligence, IJCAI'13, pp. 1310–1316 (2013)
5. Friedrich, G., Zanker, M.: A taxonomy for generating explanations in recommender systems. AI Mag. **32**(3), 90–98 (2011)
6. Herlocker, J.L., Konstan, J.A., Riedl, J.: Explaining collaborative filtering recommendations. In: Proceedings of the 2000 ACM Conference on Computer Supported Cooperative Work, CSCW'00, pp. 241–250 (2000)
7. Hu, M., Liu, B.: Mining opinion features in customer reviews. In: Proceedings of the 19th National Conference on Artificial Intelligence, pp. 755–760 (2004)
8. Justeson, J.S., Katz, S.M.: Technical terminology: some linguistic properties and an algorithm for identification in text. Nat. Lang. Eng. **1**(1), 9–27 (1995)
9. McSherry, D.: Similarity and compromise. In: Proceedings of the 5th International Conference on Case-based Reasoning, pp. 291–305 (2003)
10. Moghaddam, S., Ester, M.: Opinion digger: an unsupervised opinion miner from unstructured product reviews. In: Proceedings of the 19th ACM International Conference on Information and Knowledge Management, pp. 1825–1828 (2010)
11. Muhammad, K., Lawlor, A., Rafter, R., Smyth, B.: Great explanations: opinionated explanations for recommendation. In: Proceedings of the 23rd International Conference on Case-based Reasoning pp. 1–15 (2015)
12. Pu, P., Chen, L.: Trust-inspiring explanation interfaces for recommender systems. Knowl.-Based Syst. **20**(6), 542–556 (2007)
13. Reilly, J., McCarthy, K., McGinty, L., Smyth, B.: Explaining compound critiques. Artif. Intell. Rev. **24**(2), 199–220 (2005)
14. Symeonidis, P., Nanopoulos, A., Manolopoulos, Y.: Providing justifications in recommender systems. IEEE Trans. Syst., Man, Cybern.—Part A: Syst. Hum. **38**(6), 1262–1272 (2008)
15. Tintarev, N., Masthoff, J.: The effectiveness of personalized movie explanations: an experiment using commercial meta-data. In: Proceedings of the 5th International Conference on Adaptive Hypermedia and Adaptive Web-Based Systems, vol. 5149, pp. 204–213 (2008)
16. Vig, J., Sen, S., Riedl, J.: Tagsplanations: explaining recommendations using tags. In: Proceedings of the 13th International Conference on Intelligent User Interfaces, pp. 47–56 (2008)
17. Zhang, Y., Lai, G., Zhang, M., Zhang, Y., Liu, Y., Ma, S.: Explicit factor models for explainable recommendation based on phrase-level sentiment analysis. In: Proceedings of the 37th International ACM SIGIR Conference on Research and Development in Information Retrieval, SIGIR'14, pp. 83–92 (2014)

The Influence of Virtual Reality on the Perception of Artificial Intelligence Characters in Games

Christopher J. Headleand, Gareth Henshall, Llyr Ap Cenydd and William J. Teahan

Abstract Virtual Reality is a technology which is quickly leaving the laboratory and being placed in the hands of the consumer. With many large hardware manufacturers and games development studios investing heavily in the future of the technology, we are starting to see the first VR-based games become available. But will the consumerization of VR hardware change how games developers consider Artificial Intelligence? In this study, we begin by discussing how the perception of an AI-based character may change how a user interacts with it. Based on this survey, we ask the following question: "Do AI characters appear more or less human-like though Virtual Reality, as opposed to typical monitor-based viewing mediums?" We conduct a study where 16 participants play two games (a First Person Shooter (FPS), and a racing game), both played through VR and a monitor (4 games in total). In this study, the participants are told that they will play two games against another human participant, and two against an AI driven character and that they must make a judgement on what they are playing against for each game. However, they actually play against identical AI characters in both viewing instances of the two games. The results show a clear split in assessment for the two games; when the racing game was played through VR, the participants concluded that their opponent was Human; however, when played through the monitor, they concluded they were playing against an AI. However, the opposite trend is apparent when the participants played the FPS game. We conclude the VR does change the way we perceive AI characters; however this change in perception needs to be further investigated.

C.J. Headleand (✉) · G. Henshall · L.A. Cenydd · W.J. Teahan
School of Computer Science, Bangor University, Bangor, Wales, UK
e-mail: c.headleand@bangor.ac.uk

G. Henshall
e-mail: g.henshall@bangor.ac.uk

L.A. Cenydd
e-mail: llyr.ap.cenydd@bangor.ac.uk

W.J. Teahan
e-mail: w.j.teahan@bangor.ac.uk

M. Bramer and M. Petridis (eds.), *Research and Development in Intelligent Systems XXXII*, DOI 10.1007/978-3-319-25032-8_26

1 Motivation

The computing industry recently entered a virtual reality renaissance. The introduction of head mounted displays (HMDs) capable of immersive and comfortable experiences at affordable price points has dramatically reduced the barrier of entry for both home users and researchers.

Emerging VR HMDs are capable of producing the visceral sensation of being inside the simulated world. The depth and quality of this spacial immersion, called *presence*, is the main yardstick that all VR companies use when developing their headsets and experiences. While there are many facets to achieving a deep sense of presence, even the first generation of consumer VR hardware released in 2016 will be capable of producing this phenomenon for sustained periods.

Over the next few years, the industry is predicting an explosion of VR experiences. But what impact will this revolution have on AI research? Some are anticipating that high quality, believable AI will become increasingly important and questions have been raised as to how the appearances and behaviour of virtual characters influence peoples levels of presence, or immersion [5]. As the depth of presence becomes a commonly sought experience, players will be more critical of AI that breaks their sense of immersion. We could assume that this means 'business as usual' for AI, and we simply need to keep striving for improvement on the state of the art. But, this does raise the question of whether VR immersion in itself makes a difference to how we perceive AI characters?

A pertinent question for the gaming industry will be whether it is easier to identity an AI character through VR than a traditional monitor. For example, many modern collaborative on-line games seamlessly replace human controlled companions with AI ones if they leave the game, allowing the player to continue without breaking immersion. Virtual characters are now at a stage where it is increasingly difficult to make human/not human distinctions in virtual environments [13], and artificial intelligence is an important component of this façade. If VR had an effect on this, developers may need to reconsider their development choices, or how they implement their AI characters.

This preliminary study is motivated by two research questions: firstly,"Does virtual reality change the way we perceive non-player characters in-game?"; and secondly, "Do AI controlled characters appear more or less human-like through this viewing medium?" In essence, we will explore whether VR makes the synthetic behaviour of AI characters more noticeable.

2 Background

How we perceive AI characters within virtual worlds is a pertinent topic in both computer games research and virtual environment applications in general [18]. How AI characters are perceived could have a broad impact on believability, immersion and the usefulness of simulators, training tools and telepresence applications. There is also evidence to suggest that believing an in-game character is human could factor into enjoyment of games. Studies have shown that players will show preference towards team-mates they believe to be human controlled, even if they are actually AI characters [8]. However, what factors affect perception of an in-game character is an open question.

There is evidence to suggest that behaviour and movement are key qualities, with studies showing that people are more comfortable interacting with avatars that move like humans [22]. Studies have also demonstrated that human players will treat in-game characters more favourably if they believe that they are controlled by a human. This extends to noticing positive behaviour (such as sacrifice, or protection) more often [9, 11]. Conversely, if a player believes an in-game character to be a bot (an AI controlled character), they are more likely to assign blame to it [10]. However, an alternate theory is that participants will respond equally to human and computer controlled entities that exhibit similar social behaviour [13], a phenomenon that is sometimes attributed to the CASA (Computers Are Social Actors) theory [12].

Gender may also be a factor in perception. Avatars with gender identities have been shown to elicit specific behaviours from humans during interaction. For example, when interacting with characters with a female gender, participants exhibited stereotypical masculine behaviour, and when interacting with male avatars, the participants exhibited feminine behaviours [19]. It is important to note that the participants were not aware of their change in behaviour, meaning that we may not be conscious of how our perception of characters in virtual worlds changes our actions.

We are aware that the appearance of avatars has an impact on interaction [1]. For example, characters wearing outfits with negative social connotations [15] elicit more aggressive intentions and attitudes from participants. But how could other visual factors contribute to our perception of AI characters? To the best of our knowledge, there has not been a study which has evaluated whether changing viewing modality effects the perception of AI characters. Would the increased immersion of VR make synthetic behaviours more obvious to a human observer, or would this change in presence result in a greater perceptual inference of the agent's behaviour, making the façade easier to believe?

3 Method

To explore the research questions, 16 participants (12 Males, 4 Female) were tasked with playing two types of game: a racing game, and a first person shooter (FPS). Both game types were played through two viewing mediums, an Oculus Rift DK2 and a standard PC gaming monitor. Every person played all four games.

Games were played in a mixed order, and it was ensured that the same viewing medium wasn't repeated twice (resulting in 8 possible orders of play). Each order was therefore played by two participants.

Each participant was told that during the four games, they would play two rounds against a human, and two against an AI opponent. They were told that their task in each game was to identify whether the identity of the opponent was an AI or Human.

During the experiment, the participant was in a segregated booth, unable to see other people during the experiment. The beginning of each game included a splash screen which implied the game was connecting to a multi-player server, adding to the façade.

However, regardless of whether the person played through VR or Monitor, their opponent was the same AI (one AI for the racing game, another for the FPS). The purpose of this deception was to ensure that the players were competing against opponents of identical competence and that in-game ability was not used as a flag to differentiate between opponents. This removes one confounding variable from the experiment, and is consistent with the experimental design of studies with similar objectives [8]. By following this approach, we only identify differences in the perception of AI through the two viewing media.

At the end of the experiment, participants filled out a survey. For each of the games, they were asked to make an assessment of the identity of the opponent they competed against. This was done on a 1 to 5 scale, with 1 representing high confidence that the opponent was Human and 5 representing high confidence that the opponent was AI, a score of 3 indicated that the player was unsure either way. The player was then also asked to rate their enjoyment of the game, and a free text response provided the participant with the opportunity to provide qualitative data.

3.1 Ethical Considerations

There are two principle ethical considerations in this study. The first is that the experimental design involves deceiving the participants. However in this case, the harm caused from the deception is minimal, and it was deemed to be the only practical method of achieving the goals of the study.

The second consideration is that video games have been shown to induce motion sickness [21] and that use of the Oculus Rift compounds this nausea in users [3, 4], a condition known as cybersickness. To reduce the risk, we designed both games to adhere to VR best practices [14]. For example, low frame rates are the most common source of sickness, and so we ensured both game types were played on a machine capable of producing a constant 75fps (Frames Per Second), the native frame-rate of the Oculus Rift DK2 [20]. Efforts were also made to alleviate motion sickness due to vection [7], including limiting movement speed in the FPS game.

All participants were informed of the risk before they entered the study, and had to read the health and safety information (produced by Oculus) and sign a consent form. Participants were also told that if they felt sick, they could ask to end the study at any time. Additionally, participants was given a short break between each game, and the participants never played two VR games in a row, limiting extended exposure.

4 Test-Bed Games

In the following sections, we will explain how both games were implemented, including integration with the Oculus Rift.

4.1 Racing Game

To implement the racing game, we used two projects available freely through the Unity asset store. The first *Car Tutorial* [23] is a complete package including a track and a physics driven player car. We augmented this package with the AI car from *The Vehicle Physics Toolkit* (*VPT*) [2] package, which is also freely available. The underlying physics of both AI and player controlled cars were based on the same model [23].

The AI car follows a predetermined path along the center of the virtual racetrack. A variable in the AI car script determines how much it can deviate from that path before it needs to correct. Varying this value allows us to produce a relatively realistic driving style. The path itself is constructed from a series of game objects linked together to form a complete circuit. Within the VPT package, breaking zones are placed throughout the track so that the AI car will slow down at sharp corners.

A reset function was also added so that any AI or player controlled car crashes would result in the car resetting its position to the middle of the road after two seconds.

The majority of racing games have a natural rhythm, with players regularly changing position and overtaking each other, rather than one player dominating the race. This is a commonly sought after mechanic by game designers, and is usually achieved by applying a rubberbanding function that speeds or slows down the AI car if too far behind or ahead. The same mechanic tends to naturally arise in similarly-skilled player controlled games due to driver error etc. Rubberbanding was implemented into our game for the same reason, ensuring that the AI was constantly battling the player for position regardless of the player's ability.

The original camera position followed above and behind the car (third person perspective). This camera angle was not suitable for VR as it is both unnatural and likely lead to nausea [16]. The camera was moved to be inside the car, providing a much more natural perspective. Being surrounded by the car's interior also helps to reduce nausea as it provides a static frame of reference.

4.2 First Person Shooter

The FPS game required the player to explore a randomly generated maze and destroy an opponent. To generate the environment, we implemented the RMCM algorithm [6] in Unity. Each environment (which can be seen in Fig. 2) was 50 units square, with one unit equal to the diameter of the enemy and the player's avatar colliders (Fig. 1).

Fig. 1 Players view from inside the car during the game. The original game was played from a 3rd person perspective (outside, behind and above the car). This would have been an unnatural (non-immersed) position to play in VR, so the camera was moved inside the car

Fig. 2 The player's view within the FPS maze. This is randomly generated using the RMCM algorithm

The player and the opponent were placed at opposite corners of the environment. The player and the opponent both controlled an identical character implementation (which can be seen in Fig. 3), ensuring equivalent speed and manoeuvring ability. The two entities in the game were armed with the same simulated weapon, and had the same number of lives.

Fig. 3 The opponent character in the FPS game. The character was a simple sphere with a turret weapon. As the character moved around the scene, it *left* a decaying *blue* trail behind since in early experiments we discovered that even in small mazes, it was difficult to locate the opponent character. The player avatar also had a *blue* trail, although this was not used by the AI for tracking

The AI was controlled by a simple finite state machine operating in one of three states:

Wander In the wander state, the AI randomly explores the environment, sensing the world ahead of it with a vision cone of 120°. The vision range is limited only by occlusion from walls in the environment. As the character moves into a new room, it will detect any exits it can see, select one at random and steer towards it. If it does not see an exit directly ahead, it explores the room using a wander steering behaviour [17].

Engaged If the player enters the opponent's vision cone, the AI enters the engaged state. In this state, it will move towards the player firing its weapon. The AI stops moving forward if the player is within a 'close range' vision cone, where the length of the close range vision cone is equal to the diameter of the enemy and the player's character collider.

Seek If the AI is in the engaged state, and the player exits its vision cone, it enters a seek state. The seek state causes the AI to turn and move towards the last point at which it saw the player. Once there, if it has not seen the player again (activating the engaged state), it returns to the wander state.

Each time the player was shot, a small health bar provided the player with a visual indicator of the damage. The game was a single round ending with either the player or opponent being destroyed.

4.3 Headset and Input Device

We used the same camera position and field of view for the VR and monitor versions of both games. A standard Unity camera was used to render for the monitor versions, while the VR versions used the stereoscopic camera implementation provided by Oculus that renders a separate image for left and right eyes.

The Oculus Rift DK2 is capable of tracking the player in 3D space using an infrared camera that tracks the headset's position. This allowed the player to lean around in the VR car game, but not in 2D monitor version. While this could have a small impact on performance, we deemed positional tracking to be an integral part of VR as it has an important effect on both nausea reduction and immersion.

As half of the four experiments were run on the Oculus Rift, the games were controlled using a gamepad (specifically the Microsoft Xbox gamepad). Once the headset is on, the participant is unable to see the keys of a keyboard, so a tactile controller is more suitable.

5 Results

The results show a clear split between the monitor and VR based games. However, what makes the results particuarly interesting is that the split is inverted for the two game types.

In the VR racing game, players typically reported that the opponent was Human. The mean result for all the racing games played through VR is 2.56, placing the average player opinion between human and undecided. This is an inconclusive score by itself, but the mode score of 2 provides further insight into the perception of the players. We can also observe an obvious bias towards the players reporting the opponent as human controlled in the distribution of scores (seen in Fig. 4).

When playing the racing game through a monitor, the players typically reported that the opponent was an AI. The identification here was more statistically obvious, with a mean score of 4 and a mode of 4 (13 of the 16 participants voted that they believed the character was AI).

However, we observe an inverse trend in the FPS game. While playing the game in VR, the majority of participants reported that the opponent was AI controlled (mean score of 3.68 and a mode of 4). This falls into a similar distribution to the AI reporting in the racing game, but through the alternate viewing medium (VR rather than Monitor).

When the participants played the FPS game through the monitor, they trended towards reporting that the opponent was human controlled. As with the racing game, this was reported with significantly less confidence than the reporting of the AI character (mean score of 2.43 and a mode of 1).

It is perhaps unsurprising that where participants reported that they believed the opponent was an AI (Monitor for the racing game and VR for the FPS), they

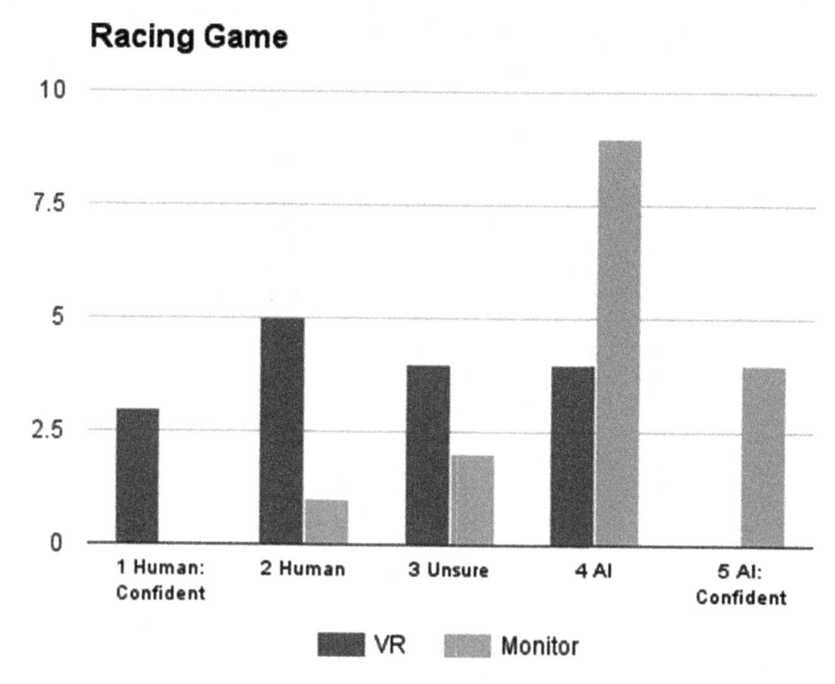

Fig. 4 The participant's assessment of their opponents identity for the First Person Shooter game. When played in VR, the majority of participants trended towards believing that their opponent was human controlled (mean score of 2.56 and a mode of 2). However, when playing the game through the monitor, the participants were more likely to report a belief that the opponent was AI controlled (mean score of 4 and a mode of 4)

responded with higher confidence. The AI designed for both the games was relatively simple and contained little sophistication or artificial stupidity to make it respond more like a human controlled character. However, as the participants voted with relatively strong conviction in these cases, it is interesting to see participants trending towards reporting that their opponent was human controlled in the alternative viewing medium (even though this was with less confidence).

We also asked the players to rate their enjoyment of each of the four games. We were expecting to see a correlation between the games where the participant believed they were playing against a human and higher enjoyment. No such correlation existed, with the games that received the highest rating being the ones played in VR. However, we do not assume that this trend necessarily means that players will enjoy VR games more than their monitor based equivalent. The majority of the participants had not played games through VR before the study. As such, the novelty of the new viewing medium likely contributed to the enjoyment results we have reported (Fig. 5).

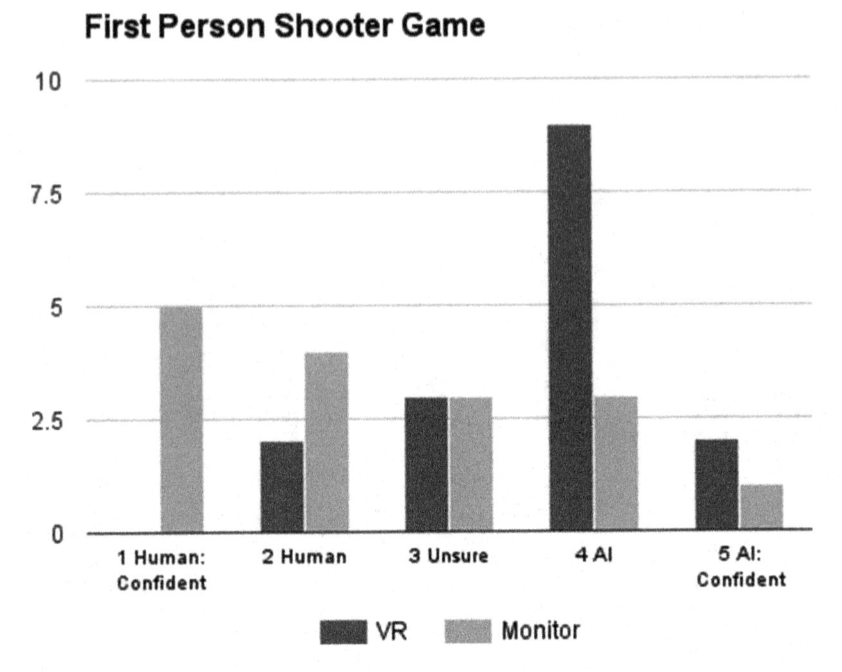

Fig. 5 The participant's assessment of their opponent's identity for the First Person Shooter game. When played in VR, the majority of participants trended towards believing that their opponent was AI controlled (mean score of 3.68 and a mode of 4). However, when playing the game through the monitor, the participants were more likely to report a belief that the opponent was human controlled (mean score of 2.43 and a mode of 1)

We also captured data regarding who won the game, the player or the opponent. In the racing game, a win was recorded if the player completed two laps in the shortest time. In the FPS, a win was recorded if the player successfully destroyed the opponent before they themselves were destroyed.

As can be seen in Fig. 6, the player was not particularly successful in either game (6 wins recorded in the racing game, 1 win in the FPS). We assume that this was because the participants were not provided with the opportunity to practice the game before the study, and conversations with participants after the experiment adds evidence towards this suspicion. However in the racing game, it did appear that the player performed moderately better through VR. We gained further insight through the free text response. In eight of the games, the player reported that the game was easier in VR, with several mentioning that the ability to look around freely was a positive experience.

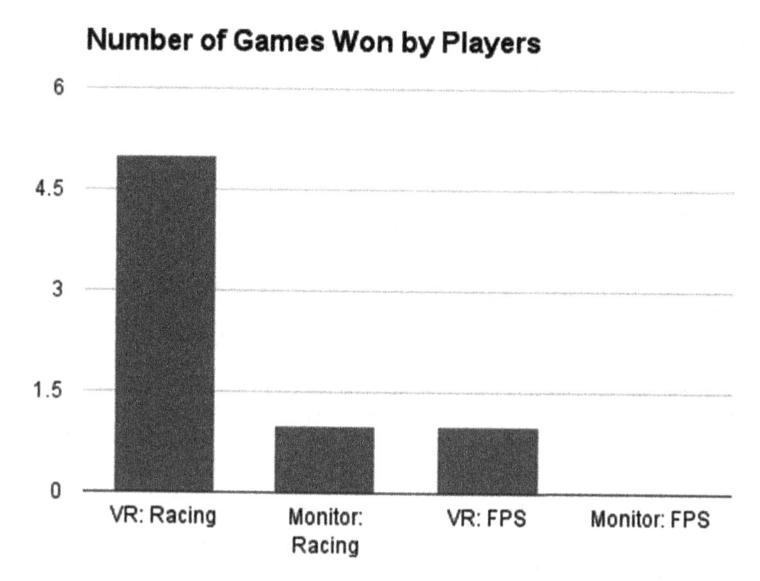

Fig. 6 Number of player wins for each of the four game instances

6 Conclusion and Future Work

The important conclusion to draw from this study is that the level of immersion provided through Virtual Reality appears to clearly impact how we perceive AI characters. Despite the study being undertaken with a relatively small number of participants, the results show a clear split in the player's perception of their in-game opponent.

We anticipated that the results for both game types would be the same, demonstrating that VR either makes AI characters more or less obvious to a human player; clearly, from our results, this is not the case. This study indicates that there is likely a link between the way a game is played, and how VR affects the player's perception of the world. Further studies need to explore this in greater detail. Future work will be to implement a much larger study with more participants engaged in a broader spectrum of games.

Perception through virtual reality could have clear implications for the future development of Artificial Intelligence in games. It appears that VR could have the effect of making AI characters more or less human-like during play, and this will impact how we design them. If prolonged presence is the ultimate goal of VR, it is clear that AI will have a significant role to play.

One further consideration is that the Rift Dk2 hardware does not invoke a sense of presence in most people beyond fleeting moments. It will be interesting to see how hardware capable of deeper and more prolonged presence (like the consumer Rift and HTC Vive) will affect the results.

Acknowledgments The authors would like to thank HPC Wales for providing their facilities and technical support during the running of the experiments described in this research. Chris Headleand would also like to thank Fujitsu for their ongoing financial support.

References

1. Banakou, D., Chorianopoulos, K.: The effects of avatars gender and appearance on social behavior in online 3D virtual worlds. J. Virtual Worlds Res. **2**(5) (2010)
2. Indie Bytes: Vehicle physics toolkit. https://www.assetstore.unity3d.com/en/#!/content/14868. September 2014
3. Davis, S., Nesbitt, K., Nalivaiko, E.: A systematic review of cybersickness. In: Proceedings of the 2014 Conference on Interactive Entertainment, pp. 1–9. ACM (2014)
4. Davis, S., Nesbitt, K., Nalivaiko, E.: Comparing the onset of cybersickness using the oculus rift and two virtual roller coasters. In: Proceedings of the 11th Australasian Conference on Interactive Entertainment (IE 2015), vol. 27, p. 30 (2015)
5. Draper, M.H., Viirre, E.S., Furness, T.A., Gawron, V.J.: Effects of image scale and system time delay on simulator sickness within head-coupled virtual environments. Hum. Factors: J. Hum. Factors Ergon. Soc. **43**(1):129–146 (2001)
6. Headleand, C.J., Henshall, G., Cenydd, L.A., Teahan, W.: Randomised multiconnected environment generator. Technical Report CS-TR-004-2014, Bangor University (2014)
7. Lawrence, J., Hettinger, L.J., Berbaum, K.S., Kennedy, R.S., Dunlap, W.P., Nolan, M.D.: Vection and simulator sickness. Mil. Psychol. **2**(3), 171 (1990)
8. Merritt, T., McGee, K., Chuah, T.L., Ong, C.: Choosing human team-mates: perceived identity as a moderator of player preference and enjoyment. In: Proceedings of the 6th International Conference on Foundations of Digital Games, pp. 196–203. ACM (2011)
9. Merritt, T., Ong, C., Chuah, T.L., McGee, K.: Did you notice? artificial team-mates take risks for players. In: *Intelligent Virtual Agents*, pp. 338–349. Springer (2011)
10. Merritt, T.R., Tan, K.B., Ong, C., Thomas, A., Chuah, T.L., McGee, K.: Are artificial team-mates scapegoats in computer games. In: Proceedings of the ACM 2011 Conference on Computer Supported Cooperative Work, pp. 685–688. ACM (2011)
11. Merritt, T.R.: A Failure of Imagination: How and Why People Respond Differently to Human and Computer Team-Mates. Ph.D. thesis, The National University of Singapore (2012)
12. Nass, C., Steuer, J., Tauber, E.R.: Computers are social actors. In: Proceedings of the SIGCHI Conference on Human Factors in Computing Systems, pp. 72–78. ACM (1994)
13. Nowak, K., Biocca, F.: The effect of the agency and anthropomorphism on users' sense of telepresence, copresence, and social presence in virtual environments. Presence **12**(5), 481–494 (2003)
14. Oculus: Vr best practices guide. http://static.oculus.com/sdk-downloads/documents/Oculus_Best_Practices_Guide.pdf. January 2015
15. Peña, J., Hancock, J.T., Merola, N.A.: The priming effects of avatars in virtual settings. Commun. Res. **36**(6), 838–856 (2009)
16. Regan, C.: An investigation into nausea and other side-effects of head-coupled immersive virtual reality. Virtual Real. **1**(1), 17–31 (1995)
17. Reynolds, C.W.: Steering behaviors for autonomous characters. In: Game Developers Conference, vol. 1999, pp. 763–782 (1999)
18. Schroeder, R.: Social interaction in virtual environments: key issues, common themes, and a framework for research. In: The Social Life of Avatars, pp. 1–18. Springer (2002)
19. Sherrick, B., Hoewe, J., Waddell, T.F.: The role of stereotypical beliefs in gender-based activation of the proteus effect. Comput. Hum. Behav. **38**, 17–24 (2014)
20. So, R.H.Y., Lo, W.T., Ho, A.T.K.: Effects of navigation speed on motion sickness caused by an immersive virtual environment. Hum. Factors: J. Hum. Factors Ergon. Soc. **43**(3), 452–461 (2001)

21. Stoffregen, T.A., Faugloire, E., Yoshida, K., Flanagan, M.B., Merhi, O.: Motion sickness and postural sway in console video games. Hum. Factors: J. Hum. Factors Ergon. Soc. **50**(2), 322–331 (2008)
22. Tanaka, K., Onoue, S., Nakanishi, H., Ishiguro, H.: Motion is enough: how real-time avatars improve distant communication. In: Collaboration Technologies and Systems (CTS), 2013 International Conference on, pp. 465–472. IEEE (2013)
23. Unity Technologies: Car tutorial. https://www.assetstore.unity3d.com/en/#!/content/10. December 2012

Short Papers

Eugene: A Generic Interactive Genetic Algorithm Controller

C. James-Reynolds and E. Currie

Abstract This paper outlines the development of an open source generic hardware-based interactive Genetic Algorithm controller (Eugene) and explores contexts in which it may be deployed. The system was first applied to the generation of synthetic sound using MIDI and a simple analogue synthesiser with 27 continuous controller values. It was then applied in the area of image evaluation using an image enhancer program with 7 continuous controller values. The system was evaluated by experimental observation of users attempting various tasks with different success criteria. This led to the identification of issues, some of which were specific to, and others divorced from the application domain. These are discussed together with areas for improvement.

1 Introduction

This paper describes the development of an embedded interactive Genetic algorithm controller (Eugene) and its application. We begin by outlining the nature of genetic algorithms and interactive genetic algorithms. We then describe the features of Eugene and finally some evaluation of the system and its interaction with users. We conclude with some discussion of the outcomes and suggestions for future work.

C. James-Reynolds (✉) · E. Currie
School of Science and Technology, Middlesex University, London NW4 4BT, UK
e-mail: C.James-Reynolds@mdx.ac.uk

E. Currie
e-mail: E.Currie@mdx.ac.uk

M. Bramer and M. Petridis (eds.), *Research and Development in Intelligent Systems XXXII*, DOI 10.1007/978-3-319-25032-8_27

2 Genetic Algorithms

Genetic algorithms (GAs) [1] are examples of evolutionary algorithms used to implement search and optimisation strategies, which loosely parallel Darwinian evolutionary theory. Some GA approaches try to accurately mirror the corresponding biological process [1]. Others have added new processes to assist the user in either reaching an optimal outcome more quickly, or to provide them with the opportunity to explore the solution space more thoroughly.

The GA process starts with a population of candidate solutions or individuals (chromosomes), randomly chosen from the solution space.

A fitness function is then used to select those individuals whose characteristics are closest to the desired outcome; these become the parents of the next generation. Each member of the next generation "inherits" variables from each of two randomly chosen parents.

The outcome is either that the system converges on a solution which scores above an agreed threshold value of the fitness function, or that a pre-set number of generations have passed. A good example of how this works in practice is the work on antenna design by [2].

3 Interactive Genetic Algorithms

Interactive Genetic Algorithms (iGAs) have been with us since [3]. They are interactive in the sense that a user's judgement replaces the fitness function used in a traditional GA [4]. The concept is appropriate where evaluation of candidate solutions requires a subjective approach that is difficult to automate; for example the generation of identikit pictures [5].

Reference [6] classifies iGA application areas as Artistic/Engineering/ Edutainment, and most published work describes applying iGAs to these areas. There has been some notable work in the area of artistic applications, which often focuses on the media seen as output from the process. [7, 8] are clearly audio focussed, whereas [5, 9] are concerned with image processing and identification. Engineering applications include interface design [10], image retrieval and robotics. Edutainment applications include composition support and games.

Although many iGAs have been used in these areas, there are interesting aspects to be explored. For example, should they be tools for those who wish to explore a solution space, or should they be used when there are clearly defined user goal states? [11] Issues such as the fitness evaluation bottleneck are discussed by [12] who also considers the impact of fatigue.

As [13] points out, there is an assumption that users are consistent in their rating across multiple generations and we do not have information on the impact of user fatigue on consistency. [12] also points out that evaluation is often poorly defined in terms of the goals. A goal state may change as the user is exposed to various individuals during the algorithm run.

4 Eugene

The concept for Eugene was one of a simple open source tool that could be easily adapted to work within a wide range of application areas and for which a simple library could be used to allow exploration of some of the issues above. Issues of particular interest to us were the understanding of goals, consistency, and tradeoffs and the ways in which users interact with an iGA interface.

Eugene was developed using the Arduino Uno. It provides six sliders attached to the analogue inputs, with which, in each successive generation, the user rates the appropriateness of each of six candidate individuals (sounds or pictures). The device has play buttons for each of the six candidates attached to digital i/o pins and an evolve button attached to a digital i/o pin, that generates the next population for consideration. One of three possible mutation values may be chosen through the use of a rotary encoder that uses two digital i/o pins. The system generates MIDI Continuous Controller (CC) data on the Arduino serial output pin which is connected to a USB interface via a standard MIDI DIN socket. All the switches are provided with pull up resistors (Fig. 1).

MIDI CC data consists of 3 bytes, where the first identifies which of the possible 16 MIDI channels are used and the type of message, the second determines the CC number (0–127) and the third the CC value (0–127). In most software (and some hardware) synthesisers, any control (for example modulation or resonance) can be mapped to any CC number.

When being used for sound synthesis, the MIDI data modifies the settings of a software synthesiser with two oscillators, envelope generators and a state variable filter built in SynthEdit. When used for image correction, the data is used to control a simple image enhancement program allowing modification of 7 parameters (red, green, blue, brightness, contrast, sharpen and blur) developed in Processing by means of the MidiBus library. This potentially allows Eugene to control any application developed in Processing.

Setting a given slider at zero causes that individual to be replaced by a new random individual (the 'randomise' enhancement described above), while setting a slider at its maximum position causes that individual to be retained, unchanged, for the next generation (elitism) [8].

Fig. 1 Eugene, showing user interface

On pressing the evolve button, the choice of parents for the next generation uses a roulette wheel approach, where the probability of a given individual being chosen as a parent is proportional to its relative rating by the user.

5 Evaluation

The evaluation was carried out using MIDI to control two very different applications. The first was a 23-parameter model of an analogue synthesiser designed in SynthEdit and the second was a simple image enhancement program built in Processing.

The initial pilot evaluation was carried out with the synthesiser and six users, all of whom had an interest in music or music technology. The users initially played with the controller by experimenting to find sounds they liked. They were then given a target sound and asked to reproduce the sound with Eugene.

One intended limitation of Eugene is that it only triggers the playback of a single note (middle C; MIDI note 60). This simplifies the selection process.

A similar pilot evaluation was performed with the image enhancer program using two users, to identify any major issues. Users were asked to experiment with the controller and then to enhance an image with poor colour, balance and contrast, which was slightly out of focus.

The iGA functioned effectively as a tool for users to explore both search spaces by experimenting with the controller. When given a specific task, users were often quick to eliminate solutions that sounded or looked different from their target. With the synthesiser, they found that it was possible to get close to the desired sounds, but more difficult to get an exact result. With images, both users felt that the initial population was often far removed from their requirements for image enhancement, with images having extreme effects. The users were using the Randomise feature extensively in order to get what they felt was a good starting population and they thought the task would have been far easier in a simple image editor. They did feel that using the system to explore the possibilities was interesting, but that it was of limited practical use without more features.

From the MIDI data logs for the users of the controller with the synthesiser, it could be seen that in their search for a specific sound, users were using Randomise to discard individuals, despite those individuals having variables that were very close to that of the target sound.

An experiment was then conducted to test observations made with subjects using Eugene. Twenty-five subjects were asked to compare six sounds to a reference sound. The sounds were chosen to provide different amounts of variation in the MIDI CC parameters used to generate them. One sound provided a control and was the same as the reference.

The subjects rated each of the sounds for similarity to the reference sound on a scale of 0–10, where 0 indicated little similarity to the reference and 10 meant that the sound was almost identical to the reference. The findings were consistent, other than three outliers that distorted the results a little.

The findings showed that modulation of the oscillators (Vibrato) was a change that resulted in very low similarity to the reference, even where other parameters were identical. Changes in waveshape (Tone) produced very high similarity to the reference where dynamics (changes in loudness) remained similar, and the substitution of one of the oscillators for white noise produced the results with the greatest variance, where users seemed to either hear the sound through the noise and rated it as highly similar to the reference, or they decided that the noise made the sound so unlike the reference that they rated it with a low similarity value.

In terms of the MIDI CC data used to create the sounds, this showed that changes in one variable (e.g. modulation depth or waveshape) can have a more significant impact on users' decisions than larger changes in other variables.

6 Discussion

When we are working with variables that are not continuous, but discrete such as waveform selection in a synthesiser, an individual candidate may be assigned a low score as a result of a single parameter being out by 1 bit. We note that the perception of 'distance' from the desired target is highly sensitive to changes in such parameters.

For control parameters such as oscillator modulation, a low value would probably be desirable if a musical context is sought. For example, vibrato requires a subtle amount of frequency modulation at a few Hertz. This would not necessarily apply if we were looking for a wider gamut of sounds; for example siren effects.

An important conclusion from the observation of the above two scenarios is that the search space and users' cognitive understanding of that search space do not necessarily match and where this occurs there will be a slower convergence on an acceptable solution.

Given that users sometimes discard candidate solutions where the difference in certain parameters is small, we could modify the encoding of the solution space to better match the users' cognitive understanding of the domain. This would allow users to make better decisions and become a more effective fitness function.

This could be resolved to a certain extent by scaling parameters. For example, using a log curve would allow more of the available range of the parameter to be used for smaller changes. This would also, in the case of the image processing software, produce an initial population more suited to the task.

There is also a good argument for limiting the solution space for certain types of application, for example cleanup of images and creative image processing have different solution spaces that may overlap.

The Eugene controller, whilst successful, has a small initial population size and the number of new children is reduced when we retain individuals or remove them (by re-randomising). This reduced amount of available offspring might limit the effectiveness of the algorithm to converge quickly, although our experiments indicate that it does seem to be effective in helping explore the solution space.

7 Conclusion

We have successfully developed a generic hardware iGA controller (Eugene) and applied it to two different domains. The controller facilitated the exploration of the solution spaces effectively, although the simple image processing application did not have as large a solution space as users would have liked. There were however, lessons to be learned from the use of the controller for solving a given task involving a reference target sound or an "ideal" target image.

Many existing iGA approaches have used additional features to enhance the iGA to make it effective for the application domain; future work will incorporate these in addition to supporting a larger initial population. This will provide a better tool for further work.

References

1. Holland, J.H.: Adaptation in Natural and Artificial Systems. University of Michigan Press, Ann Arbor (1975)
2. Hornby, G.S., Globus, A., Linden, D.S., Lohn, J.D.: Automated antenna design with evolutionary algorithms. In: AIAA Space, pp. 19–21 (2006)
3. Todd, S., Latham, W.: Evolutionary Art and Computers. Academic Press (1992)
4. Biles, J.: Genjam: a genetic algorithm for generating jazz solos. In: Proceedings of the International Computer Music Conference (1994)
5. Frowd, C.D., Hancock, P.J.B., Bruce, V., McIntyre, A.H., Pitchford, M., Atkins, R., Webster, A., Pollard, J., Hunt, B., Price, E., Morgan, S., Stoika, A., Dughila, R., Maftei, S., Sendrea, G.: Giving Crime the "evo": Catching Criminals Using EvoFIT Facial Composites, pp. 36–43. IEEE (2010). doi:10.1109/EST.2010.38
6. Cho, S.-B.: Towards creative evolutionary systems with interactive genetic algorithm. Appl. Intell. **16**, 129–138 (2002)
7. Biles, J.A., Eign, W.G.: GenJam Populi: Training an IGA via Audience-Mediated Performance, pp. 347–348. San Francisco, USA (1995)
8. Johnson, C.G.: Exploring the sound-space of synthesis algorithms using interactive genetic algorithms. In: Proceedings of the AISB'99 Symposium on Musical Creativity. Society for the Study of Artificial Intelligence and Simulation of Behaviour, pp. 20–27 (1999)
9. Takagi, H., Cho, S.-B., Noda, T.: Evaluation of an IGA-based image retrieval system using wavelet coefficients. In: 1999 IEEE International Fuzzy Systems Conference Proceedings. Presented at the Fuzzy Systems Conference Proceedings, 1999. FUZZ-IEEE'99, pp. 1775–1780. IEEE (1999)
10. Masui, T.: Graphic object layout with interactive genetic algorithms. In: Proceedings of the 1992 IEEE Workshop on Visual Languages, pp. 74–80. IEEE (1992)
11. Collomosse, J.P.: Supervised Genetic Search for Parameter Selection in Painterly Rendering. EvoWorkshops **2006**, 599–610 (2006)
12. McDermott, J., O'Neill, M., Griffith, N.J.: Interactive EC control of synthesized timbre. Evol. Comput. **18**, 277–303 (2010)
13. Bauerly, M., Liu, Y.: Evaluation and improvement of interface aesthetics with an interactive genetic algorithm. Int. J. Human-Comput. Interact. **25**, 155–166 (2009). doi:10.1080/10447310802629801

Automatically Geotagging Articles in the Welsh Newspapers Online Collection

Sean Sapstead, Illtud Daniel and Amanda Clare

Abstract The National Library of Wales' Welsh Newspapers Online collection comprises over 16 million articles from historic newspapers. It is stored in NLW's institutional repository, and is a rich source of historic text. The text of the articles has been extracted from the digitised images using OCR. This project investigates methods of determining which articles can be automatically located to places within Wales. We use machine learning, text mining and the OpenStreetMap data as a gazetteer.

1 Introduction

In 2009, the National Library of Wales started a project to digitise their collection of historic Welsh newspapers; 1.2 million pages have been digitised to date. Conceptually, the collection comprises 238 newspaper titles, each of which has an object representation in the repository and associated metadata describing the title. Each title object has a sequence of issue object children, which have an associated date. Each issue has a sequence of page objects, which have an image and OCR data. The OCR data includes the raw recognised text and layout of the page. Additional semi-automated work was outsourced to a third party to add articles, grouping together text blocks from the pages of the issue with very minimal metadata to create an article representation.

General access to the collection is via a public website with full-text and faceted search that allows users to discover articles. The data from the title, issue, page and

S. Sapstead (✉) · A. Clare
Department of Computer Science, Aberystwyth University,
Aberystwyth SY23 3DB, Wales, UK
e-mail: sean.sapstead@me.com

A. Clare
e-mail: afc@aber.ac.uk

I. Daniel
National Library of Wales, Aberystwyth SY23 3DB, Wales, UK
e-mail: illtud.daniel@llgc.org.uk

M. Bramer and M. Petridis (eds.), *Research and Development in Intelligent Systems XXXII*, DOI 10.1007/978-3-319-25032-8_28

article representations in the repository is flattened into XML records for each article which are indexed and stored in Apache Solr. The data used in this project is the XML article data as stored in Solr.

2 Metadata and Geotagging

The descriptive metadata for each article XML document is brief, some of it inherited from the title object, so may be common to many millions of articles. Essentially, all the descriptive metadata tells us about the article in particular is on what date it was published, a generalised manually-added categorisation (advertising, news, family notices or lists), the article title and the article text as extracted by OCR. An example XML document can be seen in Fig. 1.

Enriching the article metadata by extracting information from the text would enable users and researchers to discover the content they need much more efficiently, adding significantly to the value of the resource. Given that the collection currently stands at over 16 million articles, doing this kind of work by hand would require an unrealistic investment.

An example of useful information that is contained in the text of the article is the location of the events being reported, extracting that information into metadata would effectively geotag the articles, allowing not only additional weighting of results based on hits on the tag value, but location-based searching such as bounding box, proximity or geographic region searches.


```
<?xml version="1.0"?>
<doc>
  <str name="PID">llgc-id:3305977</str>
  <str name="Region">English newspapers-Wales, Mid.</str>
  <str name="ArticleID">3305977-modsarticle31-3305981-1</str>
  <str name="ArtPointer">ART31</str>
  <str name="ArticleTitle">ARTHOG.</str>
  <str name="ArticleSubject">News</str>
  <str name="ArticleAbstract">OPENING OF A NEW CHAPEL. - On Wednesday
  and Thursday, the 15th and 16th instant, the new Wesleyan Chapel in this
  place was formally opened. The services were conducted by the Revs. Samuel
  Davies, Carnarvon ;</str>
  <int name="ArticleWordCount">79</int>
  <str name="ArticleText">ARTHOG. OPENING OF A NEW CHAPEL. On Wednesday
  and Thursday, the 15th and 16th instant, the new Wesleyan Chapel in this
  place was formally opened. The services were conducted by the Revs. Samuel
  Davies, Carnarvon William Davies, Bangor; and Hugh Hughes, Cefn- mawr; and
  eloquent sermons were preached to large con- gregations. On the afternoon of
  Thursday a tea party was held, for which some hundreds of ls. tickets
  were sold. The proceeds were devoted towards the chapel fund.</str>
  <str name="PageLabel">[4]</str>
  <str name="PagePID">llgc-id:3305981</str>
  <str name="PublicationTitle">Cambrian News and Merionethshire Standard</str>
  <date name="IssueDate">1869-09-25T00:01:00Z</date>
  <str name="PublicationPID">llgc-id:3305806</str>
</doc>
```


Fig. 1 An example article in XML format (some technical fields removed for readability)

Geotagging of text has been investigated by many researchers, though generally as a problem of Named Entity Recognition (to determine how to identify place names in the text) rather than as a problem of relevance (to determine whether the place names identified are relevant as locations for the text). Sultanik and Fink [8] investigated geotagging for Twitter articles using fuzzy phonetic matching to a gazetteer to identify places and then using the geographic context of the Twitter feed to disambiguate locations. Web-a-where [1] is a solution for geotagging web pages. The Web-a-where system used a focus scoring algorithm that took into account the multiple places mentioned in the web page to come to a conclusion. Lieberman [6] used the extra context provided by comma separated lists to help to discover place names. Buscaldi and Rosso [3] use a georeferenced version of WordNet as a structured gazetteer, which provided lat/long information and part-of relationships between geographical entities (Cardiff is part of Wales, which is part of Europe).

Leidner and Lieberman [5] discuss three different methods for detecting place names in articles: gazetteer lookup, symbolic rules and a machine learning method. In all three methods they only address the question of place name detection, rather than place name relevance. As an example, an article that mentions *The Duchess of York* includes a place name but refers to a person, not a place. In our example article in Fig. 1 we can see that place name detection would extract *Arthog*, *Carnarvon*, *Cefn-mawr* and *Bangor*. The article describes events in *Arthog*. The other place names are included to identify the individuals mentioned in the article.

We use a gazetteer to discover place names, but then add a machine learning process to intelligently and automatically decide whether the place names are pertinent to the location of the article.

3 Methods

Gazetteers are lists of places, usually containing some geographical information about those places. We used an export from OpenStreetMap (OSM) [4], a collaborative project to make geographical data more accessible. They release their data under the ODbL (Open Data Commons Open Database License). We obtained a file containing the latest OSM data for Wales (*wales-latest.osm.bz2* [13/03/2015]) and extracted place names from this data.

We then need to analyse the text of the articles. The Natural Language Toolkit (NLTK) [2] is a library for Python which gives access to multiple features including part of speech tagging, sentence splitting and extracting sentence structure. We selected attributes from the text to represent the context of the place name in the sentence. We considered the part of speech tag of the words immediately before and after the place name, including verb, noun, pronoun, adjective, adverb, adposition, conjunctive, determiner and punctuation. We also recorded the presence of specific words such as *in*, *near*, *the*, *of* and *at* before the place name. Positional features of the place name were also selected, including presence at start of sentence or end of sentence, the length of the sentence, the sentence number in the article and the word number of the place name within the sentence.

We have a binary classification problem: to determine for each place name whether or not this refers to the location of the current article. The classification methods that we chose are decision trees, random forests and support vector machines, with implementations as provided in the scikit-learn library for Python [7].

We manually labelled articles to create a gold standard for evaluation. The labelling decision could be made in a number of ways, depending on the desired outcome. We might be asking one of two questions:

1. asking the user to label whether or not this sentence or article discusses this place
2. asking the user to label whether or not this article should be pinned at this location on a browsable map of articles.

The nuance of the question will lead to articles being labelled differently, and could make the machine learning task more difficult or more achievable. The article may mention a town, but be only tangentially referring to the town. For example, an article that mentions that a politician travelled from Tregaron to Cardiff in order to attend some important meeting, could be pinned at both Tregaron and Cardiff on a browsable map of goings-on in the 1800s, but is not actually discussing Tregaron.

We therefore labelled two sets of articles, one set for question 1 and one set for question 2. We shall henceforth refer to these two sets of labelled data as 'Discussed' and 'Pinned' respectively. For the *Discussed* data set we labelled a corpus containing 287 place names (97 True and 190 False). For the *Pinned* data set we labelled a corpus of 238 labelled place names (84 True, 154 False).

4 Results

Ten-fold cross validation was used to measure how accurately each of the classifiers could classify place names. To ensure it was possible to directly compare each of the classifiers the labelled data was split prior to classification and the same data split was used consistently for all three of the classifiers.

Table 1 shows the range and mean for the accuracy of each of the classifiers across each fold of the cross validation, for the two data sets. Random forest performs best across the three classifiers, and SVM worst.

Table 1 A comparison of the cross validation accuracy across the classifiers

Data	Statistic	Decision_Tree (%)	Random_Forest (%)	SVM (%)
Discussed	Range	16.6	17.8	23.4
	Mean	76.3	77.0	66.6
Pinned	Range	26.1	30.4	24.0
	Average	68.8	74.4	61.7

Table 2 The confusion matrices, showing True Negative (TP), False Positive (FP), False Negative (FN) and True Positive (TP) results, along with precision and recall

Data	Classifier	TN	FP	FN	TP	Precision	Recall
Discussed	Decision tree	166	24	44	53	0.688	0.546
	Random forest	183	7	59	38	0.844	0.392
	SVM	168	22	74	23	0.511	0.237
Pinned	Decision tree	121	33	41	43	0.566	0.512
	Random forest	143	11	50	34	0.756	0.405
	SVM	128	26	65	19	0.422	0.226

Table 2 shows the confusion matrices for the classifiers, along with precision and recall measures of their accuracy. They highlight that Random Forest has the best precision rate and Decision Tree has the best Recall rate. Random Forests label the fewest false positives. For a scenario where articles are to be pinned to a map, it is likely to be preferable that we maximise recall so as to create a map with the most potential for public browsing of the articles.

Decision trees have the benefit that their decisions can be visualized. We inspected the trees produced for each data set. Both trees use *at_before, the_before, length_of_sentence, sentence_number* and *word_number*. The tree for the *Discussed* data also uses *in_before, start_of_sentence* and *end_of_sentence* whereas the tree for the *Pinned data* also uses *adposition_before* and *of_before* instead. The root of the tree in both cases uses *at_before*, which is expected and reassuring. Decisions for the Pinned data indicate that being the first word of the first sentence is important and also that a town named very late in a sentence is unlikely to be relevant.

5 Conclusions

We have described and evaluated a method for automatically geotagging articles from the Welsh Newspapers Online collection. This combined the use of a gazetteer extracted from OpenStreetMap with machine learning based on the immediate textual context of the place name.

Random forests proved to be the better classifiers for this problem, though their results were still disappointing and show that there are more challenges to overcome. It is important to remember that random forests take far longer than decision trees during training and classification. The examples shown in this paper are only a very small subset of the 16 million articles that the Library has in its digitised collection.

There are several ways in which this work can be improved and extended:

- **Larger data sample**—The amount of data that has been labelled is a very small subset of the available data. Another source of concern is that the two classes are not evenly balanced, with twice as many negative examples as positive examples. Much more data labelling is required to be able to cover more of the possible range of articles and the variation that is present in their attributes. We now plan to investigate crowdsourcing the labels from a wide user group.
- **Alternative attributes**—the features that we use could be improved with the addition of bigrams and trigrams (multi-word combinations), for example, *of the*. We note that this was a common occurrence before a place name. Sentence structure could also be taken into account using more of the tools available in NLTK including syntactic structure and parse tree structure.
- **Better handling of OCR errors**—The OCR process has errors which mask place names or corrupt the surrounding words. Some further correction of common OCR errors could be investigated as part of a pre-processing phase.
- **Historical and variant spelling gazetteers**—Place names vary over time with the spellings of some town and city names changing. The newspapers collection dates back to the 1840 s and it is known that numerous places today had different and varied spellings during this period (e.g. the modern-day town of *Caernarfon* is referred to as *Carnarvon* 186,000 times). Building a better gazetteer would allow us to add more metadata to more articles.

Previous research has concentrated on finding place names in text with very few methods for understanding the context of the text around the place name. This work investigates methods to determine the relevance of any extracted place names from within an article. Our code can be found at https://github.com/ssapstead/geo-locating-articles. We are continuing to improve this work with the aim to use it to add metadata to the full 16 million articles that the library holds and also to any new articles as they are added to the digital collection.

Acknowledgments We would like to thank Aberystwyth University and the National Library of Wales for their support.

References

1. Amitay, E., Har'El, N., Sivan, R., Soffer, A.: Web-a-where: Geotagging web content. In: Proceedings of SIGIR'04, pp. 273–280 (2004)
2. Bird, S.: Nltk: The natural language toolkit. In: Proceedings of the COLING/ACL on Interactive Presentation Sessions, pp. 69–72 (2006)
3. Buscaldi, D., Rosso, P.: Map-based versus knowledge-based toponym disambiguation. In: Proceedings of GIR'08, pp. 19–22 (2008)
4. Haklay, M., Weber, P.: OpenStreetMap: user-generated street maps. IEEE Pervasive Comput. **7**(4), 12–18 (2008)

5. Leidner, J.L., Lieberman, M.D.: Detecting geographical references in the form of place names and associated spatial natural language. SIGSPATIAL Spec. **3**(2), 5–11 (2011)
6. Lieberman, M.D., Samet, H., Sankaranayananan, J.: Geotagging: using proximity, sibling, and prominence clues to understand comma groups. In: GIR'10, pp. 6:1–6:8 (2010)
7. Pedregosa, F., et al.: Scikit-learn: machine learning in python. J. Mach. Learn. Res. **12**, 2825–2830 (2011)
8. Sultanik, E.A., Fink, C.: Rapid geotagging and disambiguation of social media text via an indexed gazetteer. Proc. ISCRAM **12**, 1–10 (2012)

Contextual Sequential Pattern Mining in Games: Rock, Paper, Scissors, Lizard, Spock

Julien Dumartinet, Gaël Foppolo, Loïc Forthoffer, Pierre Marais, Madalina Croitoru and Julien Rabatel

Abstract Traditional sequential pattern algorithms do not usually take consideration of contextual information commonly associated with sequential data. Our project is based on the non-cooperative game *Rock Paper Scissors Lizard Spock* by associating the age and sex of the player in order to bring to light the hidden correlations in the data or general gaming trends. Considering the hypothesis that a sequential pattern is specific to a particular sex and age, we propose to extract patterns of the form: *after playing figure A followed by figure B, men aged between 18 and 22 years tend to play figure C.*

1 Introduction

Rock-paper-scissors is a game usually played between two people, in which each player simultaneously forms one of three shapes with an outstretched hand. The game is often used as a choosing method in a way similar to coin flipping, drawing straws, or throwing dice. Unlike truly random selection methods, however, rock-paper-scissors can be played with a degree of skill by recognizing and exploiting non-random behavior in opponents [7].

Many strong algorithms have emerged in recent literature addressing this game. The main strategies it employs are history matching, frequency analysis, and random guessing [7]. History matching searches for a sequence in the past that matches the last few moves in order to predict the next move of the algorithm. In frequency analysis, the program simply identifies the most frequently played move. However, none of these algorithms address the contextual information of the player such as age or sex. In this paper we propose *the first algorithm in the literature that addresses contextual information for such game.* Our algorithm has been implemented and tested in an online game where the emergent patterns were used as game strategies.

A *second contribution* of the paper deals with the investigation of an *extended rock paper scissors game* and namely rock-paper-scissors-lizard-spock. Created by

J. Dumartinet · G. Foppolo · L. Forthoffer · P. Marais · M. Croitoru (✉) · J. Rabatel
IUT Informatique University of Montpellier, Montpellier, France
e-mail: croitoru@lirmm.fr

M. Bramer and M. Petridis (eds.), *Research and Development in Intelligent Systems XXXII*, DOI 10.1007/978-3-319-25032-8_29

Sam Kass and popularized by the American series The Big Bang Theory, it adds two new possibilities to the game: the lizard and Spock. This has the effect of bringing the probability of a draw to $^1/_5$ (as opposed to $^1/_3$ in the original game). Spock smashes scissors and vaporizes rock; he is poisoned by lizard and disproven by paper. Lizard poisons Spock and eats paper; it is crushed by rock and decapitated by scissors.

This work is interested in the discovery of contextual patterns in sequences of game moves. This allows to highlight hidden correlations in the data or general game playing trends [5]. Starting from the general idea that a contextual pattern will represent expected behaviour, the two research questions we ask in this paper are:

- Can we predict the next move to be played in the game by a specific player profile?
- Can we deduce a playing trend from a set of game moves of similar profile players?

In order to answer these two research questions we implemented a web interface allowing two players to connect and play the game. This is described in Sect. 2. Based on the data collected by the means of this interface, we have used contextual pattern mining (described in Sect. 3) in order to mine frequent contextual patterns. Preliminary results are described in Sect. 4 that concludes the paper.

2 The Game Interface

The game website must provide two key features: the possibility to play against another player and play against the system. The last functionality is achieved by learning the way user profiles play and using these statistics against them. The main system must interact with the player to get the moves he plays, transmit in to the database and compute the contextual sequential patterns. It must also manage cases of errors when attempting to access an unauthorized part of the application and synchronization with the database to be always in a consistent state.

The game website is available at http://www.pfcls.me. Figure 1 displays the different choices the players have when playing against either another player or the system. At the end of the game round the users can choose to see their profile as depicted in Fig. 2.

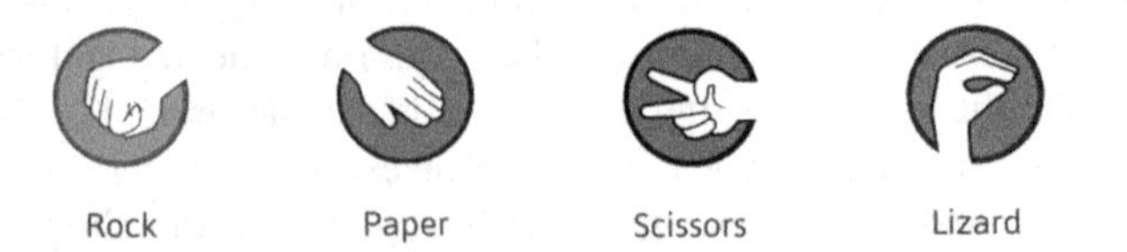

Fig. 1 Interface for playing the game

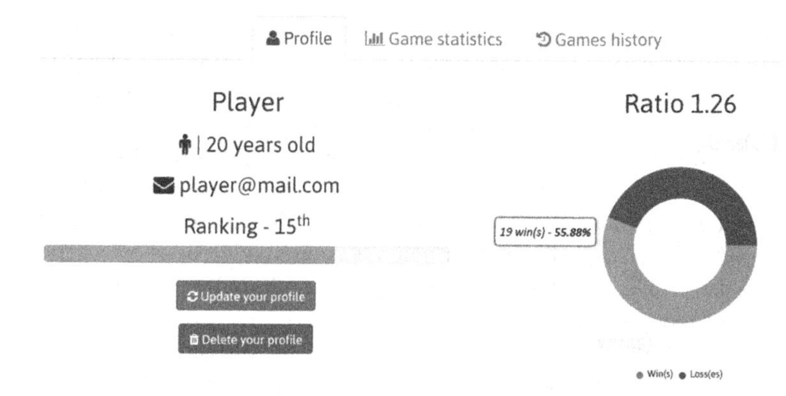

Fig. 2 Interface for visualising profile

The next section will explain the theoretical foundations of using data collected via the website in order to learn move trends in playing the game. We will introduce the notion of contextual sequential pattern mining and explain the intuition of the algorithm used in this paper.

3 Formal Background

Sequential patterns were introduced in [2] and can be considered as an extension of the concept of frequent itemset [1] by handling timestamps associated to items. Sequential pattern mining aims at extracting sets of items commonly associated over time. In the "basket market" scenario, a sequential pattern could be: "40 % of the customers buy a television, then buy later a DVD player". The problem of mining all sequential patterns in a sequence database is defined as follows. Let $\mathscr{X}$ be a set of distinct ***items***. An ***itemset*** is a subset of items, denoted by $I = (i_1 i_2 \dots i_n)$, i.e., for $1 \leq j \leq n$, $i_j \in \mathscr{X}$. A ***sequence*** is an ordered list of itemsets, denoted by $\langle I_1 I_2 \dots I_k \rangle$, where $I_i \subseteq \mathscr{X}$ for $1 \leq i \leq n$.

Let $s = \langle I_1 I_2 \dots I_m \rangle$ and $s' = \langle I'_1 I'_2 \dots I'_n \rangle$ two sequences. The sequence s is a ***subsequence*** of s', denoted by $s \sqsubseteq s'$, if $\exists i_1, i_2, \dots i_m$ with $1 \leq i_1 < i_2 < \dots < i_m \leq n$ such that $I_1 \subseteq I'_{i_1}, I_2 \subseteq I'_{i_2}, \dots, I_m \subseteq I'_{i_m}$. If $s \sqsubseteq s'$ we also say that s' ***supports*** s.

A ***sequence database*** $\mathscr{D}$ is a relation $\mathscr{R}(ID, S)$, where an element $id \in dom(ID)$ is a sequence identifier, and $dom(S)$ is a set of sequences. The ***size*** of $\mathscr{D}$, denoted by $|\mathscr{D}|$, is the number of tuples in $\mathscr{D}$.

A tuple $\prec id, s \succ$ is said to *support* a sequence α if α is a subsequence of s, i.e., $\alpha \sqsubseteq s$. The ***support*** of a sequence α in the sequence database $\mathscr{D}$ is the number of tuples in $\mathscr{D}$ supporting α, i.e., $sup_{\mathscr{D}}(\alpha) = |\{\prec id, s \succ | \, (\prec id, s \succ \in \mathscr{D}) \wedge (\alpha \sqsubseteq s)\}|$.

Given a real *minSup* such that $0 < minSup \leq 1$ as the ***minimum support threshold***, a sequence α is ***frequent*** in the sequence database $\mathscr{D}$ if the proportion of tuples in $\mathscr{D}$ supporting α is greater than or equal to *minSup*, i.e., $sup_{\mathscr{D}}(\alpha) \geq minSup \times |\mathscr{D}|$. In this case, sequence α is also called a ***sequential pattern*** in $\mathscr{D}$.

However, as also highlighted by our paper, data are very often provided with additional information, such as the age or gender. Traditional sequential patterns do not take into account this information. Having a better knowledge about the features of objects supporting a given behavior can help decision making as shown here. The set of such descriptive information about objects is referred to as contextual information and formalised as follows.

We define a ***contextual sequence database*** $\mathscr{CD}$ as a relation $\mathscr{R}(ID, S, D_1, \ldots D_n)$, where $dom(S)$ is a set of sequences and $dom(D_i)$ for $1 \leq i \leq n$ is the set of all possible values for D_i. $D_1, D_2, \ldots D_n$ are called the ***contextual dimensions*** in $\mathscr{CD}$. A *tuple* $u \in \mathscr{CD}$ is denoted by $\prec id, s, d_1, \ldots, d_n \succ$.

Values on contextual dimensions can be organized as hierarchies. For $1 \leq i \leq n$, $dom(D_i)$ can be extended to $dom'(D_i)$, where $dom(D_i) \subseteq dom'(D_i)$. Let $\subseteq_{D_i}$ be a partial order such that $dom(D_i)$ is the set of minimal elements of $dom'(D_i)$ with respect to $\subseteq_{D_i}$. Then, the partially ordered set $(dom'(D_i), \subseteq_{D_i})$ is the ***hierarchy on dimension*** D_i, denoted by $\mathscr{H}_{D_i}$.

A ***context*** c in $\mathscr{CD}$ is denoted by $[d_1, \ldots d_n]$ where $d_i \in dom'(D_i)$. If, for $1 \leq i \leq n$, $d_i \in dom(D_i)$, then c is called a ***minimal context***.

Let c_1 and c_2 be two contexts in $\mathscr{CD}$, such that $c_1 = [d_1^1, \ldots, d_n^1]$ and $c_2 = [d_1^2, \ldots d_n^2]$. Then $c_1 \leq c_2$ iff $\forall i$ with $1 \leq i \leq n, d_i^1 \subseteq_{D_i} d_i^2$. Moreover, if $\exists i$ with $1 \leq i \leq n$ such that $d_i^1 \subset_{D_i} d_i^2$, then $c_1 < c_2$. In this case, c_1 is said then to be ***more specific*** than c_2, and c_2 is ***more general*** than c_1. The set of all contexts associated with the partial order $\leq$ is called the ***context hierarchy*** and denoted by $\mathscr{H}$.

Let us now consider the tuples $u = \prec id, s, d_1, \ldots d_n \succ$ of $\mathscr{CD}$ according to contexts defined above. The context $c = [d_1, \ldots d_n]$ is called the ***context of*** u. Note that the context of u is minimal ($\forall i$ with $1 \leq i \leq n$, $d_i \in dom(D_i)$). Let u be a tuple in $\mathscr{CD}$ and c the context of u. For all contexts c' such that $c' \geq c$ we say that c' ***contains*** u (and u is contained by c'). Let c be a context (not necessarily minimal) in $\mathscr{CD}$. The ***sequence database of*** c, denoted by $\mathscr{D}(c)$, is the set of tuples contained by c.

In the following, we consider the context c and the sequence s. Sequence s is **frequent in** c (c-frequent) iff s is frequent in $\mathscr{D}(c)$, i.e., if $sup_{\mathscr{D}(c)}(s) \geq minSup \times |c|$. We also say that s is a **sequential pattern in the context** c.

We can now focus on mining sequential patterns that, regarding contextual information, are of interest in a given context. The algorithm employed in this paper consists of mining several contextual levels: search in the player's own game data, search the player dataset with the same sex and close age (with a margin of plus or minus two years), search the players of close age but of the opposite sex, search in the players dataset with the same sex and not so close age (with a margin more or less five years), search in the dataset players of age with a margin of plus or minus five years but of opposite sex. Preliminary results are explained in the next section.

4 Discussion

4.1 Preliminary Results

Having chosen the age and gender as contextual data here are our results. We take as an example a man between 18 and 22 years old (see Fig. 3).

Note that after playing a lizard young men tend to replay again the move lizard. However, after playing Spock, each move is played roughly equal. Our plan results go even further and we can extract more complicated patterns. Taking as a criterion of age and sex, we are able to define a move trend, whether individual or general.

Of course, after a while the system always wins (due to extensive training on the data). While this makes the game less fun for human users after a certain while, it achieves the purpose of validating the research question we had and namely to learn how human users play. It could be used, for instance in the famous auction dilemma where Takashi Hashiyama, president of Maspro Denkoh Corporation, an electronics company based outside of Nagoya, Japan, could not decide whether Christie's or Sotheby's should sell the company's art collection, which is worth more than 20 million dollars. The two auction houses were informed that they will be playing rock paper scissors to decide the winner. Such website could provide interesting insights into the game for further such auctions (and not only).

4.2 Related Work

The first approach that aims at extracting sequential patterns being dependent on more or less general contexts has been proposed in [6]. However, this work focuses on mining sequential patterns in specific case of an online game. The contribution of the paper lies in the use of such algorithms in order to address the game playing problem and not in improving the theoretical foundations of such patterns. The

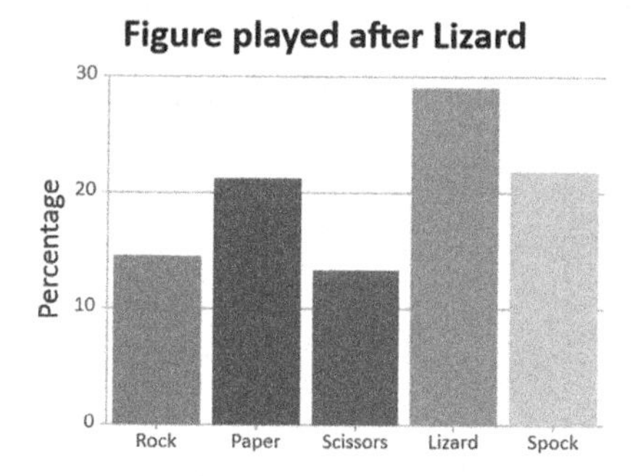

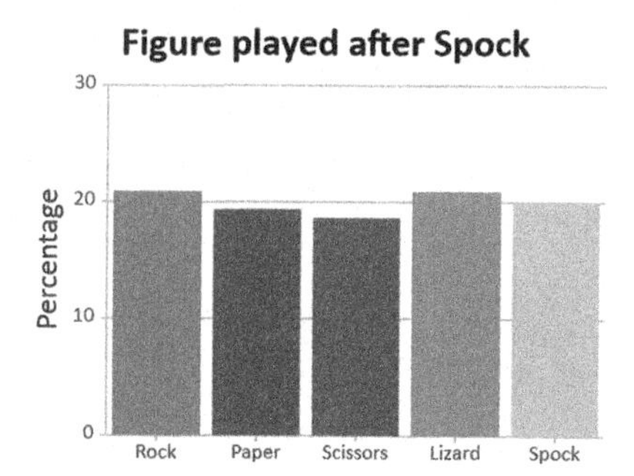

Fig. 3 Figures played after lizard and Spock by men between 18 and 22 years old

choice of sequential pattern mining with contexts is motivated by the nature of the data the game naturally offers. Of course alternative solutions can be investigated from the literature. For instance, a closely related problem is the problem of mining multidimensional sequential patterns, i.e., extracting sequential patterns dealing with several dimensions. The first proposition is described in [3], where sequential patterns are extracted in data similar to the contextual data considered in our approach.

However, while multidimensional sequential patterns consider contextual information, they associate a context to a sequential pattern only if this association is globally frequent in the whole database. Moreover, multidimensional sequential patterns do not consider the representativity in their corresponding context. As a consequence, a sequential pattern which is not frequent in the whole database can not be extracted, even if it is very frequent in a sub-category only. Other approaches have considered more complex multidimensional sequence databases, e.g., where items are also described over several dimensions [4], but the same principle is used to extract such patterns. The same remark can also be mentioned about [8], in spite of the rich definition of context proposed.

References

1. Agrawal, R., Imieliński, T., Swami, A.: Mining association rules between sets of items in large databases. SIGMOD Rec. 22(2) (1993)
2. Agrawal, R., Srikant, R.: Mining sequential patterns. In: Yu, P.S., Chen, A.S.P. (eds.) Eleventh International Conference on Data Engineering. IEEE Computer Society Press (1995)
3. Pinto, H., Han, J., Pei, J., Wang, K., Chen, Q., Dayal, U.: Multi-dimensional sequential pattern mining. In; Proceedings of the Tenth International Conference on Information and Knowledge Management. ACM (2001)
4. Plantevit, M., Choong, Y.W., Laurent, A., Laurent, D., Teisseire, M.: M^2SP: mining sequential patterns among several dimensions. In: Jorge, A., Torgo, L., Brazdil, P., Camacho, R., Gama, J. (eds.) PKDD. Lecture Notes in Computer Science, vol. 3721. Springer (2005)
5. Rabatel, J.: Extraction de motifs contextuels: enjeux et applications dans les donnes squentielles. Ph.D. thesis, Universit Montpellier II (2011)
6. Rabatel, J., Bringay, S., Poncelet, P.: Contextual sequential pattern mining. In: 2010 IEEE International Conference on Data Mining Workshops, pp. 981–988. IEEE (2010)
7. Wang, Z., Xu, B., Zhou, H.-J.: Social cycling and conditional responses in the rock-paper-scissors game. Nature (2014)
8. Ziembiński, R.: Algorithms for context based sequential pattern mining. Fundamenta Informaticae **76**(4), 495–510 (2007)

On the Way Towards Automated Planning and Budgeting: Combining Fuzzy Planning with CBR Learning Systems

Miriam Gebauer, Peter Rausch and Michael Stumpf

Abstract In this paper we present a novel hybrid approach to improve the complex process of planning and budgeting (P&B). Especially in the field of fuzzy P&B a lot of expert knowledge is needed to find an appropriate solution for planning problems. To support decision makers we consider a combination of a fuzzy P&B tool with a case-based reasoning (CBR) system. The novel hybrid approach will be presented, and it will be analysed in which way decision making and learning processes can be improved. Due to the promising results, we also discuss possible extensions to automate P&B processes completely.

1 Issues and Challenges of Planning and Budgeting

In this research we present a novel hybrid approach to combine the strengths of fuzzy planning with the benefits of CBR learning systems. Recent research indicates that planning and budgeting (P&B) causes a lot of effort and is a challenging task for many organisations. Due to lengthy planning cycles final budgets often fall out of step with quickly evolving business conditions [5]. Although fuzzy approaches reduce these issues a lot, expert knowledge is still needed to understand the impact of possible P&B parameter configurations, and gathering all relevant information is a challenge. At the focussed organisation an interactive fuzzy optimisation approach which is described in [6, 7] is applied. So, the idea came up to create a learning system. A well established approach for learning systems is case-based reasoning (CBR). It is used in various domains, e.g. medical research [4]. The related available technologies are mature, and even free software tools are available. For these

M. Gebauer · P. Rausch (✉) · M. Stumpf
Technische Hochschule Nürnberg Georg Simon Ohm, Keßlerplatz 12, 90489 Nuremberg, Germany
e-mail: peter.rausch@th-nuernberg.de

M. Gebauer
e-mail: gebauermi52750@th-nuernberg.de

M. Stumpf
e-mail: michael.stumpf@th-nuernberg.de

M. Bramer and M. Petridis (eds.), *Research and Development in Intelligent Systems XXXII*, DOI 10.1007/978-3-319-25032-8_30

reasons, a project to combine the fuzzy P&B with a CBR approach helping novices to learn about parameter configurations and to evaluate solutions was started. The first promising results of this research will be presented here. Also possible steps towards fully automated P&B will be discussed. The paper is organised as follows: After this introduction we provide an overview of recent research in the related fields of fuzzy P&B and CBR. Subsequently, the combination of the fuzzy P&B approach with a CBR system is presented and evaluated. This is followed by a brief discussion of further steps to achieve a fully automated P&B. Finally, the paper is concluded by a short summary and an outlook on future work.

2 Recent Research

2.1 FULPAL—A Fuzzy P&B Approach

To analyse the potential benefits of our hybrid approach it is necessary to summarise a few basics of the fuzzy P&B approach. FULPAL (FUzzy Linear Programming based on Aspiration Levels) is an iterative approach for finding compromise solutions of linear problems with multiple objectives using fuzzy sets [6]. It is a generic approach. Model parameters can be specified using fuzzy intervals and fuzzy constraint borders.

FULPAL has been applied to P&B in [6, 7]. In the specific case the objective function $z = \sum_{i=1}^{n} w_i x_i$ was used, where n denotes the number of budgetary items x_i, such as staff expenses, budget for assistants, budget for mentoring programs, budget for the library, etc. These budgetary items are multiplied by independent weights w_i to influence the overall solution. The objective function $z \rightarrow max$ is combined with an equal constraint to limit the total spending. This part of the model is needed to satisfy the requirement that all available resources have to be budgeted completely. Additionally, constraints with fuzzy borders and aspiration levels for each budgetary item are used. Individual budget limits are denoted by the fuzzy constraint borders which indicate degrees ranging from very bad to very good representations. Whereas aspiration levels are used to describe the expectations of the decision maker (DM). For instance, the library should not spend more than €10,000.

At the start of a planning iteration the DM has to specify aspiration levels for all fuzzy constraint borders. If the problem is feasible an initial solution can be computed. This solution consists of values for all budgetary items as well as indicators λ for each fuzzy border, which represent a measure on how well the targeted aspiration levels were reached. If one or more aspiration levels could not be fulfilled, at least one aspiration level has to be relaxed or no valid solution can be found. In case all aspiration levels can be achieved, an acceptable solution is found or at least one aspiration level can be tightened. Further details can be found in [7].

Experienced decision makers know their planning model and the interplay of budgetary items from past planning periods and have developed their own best practices to find good compromise solutions. However, this knowledge is not embedded into the planning models. Young accountants and novices are left to repeat this learning process from scratch. So, options to improve this learning process and to be able to retain the knowledge of expert DMs should be explored.

2.2 CBR Learning Systems

CBR uses the knowledge of familiar situations to solve a new problem by finding comparable cases, reusing the solution of the most adequate case as a starting point and if possible adapting it to fit the new situation [1]. This principle implies that similar problems often have comparable solutions. By storing problems and their respective solutions in a case base, the CBR system is constantly learning. In contrast to other approaches, the CBR methodology does not require a complex formalisation or a large case base to handle problems [8].

A CBR system is represented as a cyclic process which comprises four steps, see Fig. 1: A problem is formalised to fulfil the demands of the CBR system. The formalised problem represents a new case which is used to retrieve the most similar case or set of cases to the problem. The retrieved case set serves as a base for the problem solving process. Subsequently, the retrieved set is reused by combining it with the new case solving the initial problem. As problems are usually similar but almost never equal the available solutions can be adapted to fit all criteria. In the revise phase the proposed solution has to be evaluated. Retaining is the last step of the cycle. The solved case is stored in the database after being assessed. It is also possible to store only the parts that are likely to be useful for future problems [1].

In conclusion, CBR is advantageous, because it can be used with only a small amount of cases and deal with a wide range of problems, especially those where human experiences play a significant role. So obviously, an enhancement of FULPAL by CBR can contribute to solve the mentioned issues related with P&B.

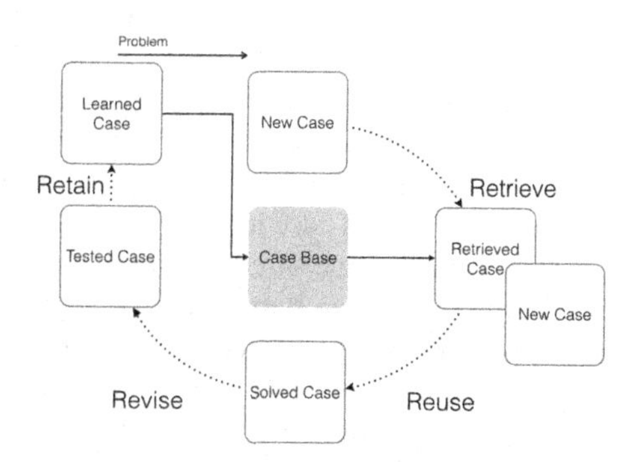

Fig. 1 Case-based reasoning cycle (adapted from [1])

3 Combing FULPAL with CBR

The following section describes the characteristics of P&B problems in a CBR context and how FULPAL is integrated with jCOLIBRI, a CBR software engineering tool. jCOLIBRI is an open source platform for developing CBR applications. The platform has been chosen since it provides an efficient development process [2]. It can easily be integrated with the existing P&B software. This is advantageous because users can access the knowledge contained in the CBR system directly.

At first, the representation form of cases had to be fixed. In the P&B domain, a model is used to describe the planning problem. A configuration of planning parameters is called an iteration. The computed iteration may have a solution for the specific configuration of planning parameters. From the CBR perspective model, iteration, and a solution vector for all unknowns are case components. They also include an overall λ^*-value, representing the total level of satisfaction with the solution, and individual λ-values for all constraints and the objective function denoting the degree of the fulfilment of the aspirations levels. Furthermore, a case solution is required to complement a CBR case, and its representation form has to be considered. When the case base was built independently from the P&B-model very few cases could be stored or created. Therefore, a specific model was used to build the case base. Based on this model, individual budgets are known as own concepts and are addressed as named components in the CBR system (e.g. "staff expenses" could be used to address a specific budget).

After this preliminary work the generic CBR cycle of Sect. 2.2 was applied to our P&B problem. As a result, CBR steps were integrated into the FULPAL algorithm, see Fig. 2. As stated previously, the first step of a CBR cycle is to retrieve a prior case or multiple prior cases to solve a problem. Although CBR systems tend to function with rather limited case bases, the quality is a critical factor in successfully recalling suitable solutions. On this account, the focus was initially on building a solid case base. Usually, planning is done periodically at fixed time intervals and is case-specific. In contrast to domains where mass data or data with statistical accumulations can be collected, the case base of planning problems is rather small. Due to these circumstances, various alternative cases were retrieved by similarity comparison from the case base and were presented to the decision maker who had to decide which cases are relevant for the problem solution.

The reuse-step which proposes a solution for the new case can be executed in different ways depending on the nature of the initial problem. One possibility would be to reuse a solution unaltered. For highly complex problems, such as planning, it is however unlikely that a case solution perfectly matches the current case [3]. Hence, the retrieved case has to be combined with the new case in order to adapt it. There is a wide range of adaptation techniques [3]. In our case generative adaptation which replays the steps to reproduce the retrieved solution on the new problem was appropriate. For instance, with regard to FULPAL and aspiration levels as control mechanism, an instruction to adapt the solution could be "Increase aspiration level of library budget by 20 %". If the representation of a solution is merely a text instruction a fully automated process cannot be supported easily. However, if there

```
LEARNFULPALWITHCBR(fulpalModel)
1  repeat
2      // DM denotes human decision maker.
3      DM.MODIFYORACCEPTASPIRATIONLEVELS(fulpalModel)
4      fulpalSolution = fulpalModel.COMPUTE()
5      selectedCase = DM.SELECTCASE(CBR.RETRIEVE(fulpalSolution))
6      acceptSolution, fulpalModel = CBR.REUSE(selectedCase, fulpalModel)
7      // re-compute after changes of aspiration levels and/or model
8      fulpalSolution = fulpalModel.COMPUTE()
9      CBR.REVISE(selectedCase)
10     if (DM.WASADAPTIONAPPROPRIATE() == TRUE)
11         CBR.RETAIN(selectedCase)
12 until acceptSolution == TRUE
```

Fig. 2 Pseudocode of CBR integration in FULPAL

are dedicated data structures which represent the changes applied to aspiration levels, this instruction could be replayed automatically on the current case. Adjustments to the FULPAL model or aspiration levels by the DM result in a recalculation, see Fig. 2.

The revise-step involves evaluating and quantifying the generated solution. The automation of this step is limited, because it has to be kept in mind that P&B problems also tend to deal with political issues and relationships within an organisation. Our model abstracts reality and, thus, can not indicate all relationships a DM knows. Hence, there is no guarantee that a mathematical optimal and economically viable plan is reasonable in practice. Similar to the retrieve-step, a DM has to evaluate the solution by taking soft factors and individual experiences into account.

Retaining the new solution is the last step of the CBR cycle. Due to the fact that the case base was small at the beginning of this research, all appropriate CBR iterations were stored in the case base.

4 Evaluation and Ongoing Work

The presented approach was successfully tested at our university. As described in Sect. 3 an initial case base was implemented. A group of students was requested to perform a few planning iterations which are usually done by an expert. They got only a brief introduction into the P&B subject and had just a very vague idea what would be a good configuration of planning parameters and how to evaluate the results. Assisted by the CBR system they could find appropriate solutions indicating that internal knowledge of the university's expert could be captured. For the next step, tests with other planning configurations are intended. Based on the results the further automation of steps, like the retrieve- and the revise-step, can be explored. At the focussed organisation professors executing P&B tasks in terms of institutional autonomy as well as employees in the administration department with work overload might be quite excited about the idea of automated P&B. However, it can not be denied that other employees might fear job losses. So, it is highly recommended to combine the idea of automating P&B steps with the introduction of incentives and/or a reassignment of tasks. In any case some effort has to be made to deploy the presented hybrid approach. For instance, expert knowledge is needed to set up the technical configuration, to adapt the CBR system to other planning models, and to identify relevant cases for the case base. On the other hand, time consuming P&B learning processes can be shortened and expensive staff resources can be used more efficiently. So, we expect that the benefits surplus the efforts.

5 Conclusions and Future Work

In this paper, we presented a novel hybrid approach to improve the efficiency of P&B processes and to master the related challenges. The technical issues of the realised prototype were all solvable with reasonable effort. It could be shown that the combination of the strengths of fuzzy planning and CBR contributes to save human and financial resources. Besides, the CBR component improves the transparency of knowledge in the field of P&B. Furthermore, P&B processes become more robust, for instance, in case of unexpected staff absence. For the next steps it is intended to test other use cases, to compare our approach with other methods and to start experiments to analyse further automation steps.

References

1. Aamodt, A., Plaza, E.: Case-based reasoning: foundational issues, methodological variations, and system approaches. AI Commun. **7**(1), 39–59 (1994)
2. Díaz-Agudo, B., González-Calero, P.A., Recio-García, J.A.,Sánchez-Ruiz-Granados, A.A.: Building CBR systems with jCOLIBRI.Sc. Comput. Program. **69**(1–3), 68–75 (2007)
3. Lopez de Mantaras, R., McSherry, D., Bridge, D., Leake, D., Smyth, B., Craw, S., Faltings, B., Maher, M.L., Cox, M.T., Forbus, K., Keane, M., Aamodt, A., Watson, I.: Retrieval, reuse, revision and retention in case-based reasoning. The Knowl. Eng. Rev. **20**, 215–240 (2005)
4. Ping, X., Tseng, Y., Lin, Y., Chiu, H., Lai, F., Liang, J., Huang, G., Yang, P.: A multiple measurements case-based reasoning method for predicting recurrent status of liver cancer patients. Computers in Industry **69**, 12–21 (2015)
5. PwC: Financial planning: Realizing the value of budgeting and forecasting (2011). https://www.pwc.ch/user_content/editor/files/publ_adv/pwc_financial_planning_realizing_value_of_budgeting_11.pdf. Accessed 26 May 2015
6. Rausch, P., Rommelfanger, H.J., Stumpf, M., Jehle, B.: Managing uncertainties in the field of planning and budgeting—an interactive fuzzy approach. In: Bramer, M., Petridis, M. (eds.) Research and Development in Intelligent Systems XXIX, pp. 375–388. Springer, London (2012)
7. Rausch, P., Stahl, F., Stumpf, M.: Efficient interactive budget planning and adjusting under financial stress. In: Bramer, M., Petridis, M. (eds.) Research and Development in Intelligent Systems XXX, pp. 375–388. Springer, London (2013)
8. Richter, M., Weber, R.: Case-Based Reasoning: A Textbook. Springer, Berlin (2013)

A Comparative Analysis of Ranking Methods in a Hybrid Filter-Wrapper Model for Feature Selection in DNA Microarrays

Conor Fahy, Samad Ahmadi and Arlene Casey

Abstract Feature selection in DNA microarrays aims to identify compact subsets of the most relevant, descriptive genes and to eliminate noisy, redundant genes in order to construct better models for classification and to identify potential drug leads. Of the two broad methods for feature selection; filters and wrappers, wrapper methods tend to yield better results but are hindered by high computational complexity and the curse of dimensionality. Filter methods are often used as a pre-processing step for wrapper methods to reduce this complexity. This paper presents a comparative analysis of a univariate filter and a multivariate filter method. Both approaches are evaluated as filters for a wrapper method which consists of a genetic algorithm and a support vector machine. The methods are tested on benchmark DNA microarray datasets. The main contribution from this study is the observation that a multivariate filter outperforms a univariate filter in filter–wrapper hybrid. The study also shows that in some cases a combination of both filters can give better results than each one taken individually.

1 Introduction

The development of DNA microarrays in the mid-nineties has made it possible to examine thousands of gene expressions simultaneously. Researchers have placed a particular focus on determining which genes are expressed differently in cancerous and normal cells [4]. One of the biggest problems in analysing gene expression microarrays is that the number of patients (samples) is often very small compared to the number of genes (features). This huge sample-to-feature imbalance makes useful information extraction difficult due to high computational cost and the curse of dimensionality. Feature selection (FS) is a method which aims at reducing the dimensionality of each sample. The underlying assumption of FS is that each sample contains both relevant and redundant features. By selecting a subset of highly relevant features and ignoring the redundant features better models can be constructed for

C. Fahy (✉) · S. Ahmadi · A. Casey
De Montfort University, Leicester, UK
e-mail: conorfahy@hotmail.com

M. Bramer and M. Petridis (eds.), *Research and Development in Intelligent Systems XXXII*, DOI 10.1007/978-3-319-25032-8_31

classification. FS does not change the original features (as opposed to feature extraction methods like principal component analysis) but rather selects a subset of them which leaves the results open to interpretation by a domain expert and can provide meaningful biological insights and potential drug leads.

FS techniques can be broken into two broad categories: filters and wrappers. Filter techniques asses the importance of a feature based on the intrinsic properties of the data regardless of the model or classifier used. They usually assign a score to each feature based on the perceived importance of the feature. Whereas filter methods evaluate features independently of the classification model, wrapper methods embed the model within the feature subset search. A search algorithm is 'wrapped' around a classification algorithm—feature subsets are identified, and a classifier is trained and tested using these subsets. An exhaustive search is not feasible in gene expression microarrays as the problem is known to be NP-hard and most wrapper methods use population based search heuristics. Wrapper methods can model the interaction between features and identify feature dependencies leading to better biological insights and greater accuracy in classification. They are however, prone to over-fitting the data and can be computationally very expensive [2]. A common approach to reduce this complexity is to combine a filter method with a wrapper method [1, 6]. A filter method is used to rank features as a pre-processing step and the top n features is selected as the search space for a subsequent combinatorial wrapper method. This hybrid approach filters out the least promising features but can still potentially identify variables that are apparently redundant in isolation but relevant in the presence of others.

2 Filter Methods for Variable Ranking

Filters can be either univariate or multivariate.Univariate filters, for example mutual information or coefficient correlation, consider each feature separately, one at a time. This is computationally simple, fast and scalable but it ignores dependency between features. Certain features might seem redundant or irrelevant when considered individually but can be relevant when considered in the presence of others. Ignoring this dependency can lead to a poorer classification accuracy. Multivariate filter techniques, for example recursive feature elimination and top scoring pair consider features as group. They can model feature-dependency to some degree but are slower and less-scalable. A study in [5] compared the performance of univariate and multivariate filter techniques across seven different types of cancer gene expression datasets. Univariate techniques performed better in five of the datasets. The authors argue that there are not enough samples available in order to detect and extract correlation structures in each sequence. However, subsequent studies using more complex combinatorial methods have obtained better results on the same datasets [6, 7, 9] suggesting that these correlation structures do exist and can be extracted but not by using a filter method by itself. The studies which attained better results each use a hybrid method with a filter as a preprocessing,variable ranking step. In [8] it is advised that a univariate filter method is used as this pre-processing step for the subsequent wrapper method, however there is little evidence in the literature to suggest that one type of filter provides a better subset than the other.

2.1 Univariate Method: Mutual Information

In Information theory mutual information (MI) is a method used to measure dependencies between features. Given two variables X and Y, MI quantifies the reduction of uncertainty in X by having knowledge about Y. To describe MI it is necessary to start with Shannon's definition of entropy:

$$H(Y) = -\sum_{y} p(y) log(p(y)) \tag{1}$$

This equation represents the uncertainty in Y. If a variable X is observed then the conditional entropy is given by [1]:

$$H(Y \| X) = -\sum_{x} \sum_{y} p(x, y) log(p(y|x)) \tag{2}$$

This implies that by observing a variable X, the uncertainty of the variable Y is reduced. This reduction in uncertainty is given as:

$$MI(Y, X) = H(Y) H(Y \| X) \tag{3}$$

This gives the MI between X and Y. The MI between each feature and the output class is calculated and the features are ranked based on this score. If a feature and the output class are completely independent then the MI is zero.

2.2 Multivariate Method: Recursive Feature Elimination

RFE uses a classifier to assign a weight to each feature in the sample. It sorts these weights and removes the smallest, the assumption being that the smaller the weight, the less informative the feature. RFE is a greedy, backward elimination method, it begins with the full set of features and progressively removes the least promising ones. Typically, a support vector machine (SVM) is used as the classifier. The hyperplane which linearly separates the data is described by a set of coefficient weights and each correspond to a feature. The larger weights are more necessary for a good hyperplane and the smallest ones are iteratively removed until either (a) a certain threshold is reached or (b) until there are not enough features left to carry out classification.The SVM used in this study follows the standard RFE algorithm originally outlined in [3].

3 Wrapper Method: Genetic Algorithm and SVM

The previous filter methods ranked the features but the top ranked features are not necessarily the most relevant or important. As succinctly stated in [2] *"A good feature ranking criterion is not necessarily a good feature subset ranking criterion"* The top ranked list of n features Fn is less likely to be optimal than a subset of Fn. This subset will likely be more compact and have redundant genes removed. A genetic algorithm (GA) is used to search for optimal subsets of Fn. Individuals in the GA are represented as binary strings of length n whereby a 1 at position i in the string means the ith feature is selected and a 0 means that feature is omitted. The resulting subsets are evaluated using a SVM. The fitness function consists of two competing objectives: (1) maximising the classification accuracy and (2) minimising the number of features. The fitness function used for the experiments in this study is:

$$Fit(x) = 100 * accuracy(x) + 1 * size(x) \tag{4}$$

The fitness objective is multi-modal and the weights, $w1 = 100$ and $w2 = 1$, were chosen to put pressure on the selection process to generate highly accurate solutions as the primary objective with the number of genes being less important.The population is updated at each generation using tournament selection with replacement. Random point crossover is used with a probability of 1 and children are randomly mutated using a random, single bit-flip with a low probability (0.04).

4 Expermiments and Results

Two benchmark gene expression datasets were used for testing, both obtained from the Gene Expression Model Selector website. The first is the diffuse large b-cell lymphomas (DLBCL) and follicular lymphomas dataset consisting of 77 samples described by 5,470 genes. The data describes two classes of blood cell tumours. The second dataset is a prostate tumour dataset. This consists of 102 samples (50 samples of normal tissue, 52 samples of tumours) described by 10,510 genes. The results are validated using 10-fold cross validation.To evaluate the univariate and multivariate filter methods the 30 top ranked genes in each filter were selected. These 30 features formed the candidate pool for the genetic algorithm. The GA had a population of 50 and ran for only 25 epochs. While this is a relatively small GA it gives good results. Tables 1 and 2 show the results of each method over 5 runs. The best performance of each method, while giving relatively similar results, were obtained using different genes; completely different in the lymphomas experiments and sharing only one common gene in the prostate data. These 'best' gene combinations from each method were pooled together for another GA (MI-RFE-GA) using the same parameters. This combination of results yielded a better performance than each one individually.

Table 1 Results using lymphomas dataset

Algorithm	Average	Best	Genes selected
MI-GA	g[a]: 5.8 (σ 1.4)	6	[1670] [2164] [409] [3264] [1797] [1303]
	%: 94.1 (σ 3.4)	100	
RFE-GA	g: 6.2 (σ .8)	5	[453] [893] [2971] [3257] [4053]
	%: 99 (σ .8)	100	
MI-RFE-GA	g: 3.4 (σ .5)	3	[453] [409] [3264]
	% 99.2 (σ .2)	100	

[a] g = number of genes selected,
% = prediction accuracy

Table 2 Results using prostate dataset

Algorithm	Average	Best	Genes selected
MI-GA	g: 5.6 (σ 1.6)	4	[4823] [7652] [7451] [1943]
	%: 91 (σ .4)	94	
RFE-GA	g: 7.4 (σ .8)	9	[2569] [4760] [4823] [5674] [6087] [6168] [6773] [7574] [8545]
	%: 98 (σ 1.5)	100	
MI-RFE-GA	g: 4.2 (σ .4)	5	[5674] [6087] [6168] [7574] [4823]
	% 98.7 (σ .2)	100	

Table 3 Comparative results using prostate dataset

Method	% Accuracy	Genes	Reference
MI-RFE-GA	100	5	This Study
SVM-RFE-GA	100	5	[6]
SVM-RFE	99	9	[9]
SVM-RFE-MRMR	98	10	[7]
LS-Bound+SVM	96	4	[7]
Bayes+SVM	95	3	[7]

Table 3 presents a comparison of the results obtained on the prostate dataset with other methods found in the literature. There are not as many comparative results for the lymphomas dataset so that comparison was omitted.

5 Conclusion

The aim of this research was to establish whether a univariate filter method outperforms a multivariate filter ranking method as the preprocessing stage in a filter–wrapper hybrid model for feature selection in DNA microarrays. Mutual Information was selected as a univariate method and recursive feature elimination was used as the multivariate method, these techniques being among the most widely used in the literature. Both techniques were used to rank features in two benchmark gene expression microarray datasets. The 30 top ranked features from each method were used as the search space for a subsequent wrapper method to select the most compact, relevant subset of the top ranked genes. The results show that recursive feature elimination—a multivariate method outperforms mutual information—a univariate method across both datasets. Because only one method from each class of filter was tested we can't conclude that multivariate methods, on the whole, outperform univariate methods but this study is part of ongoing research and future experiments will evaluate a greater number of filtering techniques from each class across a wider range of benchmark datasets. Interestingly, in one of the experiments in this study a combination of univariate and multivariate variables outperformed each set taken individually suggesting that a hybrid, combinatorial approach could lead to a better ranking of relevant features. This too will be investigated in future research.

References

1. Estévez, P.A., Tesmer, M., Perez, C.A., Zurada, J.M.: Normalized mutual information feature selection. IEEE Trans. Neural Netw. **20**(2), 189–201 (2009)
2. Guyon, I., Elisseeff, A.: An introduction to variable and feature selection. J. Mach. Learn. Res. **3**, 1157–1182 (2003)
3. Guyon, I., Weston, J., Barnhill, S., Vapnik, V.: Gene selection for cancer classification using support vector machines. Mach. Learn. **46**(1–3), 389–422 (2002)
4. Hoopes, L.: Genetic diagnosis: DNA microarrays and cancer. Nat. Educ. **1**(1), 3 (2008)
5. Lai, C., Reinders, M.J., van't Veer, L.J., Wessels, L.F.: A comparison of univariate and multivariate gene selection techniques for classification of cancer datasets. BMC Bioinform. **7**(1), 235 (2006)
6. Li, E.: Gene selection for cancer classification using the combination of SVM-RFE and GA (2014)
7. Mundra, P.A., Rajapakse, J.C.: SVM-RFE with mrmr filter for gene selection. IEEE Trans. Nanobiosci. **9**(1), 31–37 (2010)
8. Saeys, Y., Inza, I., Larrañaga, P.: A review of feature selection techniques in bioinformatics. Bioinformatics **23**(19), 2507–2517 (2007)
9. Tang, Y., Zhang, Y.Q., Huang, Z., Hu, X.: Granular SVM-RFE gene selection algorithm for reliable prostate cancer classification on microarray expression data. In: Fifth IEEE Symposium on Bioinformatics and Bioengineering, 2005. BIBE 2005, pp. 290–293. IEEE (2005)

A New Content-Based Recommendation Algorithm for Job Recruiting

N. Almalis, G. Tsihrintzis and N. Karagiannis

Abstract In this paper, we propose a *Content-Based Recommendation Algorithm* that takes into consideration the organization needs and the skills of candidate employees in order to quantify the suitability of a candidate employee for a specific job position. The proposed algorithm extends the *Minkowski* distance to perform a primary study in order to investigate how the *Job Recruiting* field could benefit further. Also we conduct an experimental evaluation with the objective of checking the quality and the effectiveness of the proposed algorithm. Our primary study produces promising results and shows that this algorithm can play an important role in the area of *Job Recruiting*.

1 Introduction

A main characteristic of our age is the rapid development of technology which increased the amount of available information on every sector of our life. The above problem -known as *Information Overload*- along with the increasing demand for specialized and personalized customization, led to a new era of information filtering and to the emergence of new technologies to assist human decisions [1]. One field of research in this direction is that one of *Recommender Systems* (RSs) [2].

Although RSs are based on other common scientific domains such as artificial intelligence, data mining, pattern recognition etc., they have some requirements that

N. Almalis (✉) · G. Tsihrintzis · N. Karagiannis
University of Piraeus, Karaoli & Dimitriou 80, 18534 Piraeus, Greece
e-mail: nikosalmalis@unipi.gr

G. Tsihrintzis
e-mail: geoatsi@unipi.gr

N. Karagiannis
e-mail: nkaragiannis87@unipi.gr

M. Bramer and M. Petridis (eds.), *Research and Development in Intelligent Systems XXXII*, DOI 10.1007/978-3-319-25032-8_32

differentiate them. From this point of view, Burke *et al.* [3] consider two basic principles which distinguish RS research. A RS must be: (i) personalized and (ii) intended to help the user select among discrete options. Thus, RSs typically use one of the next four basic techniques [4], in order to achieve technical adjustment of the aforementioned two principles: (i) *Collaborative Filtering Recommenders* (ii) *Content-Based Recommenders (CBRs)* (iii) *Knowledge-Based Recommenders* and (iv) *Hybrid Recommenders*.

In the area of *Job Seeking and Recruiting*, we have really remarkable results with 73% of recruiters to have made successful hires through social media and 59% of recruiters to declare that candidates who were recommended through social networks are of the highest quality [5]. Thus, online or offline *Job Recommender Systems* (JRSs) seem to be an unquestionably strong necessity that attracts significant research efforts. Singh *et al.* have presented PROSECT [6], a decision support tool designed to aid in the short listing of candidates for jobs by extracting several pieces of information from the unstructured resumes with the help of statistical data-driven techniques. Drigas et al. [7] have presented an expert system which evaluates the unemployed at certain offered posts using neuro-fuzzy techniques. An ontology-based hybrid approach has been presented by Fazel-Zarandi [8] which effectively matches job seekers and job postings. The approach uses a deductive model to determine the type of match between a posting and a job seeker by applying a similarity-based approach to rank applicants. In our previous work [9], we suggested a content-based job recommendation approach which used similarity measure to produce recommendation and took under consideration that the candidate with the *closer or even higher* of the respective values of the specific job requirements had higher probability to be recruited.

The above works do not use all the dimensions of qualifications and so they fail to produce accurate and targeted recommendations. Qualifications for a job can be modeled in several ways, that is an attribute can be qualified if it is below a certain threshold (e.g. a job requires the candidate to be at most 40 years old) or in between a predefined range (e.g. qualified candidates are between 18 and 40 years old) and not only to have an exact value. The contribution of this paper is to introduce a new approach based on CBR, overcoming the limitations of the existing techniques and taking under consideration four different categories of the measurable attributes/skills which depend on the values range and providing a final score which shows the suitability of the candidate for a specific job.

The remainder of the paper is structured as follows: In Sect. 2, we analyze our suggested algorithm. In Sect. 3, we perform our experimental evaluation presenting the methodology and results. Finally, we conclude the paper in Sect. 4 providing a brief summary of our current work and discussing some of its possible future extensions.

2 The Proposed Algorithm

The proposed algorithm extends the Minkowski distance [10] and quantifies the suitability of a candidate employee for a job position. The algorithm takes into consideration four different categories of measurable attributes/skills which depend on the range of values: i.e. an exact value (E), a range with lower limit (L), a range with upper limit (U) and a range with both lower and upper limit (LU), providing a final score which shows the suitability of every candidate for the specific job. The algorithm is presented bellow:

Consider a vector J to be a specific job of the form $J = (j_1, j_2, \ldots, j_n)$ with attributes $j_1, j_2, \ldots, j_n, n \in N^+$ and each attribute belongs to one of the four referred types: i.e. E, L, U, LU. Also, let vector $C = (c_1, c_2, \ldots, c_n)$ with attributes $c_1, c_2, \ldots, c_n, n \in N^+$ be a candidate employee that will be compared with the job J (vector J and C have the same structured form). We follow the next three steps, in order to find the suitability of candidate C for job J.

Step 1
We construct four vectors $J_x, x = 1, 2, 3, 4$ where their attributes belong to the initial vector of job J and their classification is done based on the type of their attributes. It is underlined that each attribute of the job J belongs *to one and only one* vector J_x.

Step 2
Case 1—*Exact value*- The suitability S_E will be given by the metric:

$$S_E = -\left(\sum_{i=1}^{n} |j_i - c_i|^p\right)^{1/p}, p \in R^* \tag{1}$$

Case 2—*Lower limit*- The suitability S_L will be given by the sum of the following two metrics S_{L+} and S_{L-}, where:

$$S_{L+} = +\left(\sum_{k=1}^{n} |j_k - c_k|^p\right)^{1/p}, p \in R^* \tag{2}$$

$$S_{L-} = -\left(\sum_{l=1}^{n} |j_l - c_l|^p\right)^{1/p}, p \in R^* \tag{3}$$

$$S_L = S_{L+} + S_{L-} \tag{4}$$

with $c_k \geq j_k$ for (2), $c_l < j_l$ for (3) and $k \neq l$

Case 3—*Upper limit*- The suitability S_U will be given by the sum of the following two metrics S_{U+} and S_{U-}, where:

$$S_{U-} = -\left(\sum_{u=1}^{n} |j_u - c_u|^p\right)^{1/p}, p \in R^* \tag{5}$$

$$S_{U+} = +\left(\sum_{v=1}^{n} |j_v - c_v|^p\right)^{1/p}, p \in R^* \tag{6}$$

$$S_U = S_{U+} + S_{U-} \tag{7}$$

with $c_u \geq j_u$ for (5), $c_v < j_v$ for (6) and $u \neq v$.

Case 4 -*Lower and upper limit*- The suitability S_{LU} will be given by the metric:

$$S_{LU} = -\left(\sum_{m=1}^{n} |j_m - c_m|^p\right)^{1/p}, p \in R^* \quad (8)$$

where j_m is the lower limit if $c_m \leq j_m$ and j_m is the upper limit if $c_m \geq j_m$ (for $j_{mMax} \leq c_m \leq j_{mMin}$ the $S_{LU} = 0$ as the value is within the desirable range)

Step 3
Overall, the suitability S_{CJ} of the candidate employee C for the job J is defined as follows:

$$S_{CJ} = S_E + S_L + S_U + S_{LU} \quad (9)$$

3 Experimental Evaluation

We conducted the following *three-step* experiment (i.e. *Step 1: Content analysis, Step 2: Adjustment of the algorithm, Step 3: Execution and result*) in order to check the quality of the proposed algorithm. Our suggested algorithm is executed for a real world dataset of 52 CVs. These CVs are available at the online website www.kaggle.com/c/job-recommendation/data and the job position is opened by an IT Company with the title "*Web Designer*". The three steps have as follow:

Step 1: Content analysis: In the first step, the personnel director of the IT Company creates the job description in a structure way, through an application which stores: the desirable skills, the scoring table for each one of the skills, the level of skills according the scoring table and finally he discriminates the skills into the four types: Exact value (E), Lower limit (L), Upper limit (U), both Lower and Upper limit (LU), e.g. the desirable skill is *education level*, the scoring table is: *BSc-20pts, Master-40pts, PhD-60 pts, postdoc-80 pts* and the level of this skill for the job position is *Master-40pts at least (L)*. Also each candidate inserts his/her skills and the application computes the respective points for each skill. In detail the final product of the first step is: *(A)Web development experience-260pts-L, (B) Education level-60pts-U, (C)Language skill-50pts-U, (D)Programming knowledge-25pts-L, (E)Network knowledge-10pts-(E, F)Technical support knowledge-20pts-(E, G)Project management experience-5pts-L and (H)Age-25 till 40pts-LU* and the skills values of the Candidates. Table 2 shows these values for the top 5.

Step 2: Adjustment of the algorithm: In the second step, the output of the content analysis is mapped into a manageable and flexible shape which is proper for our CBR algorithm to manipulate it. More specifically, an 8-dimensional vector is used to represent the job position (J). Each element of the vector is mapped to an attribute and each attribute has a value. Respectively, the candidates profile is represented with another 8-dimensional vector (C) in which each element is mapped into the same attribute as the job position needs. The values of the attributes use the same

Table 1 The attributes values of the top 5 candidates

Candidate ID	A (L)	B (U)	C (U)	D (L)	E (E)	F (E)	G (L)	H (LU)
27	290	124	0	10	20	0	0	30
11	300	131	0	9	20	10	50	28
12	300	75	0	4	25	15	10	33
32	280	121	0	31	20	10	5	24
41	295	71	30	0	15	0	25	38

Table 2 The top 5 candidates after the algorithm execution for different values of p variable

p = 1[a]		p = 2[b]	
Candidate ID	Score	Candidate ID	Score
12	+39	12	+38.5
11	+28	11	+9.06
41	+14	41	+3.7
32	−6	32	−5.26
27	−34	27	−22.17

[a]z = 1 is Manhattan distance
[b]z = 2 is the Euclidian distance

scoring table. Finals the two vectors J and C which produced in step 2 are the input of our algorithm.

Step 3: Execution and result: The algorithm executed for two different values of the *p* variable namely 1 (*Manhattan distance*) and 2 (*Euclidian distance*). Two sorted lists are produced where the top 5 Candidates are retrieved for further processing (see Table 2). According to the personnel director the sorted list for $p = 2$ was close to the list which produced from his/her experienced choice. Of course this primary research must be conducted by other experts on job recruiting. The p variable impacts the suitability of the personnel and is delineated through two different meanings: (i) the physical meaning, which defines the candidates which have the same degree of suitability although their qualifications may be divergent and (ii) the logical meaning, which is the similarity measure (Table 1).

4 Conclusion and Future Work

We can conclude that there is a great opportunity for applying RSs in the area of *Job Recruiting* in order to improve the matching quality. The results of the conducted experiment show exactly this perspective. The companies are navigated through a coherent procedure to use effectively our suggested algorithm as a part of a JRS and to take advantage of these benefits: (i) less candidates for in depth evaluation (ii) increase in the internal operation level of companies (iii) improvement of job position description (iv) results can be utilized for internal personnel evaluation, etc.

Our proposed methodology presents promising performance and encourages us to investigate and improve further its extensions. Consequently, we aim at conducting future research in order to improve the performance, in terms of time response and reliability. The algorithm must be tested on other datasets retrieved from various branches of the enterprise sector. Furthermore, different values of the p variable should be examined. Guidelines for further research also include the reverse procedure i.e. the modification of the job description based on the production level of the selected candidates and the scalability of our suggested algorithm on other fields of common interest. This and other relative research is currently under consideration and investigation and will be published elsewhere in the near future.

References

1. Linden, G., Smith, B., York, J.: Amazon.com recommendations: item-to-item collaborative filtering. In: Published by the IEEE Computer Society, IEEE Internet Computing, vol. 7(1), pp. 76–80 (2003)
2. Resnick, P., Varian, H.: Recommender systems. Commun. ACM **30**(3), 56–58 (1997)
3. Burke, R., Felfernig, A., Goker, M.H.: Recommender systems: an overview. AI Mag. **32**(3), 13–18 (2011)
4. Ramezani, M., Bergman, L., Thompson, R., Burke, R., Mobasher, B.: Selecting and applying recommendation technology. In: Proceeding of International Workshop on Recommendation and Collaboration, in Conjunction with 2008 International ACM Conference on Intelligent User Interfaces, Canaria, Canary Islands, Spain (2008)
5. Jobvite.: Social Recruitment Survey. http://recruiting.jobvite.com (2014).
6. Singh, A., Rose, C., Visweswariah, K., Chenthamarakshan, V., Kambhatla, N.: PROSPECT: a system for screening candidates for recruitment. In: Proceedings of the 19th ACM International Conference on Information and Knowledge Management, pp. 659–668, Toronto, Canada (2010)
7. Drigas, A., Kouremenos, S., Vrettos, S., Vrettatos, J., Kouremenos, D.: An expert system for job matching of the unemployed. Expert Syst. Appl. **26**, 217–224 (2004)
8. Fazel-Zarandi, M., Fox, M.S.: Semantic matchmaking for job recruitment an ontology based hybrid approach. In: proceedings of the 3rd International Workshop on Service Matchmaking and Resource Retrieval in the Semantic Web at the 8th International Semantic Web Conference, Washington D.C., USA (2010)
9. Almalis, N.D., Tsihrintzis, G.A., Karagiannis, N.: A content based approach for recommending personnel for job positions. In: Proceedings of the 5th International Conference on Information, Intelligence, Systems and Applications, IISA 2014, pp. 45–49. Chania, Greece (2014)
10. Minkowski, H.: Allgemeine Lehrsatze uber die konvexe Polyeder, Nachr. Ges. Wiss., Gottingen, 1897, 198–219 (=Ges. Abh., vol 2, pp. 103-121, Leipzig-Berlin, 1911)

Mining Fuzzy Time-Interval Patterns in Clinical Databases

A. Mills-Mullett and J. Lu

Abstract Knowledge discovery within the health informatics domain provides healthcare professionals with the ability to further inform their decision-making regarding patient treatment. Discovering patterns within this domain has however proven difficult for data scientists; one of the main challenges of knowledge extraction is the ability to extract meaningful patterns, which can be used as effective support alongside specialist knowledge. This paper demonstrates the value behind extracting temporal patterns from live datasets. In particular it shows how fuzzy logic and divisive hierarchical clustering can be used to extract frequent sequential patterns with time-intervals from a series of breast cancer diagnoses. A discussion regarding the link between time-intervals and cancer treatment is explored through the use of multi-dimensional sequential pattern mining with fuzzy time-intervals.

1 Introduction

Within the health informatics domain a vast amount of data has been generated and developed over a number of decades. In recent years, this data has played a vital role in the enabling of support for decision-making [1, 2]. In particular there has been a vast amount of research into the causes of breast cancer, but there has been little analysis with regard to the development of a temporal link between breast cancer treatments and the time between cancer episodes [3].

This paper involves the application of an existing data mining technique used to extract sequential pattern mining with time-intervals to a case study that deals explicitly with breast cancer datasets. This case study involves the use of The University Hospital Southampton (UHS) Southampton Breast Cancer Data System (SBCDS)

A. Mills-Mullett (✉) · J. Lu
Southampton Solent University, Southampton SO14 0YN, UK
e-mail: a.m.mullett@gmail.com

J. Lu
e-mail: jing.lu@solent.ac.uk

M. Bramer and M. Petridis (eds.), *Research and Development in Intelligent Systems XXXII*, DOI 10.1007/978-3-319-25032-8_33

and its anonymised datasets, as there is a current requirement within the hospital to implement a data mining method that analyses the temporal structure of the current datasets.

2 Sequential Patterns Mining with Fuzzy Time-Intervals

The analysis of temporality in transactional databases is not without its difficulties. This is due in part to the complexity of grouping distinct time-intervals together in an efficient and accurate fashion. Generating time-clusters and determining a metric, which denotes multiple time-intervals as being similar can be computationally expensive.

The Temporally-Annotated Sequence (TAS) algorithm introduced in [4] uses probability distribution estimation to determine the similarity between a clustered group of time-intervals, whereas the Integrated Sequential Patterns Mining with Fuzzy Time-Intervals (ISPFTI) algorithm introduced in [5] uses fuzzy set membership theory to determine similarity. Divisive hierarchical clustering is used to generate a maximum of two time-clusters (for each frequent sequence.) The ISPFTI algorithm has been selected over TAS in this paper, as the TAS algorithm uses a hypercube Parzen window, which becomes computationally expensive when the lengths of the frequent sequential patterns to be extracted increases.

Within the ISPFTI algorithm divisive hierarchical clustering is used to split a set of time-intervals at the maximum time-interval to create two time-clusters for each frequent sequential pattern. Fuzzy logic, fuzzy sets, and in particular fuzzy set membership are used to determine the set membership value of a time-interval for a particular time-cluster.

3 Case Study in Clinical Databases

There is an opportunity to apply the ISPFTI algorithm to a domain that benefits from a temporal analysis of patterns. The algorithm will be used in conjunction with the SBCDS in an attempt to extract meaningful patterns for decision support.

3.1 Breast Cancer Data System

At the time of writing the SBCDS system consists of 15,500 patient records. The first records were obtained in the 1940s, with the most recent records dating from 2014. The vast amount of sequence data present within the system leads to an exploitative amount of information to be extracted in the form of temporal patterns.

Table 1 UHS sample sequence database

ID	Sequence
1	('PBC', 0), ('MSD', 1370), ('LRR', 1400)
2	('PBC', 0), ('OCD', 2343), ('MSD', 2861)
3	('PBC', 0), ('LRR', 153), ('MSD', 215)
4	('PBC', 0), ('LRR', 791), ('MSD', 1096)
5	('PBC', 0), ('PBC', 61), ('LRR', 276)
6	('PBC', 0), ('LRR', 2223), ('PBC', 2496)
7	('PBC', 0), ('PBC', 1219), ('MSD', 3014)
8	('PBC', 0), ('LRR', 1125), ('MSD', 1795)
9	('PBC', 0), ('LRR', 3957), ('PBC', 6756)
10	('PBC', 0), ('LRR', 8370), ('MSD', 8401)

PBC: Primary breast cancer; LRR: Loco-regional recurrence; MSD: Metastatic disease; OCD: Other cancer diagnoses

A sample database is shown in Table 1 and this will be used as an example for the mining results illustration. Each entry in the sample sequence database represents a patient and is denoted by a unique identifier. The tuples within the 'Sequence' column represent diagnosis episodes with corresponding time-stamps in days.

3.2 Mining Results

After running the algorithm on the patient record sample shown in Table 1 (with a minimum support of 0.3—a sequence to must appear within in the dataset with a frequency of 30% or more to be frequent) several frequent sequential patterns with fuzzy time-intervals were found.

The initial stages involve following steps 1 and 2 of the ISPFTI algorithm denoted in Fig. 1. From these steps frequent sequential patterns L_1 and L_2 are extracted. A subset of extracted patterns and their corresponding supports are shown in Table 2. A L_3 pattern (the maximum length for this sample) is also shown to highlight the fact that a traditional sequential pattern mining algorithm would complete frequent sequential pattern extraction at this stage. As noted this is due to the fact that time-intervals between events are traditionally ignored.

The L_2 patterns *PBC*→ *LRR* and *LRR*→ *MSD* will be used throughout this section to demonstrate the existence of multiple frequent time-ranges between cancer diagnoses in the sample. Returning to the ISPFTI algorithm, Step 3.1 involves the obtaining of time-clusters (ordered time-intervals between events) for each L_2 pattern. The results are shown in Table 3.

The next stages denoted by steps 3.2, 3.3 and 3.4 involve the obtaining of L_2' by determining fuzzy numbers for each time-cluster (generated by identifying the maximum interval for each time-cluster from Table 3) and a respective fuzzy support.

```
ISPFTI Algorithm Pseudocode:

1.  Find all frequent 1-sequences L1
2.  Determine frequent 2-sequences L2
3.  Find frequent fuzzy time 2-sequences L'2
    3.1. Record all time-intervals increasing order with their
         frequencies
    3.2. Divide each time-interval into clusters
    3.3. Define the trapezoid fuzzy number to each obtained
         cluster
    3.4. Compute the fuzzy support for each contained cluster
4.  Determine the clusters to be preseved
5.  Determine L'2 the set of frequent fuzzy time 2-sequences
6.  For k >= 2 Find L'k+1
    6.1. Generate C'k+1 by joining L'k and L'k
    6.2. Compute fuzzy support of each sequnce in C'k+1
    6.3. Generate frequent fuzzy time (k+1)-sequences
```

Fig. 1 Algorithm pseudocode—integrated sequential patterns mining with fuzzy time-intervals

Table 2 L_1, L_2 and L_3 frequent sequence patterns

L_1	Support	L_2	Support	L_3	Support
PBC	1.0	*PBC→ LRR*	0.6	*PBC→ LRR→ MSD*	0.4
LRR	0.8	*LRR→ MSD*	0.4		
MSD	0.7				

For the sake of brevity several L_1, L_2 and L_3 patterns are omitted

Table 3 L_2 time clusters

L_2	Time clusters
PBC→ LRR	(153, 1), (276, 1), (791, 1), (1125, 1), (1400, 1), (2223, 1), (3957, 1), (8370, 1)
LRR→ MSD	(31, 1), (62, 1), (305, 1), (670, 1)

For the sake of brevity several frequent sequence patterns are their respective time-clusters are omitted

If the fuzzy support is more than or equal to 0.3 the time-cluster is preserved (steps 4 to 5). The preserved time-clusters are shown in Table 4. Step 6 involves determining L'_3 by combining L'_2 and L'_2 together.

Step 6.2 requires determining the fuzzy support of each candidate generated (C'_3.) The preserved frequent sequential patterns with fuzzy time-intervals are shown in L'_3 in Table 5.

There is a clear difference in time-intervals between $PBC \xrightarrow{153\sim8370} LRR \xrightarrow{31\sim670} MSD$ and $PBC \xrightarrow{153\sim3957} LRR \xrightarrow{31\sim670} MSD$ with regard to the *PBC→ LRR* aspect

Table 4 Preserved time-clusters

L'_2	Fuzzy number	Fuzzy support
$PBC \xrightarrow{153\sim8370} LRR$	(153, 153, 8370, 8370)	0.8
$PBC \xrightarrow{153\sim3957} LRR$	(153, 153, 3957, 8370)	0.7
$PBC \xrightarrow{3957\sim8370} LRR$	(153, 3957, 8370, 8370)	0.333
$LRR \xrightarrow{31\sim670} MSD$	(31, 31, 670, 670)	0.4
$LRR \xrightarrow{31\sim305} MSD$	(31, 31, 305, 670)	0.3

Each preserved L'_2 contains a corresponding time-cluster in the form of an approximate time-range measured in days above its respective right-hand arrow

Table 5 L'_3—preserved sequential patterns with fuzzy time-intervals

	L'_3		L'_3
1.1	$PBC \xrightarrow{153\sim8370} LRR \xrightarrow{31\sim670} MSD$	2.1	$PBC \xrightarrow{153\sim3957} LRR \xrightarrow{31\sim670} MSD$
1.2	$PBC \xrightarrow{153\sim8370} LRR \xrightarrow{31\sim305} MSD$		

Each preserved L'_3 contains a corresponding time-cluster in the form of an approximate time-range measured in days above its respective right-hand arrow

of the frequent sequences. This distinction is denoted by a ~4000 day difference between the two patterns. This amongst other observable time differences in Table 5 highlights the value behind performing a temporal analysis on a time-series dataset.

3.3 Evaluation

By considering time-intervals between cancer diagnoses, in addition to the sequence order of the cancer diagnoses, further granularity is added to the data mining methodology. The results show that the sequential patterns can be split into different groups based on similar time-intervals. The next stage is to add further dimensions, such as the tumour size of the breast cancer; this would provide greater insight into what types of patient attributes contribute to the differing frequent time-intervals.

An efficient algorithm for mining frequent multi-dimensional sequential patterns was introduced by Boonjing and Songram in [6]. The dimensions are mined using a closed frequent item set algorithm, while the sequence patterns are mined using a closed sequential patterns mining algorithm. The product of the two mined results is generated and the support for each generated sequence is determined. It is possible to modify the algorithm to accommodate the use of fuzzy set membership, which would be used to determine frequent time-intervals between frequent diagnoses. This will be explored in a future paper.

4 Conclusion

The ISPFTI algorithm provides a method to solve the problem of extracting temporal sequential patterns from within the UHS breast cancer dataset. Fuzzy logic is a significant method to group similar time-intervals to create frequent sequential patterns with time-intervals, along with hierarchical clustering for generating time-clusters. From this analysis it is useful to point out that the ISPFTI algorithm can also be applied to other domains that involve time-series analysis, such as market basket analysis, weather prediction, web page navigation and so on.

Future work will involve the inclusion of additional dimensions to add further granularity to the patterns returned. A worthwhile hypothesis should be established with regard to time-intervals between cancer episodes, such as determining the correlation between a specific treatment package and the survival rate of a grouping of patients. Little research has been conducted in the area of combining multi-dimensional sequential pattern mining with the temporal analysis of time-intervals. This area should be further explored therefore, due to the potential value obtained when determining correlations between frequent time-intervals and additional dimensions, as this leads to greater granularity for improved decision support.

References

1. Milovic, B., Milovic, M.: Prediction and decision making in health care using data mining. Kuwait Chapter Arabian J. Bus. Manag. Rev. **12**, 126–136 (2012)
2. Thangarsu, G., Dominic, P.D.D.: Prediction of hidden knowledge from clinical database using data mining techniques. In: 2014 International Conference on Computer and Information Sciences (ICCOINS), pp. 1–5 (2014)
3. Lu, J., Hales, A., Rew, D., Keech, M., Frhlingsdorf, C., Mills-Mullett, A., and Wette, C.: Data mining techniques in health informatics: a case study from breast cancer research. In: 6th International Conference on IT in Bio- and Medical Informatics, to be published in LNCS by Springer (2015)
4. Giannotti F, Nanni, N., Pedreschi, D.: Efcient mining of temporally annotated sequences. In: Proceedings of the Sixth SIAM International Conference on Data Mining, vol. 124, p. 348. SIAM (2006)
5. Chung-I, C., Hao-En, C., Yu-Chun, L.: An integrated sequential patterns mining with fuzzy time-interval. In: International Conference on Systems and Informatics (ICSAI), pp.2294–2298 (2012)
6. Boonjing, V., Songram, P.: Efficient algorithms for mining closed multidimensional sequential patterns. In: IEEE Computer Society. Fourth International Conference on Fuzzy Systems and Knowledge Discovery (2007)

A Sense-Think-Act Architecture for Low-Cost Mobile Robotics

Liam Chapman, Cameron Gray and Chris Headleand

Abstract The use of low cost devices to build autonomous robotic systems has gown significantly over recent years. The availability of high quality, low cost micro processors has only furthered this, and the subsequent development in programming paradigms. Development boards, such as the Arduino and Raspberry Pi products, have become a standard in a wide range of these robotic projects. However, there is no universally accepted architecture for using these boards for autonomous robotics. As these are commodity components, any solution must be resilient to any potential failure. This paper investigates the modulation of a robotic platform, to produce a stable and reliable base on which to build automated devices. A further, and critical, motivation is to create a fail safe system in which the loss of a module does not affect performance.

1 Background and Motivation

The availability of budget development boards have led to their use in a wide range of applications; from educational tools in university [9], through low cost computing in developing countries [4] and use as laboratory apparatus [2], to low cost autonomous robotics—the subject of this paper.

The most popular of these type of boards are the Arduino and the Raspberry Pi. Both have their own advantages, the Raspberry Pi provides more computational power with a clock speed of 700 MHz [5] compared to the 16 MHz available with the Arduino Uno [1]. The limited processing power therefore restricts the use of an Arduino in complex projects. I/O connectivity, however, allows the Arduino to excel providing both digital and analogue streams [1].

L. Chapman · C. Gray · C. Headleand (✉)
School of Computer Science, Bangor University, Bangor, Wales, UK
e-mail: c.headleand@bangor.ac.uk

L. Chapman
e-mail: eeu239@bangor.ac.uk

C. Gray
e-mail: c.gray@bangor.ac.uk

M. Bramer and M. Petridis (eds.), *Research and Development in Intelligent Systems XXXII*, DOI 10.1007/978-3-319-25032-8_34

The use of development boards to produce low-cost autonomous robots has wide ranging impact outside of the laboratory. The availability of parts, and ease of development has the possibility of democratising the field. Allowing access to autonomous robotics by parties who would have previously lacked the funds or technical abilities to exploit the opportunity. We are primarily motivated by possible applications in high risk environments. In which expensive robotics may be undesirable due to the possibility of losing a high-value asset. While deploying high-cost systems my be undesirable, the nature of these environments means that the guiding AI must be robust, and capable to recovering from hardware failure. Unfortunately, the limitations of individual boards make them generally unsuitable for many of these applications.

We, therefore, propose the use of Arduino and Raspberry Pi boards together in a single prototype architecture. This opens several opportunities with regards to accessibility and usability. The use of multiple independent boards could allow for real-time operation of a variety of tasks. Made possible by separating processing of sensory data, action selection and actuator output.

Raspberry Pi and Arduino boards have be used together before, in a range of different projects (for examples see [7]). These projects appear to use a single connection between a Raspberry Pi and an Arduino. The single Arduino board is responsible for both processing of information collected, and motor control.

The use of a three-layered architecture is commonly associated with the 'sense-think-act' artificial intelligence paradigm [3]. However, in these implementations a single microprocessor handles all the execution of all three layers. Expanding to a two board setup does not add resilience; a system crash on a single board is significant as two thirds of the platform is reliant on a single piece of hardware. One solution would be to separate the sense and act components onto their own boards/processors, to produce a more resilient system. In the event of a crash on any of the three boards, the robot can continue to function albeit in a limited or reduced capacity. This decreases the impact of any failure, and may allow recovery of the failed component as it is no longer actively relied on for operations [8].

2 Prototype Robot Platform

Development of our prototype platform tests the practicality of embodying the three layer model in a simple robot. This prototype is just one example of our proposed architecture, providing an elegant test-bed for future projects.

2.1 Hardware

Our processing architecture utilised one Raspberry Pi, powered by a 5 v DC supply (4 × AA 1.5 v cells). Two Arduino boards connected to the Pi via USB,supplying both a control channel and power. The robotics platform was an off-the-shelf 'YoBot!'—a small battery powered tracked robot. Physically carrying the boards, and three (HC-SR04) ultrasonic sensors. The YoBot! is controlled by four internal light sensors. These are normally activated by a mobile phone app, but any light source can be used. Actuating different combinations of these sensors select which motor(s), left, right or both, are enabled and in which direction. Our prototype used four 3 mm LEDs to interface with the YoBot!'s control interface. The Raspberry Pi requires a standard SD card to store and run the Python control script.

2.2 Implementation

The first Arduino board (providing SENSE services) processes incoming sensor data. Our implementation receives three distances from the ultrasonic sensors. These readings are collated into a single packet/string and sent via a serial USB interface to the Raspberry Pi.

The Raspberry Pi, (the THINK component) the central node in the architecture, manages the logic and performs action selection. The selection is based on the data packet received from the SENSE Arduino. At the beginning of each operational cycle, the control script checks for a new packet. If one is available, the contents are unpacked into three variables and the values are used by the algorithm to determine the next action. In the event that a new packet isn't available or only a partial packet is received, then action selection is based on the last received values. A timer is maintained that tracks when the control variables were last updated. An increasing value of this timer indicates likely hardware failure is the SENSE component.

The second Arduino (ACT implementation) receives commands (such as moveForward) from the Raspberry Pi. These commands are interpreted into an appropriate actions. This step can be compared with the locomotion layer of a Reynolds-Boids animation system [6]. In our prototype implementation the appropriate actions are triggering one or more of the connected LEDs to control the YoBot!. As with the THINK layer, each loop checks for a new command. If one is available, the current command is updated. Otherwise, the action continues acting under the last received instruction. A timer tracks the last update. As in the THINK component, an ever increasing timer indicates a likely hardware fault.

As both the THINK and ACT layers are loosely coupled, they will continue operating under if new data is temporarily not available. Crucially, the validity of this data decays over time (i.e. as it becomes less and less current). After a preset time-frame, subsequent layers will pause rather than continue blindly. If the delay continues, the receiving board will attempt to send a reboot command to the sending board. This two-tier response attempts to both limit the effect of a transient failure and prevent unacceptable (and potentially critical) damage to the platform if it were to continue without guiding sensor data. Simple wiring diagrams for the SENSE and ACT layers are presented in Figs. 1 and 2 respectively.

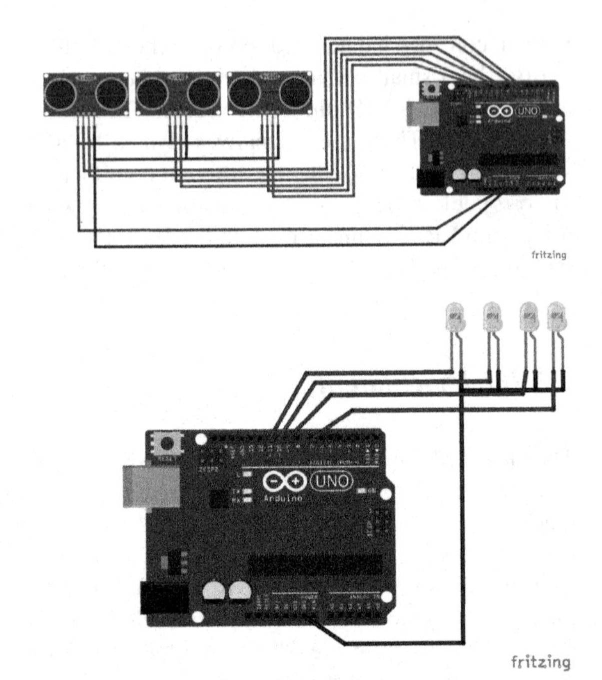

Fig. 1 Wiring diagram for the SENSE layer, indicating how the three HC-SR04 ultrasonic sensors were connected to the board

Fig. 2 Wiring diagram for the ACT component. Four LEDs allow the board to interface with, and control the YoBot! platform. Two LEDs activate each motor; one representing forward, the other reverse. For example, triggering the reverse LED for the left motor and the forward LED for the right motor would make the YoBot! rotate anticlockwise

The fundamental limiting factor in this design is the USB command pathway. The speed at which data can be exchanged on a serial interface produces a bottleneck. For our prototype packets this bottleneck was not significant enough to produce any malfunction, but this situation must be taken into account in more complex implementations.

3 Testing

In order to evaluate the architecture, we equipped the THINK component with a simple obstacle avoidance algorithm (shown in Fig. 3).

As well as testing the overall architecture, our test protocol specifically targeted failure resolution (ensuring the system is robust to the temporary failure of a single board).

The task involved placing an object in the robot's path, illustrated in Fig. 4. This task was repeated 40 times, both the obstacle and robot were placed back at the starting positions each time. We ended each cycle once the robot had negotiated the obstacle. During the experiment, the THINK component received 14 invalid (or incorrect) packets of distance information. The algorithm successfully identified the invalid input, and continued with the last valid data received. At no point did enough decay occur to exercise the reboot protocol.

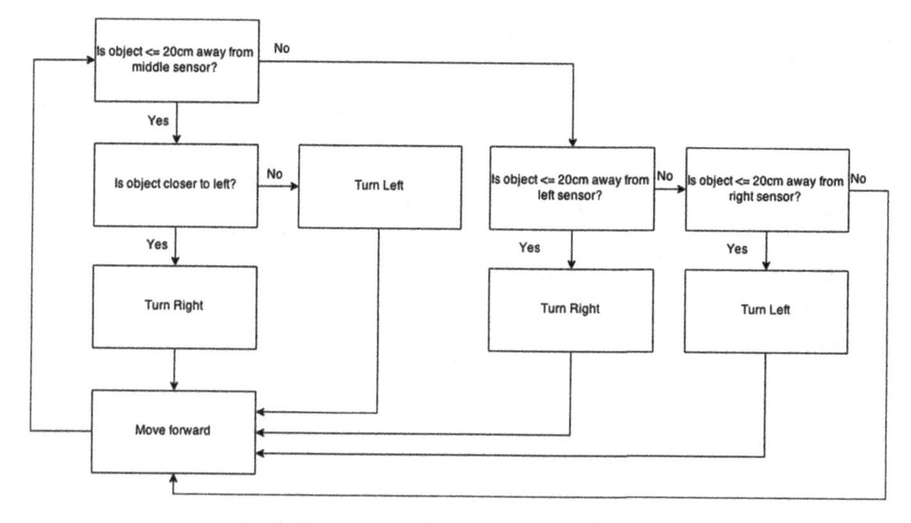

Fig. 3 Flow chart outlining the simple object avoidance algorithm that was implemented

(a) (b) (c)

Fig. 4 Three frames taken from a video of the obstacle avoidance experiment. **a** The robot moving forward detects an object **b** the robot turns towards the closest unobstructed path, **c** once the path ahead is clear the robot continues to move forwards

A second test protocol tested the hardware failure functionality. One of the ultrasonic sensors were manually disconnected and the resulting behaviour observed before reconnecting the sensor. As expected, the robot maintained the last given command and continues seamlessly once the sensor was reconnected. This interruption to the sensing input did not require a reboot of any component.

Combined these two, admittedly simple, test protocols have shown that the proposed architecture; (a) completes the task set correctly, (b) is resilient to external hardware removal/failure and (c) handles internal (unpredictable) delays. A third test protocol would need to be devised to artificially introduce enough delay to test the reboot functionality. However, we believe that the hardware removal test is more likely to occur, making this test academic at best.

4 Conclusions and Future Work

We believe that we have successfully demonstrated that a Sense-Think-Act architecture using loosely coupled hardware is an appropriate platform for autonomous robotics. Our experiment has shown that this architecture is resilient, being able to respond to both internal delay or failure and external failures. We have also demonstrated that the use of commodity hardware platforms, such as the Arduino and Raspberry Pi boards, is not only viable but suited to the task at a fraction of the cost of bespoke electronics.

The system has short term fault tolerance. This is achieved by each layer acting upon the most recently received 'good' data. If no new (or invalid) data is received, the confidence in the previous values are decayed over time. The decay function ensures change, over time, is accounted for. Therefore; the robot will pause to wait for newer data, if the last recorded value becomes unusable.

As a result, we believe this architecture could be implemented in a wide variety of robotic projects, with a strong potential for further development. Future projects could seek to expand on our work, to use the architecture to create low-cost sense-think-act search and rescue robots. Such a platform would need to meet or exceed expected durability and stability of more conventional systems but at a reduced cost in both terms of materials and maintenance.

References

1. Arduino: Arduino uno (2015)
2. DAusilio, A.: Arduino: a low-cost multipurpose lab equipment. Behav. Res. Methods **44**(2), 305–313 (2012)
3. Gat, E., et al.: On three-layer architectures. Artif. Int. Mob. Robots **195**, 210 (1998)
4. Heeks, R., Robinson, A.: Ultra-low-cost computing and developing countries. Commun. ACM **56**(8), 22–24 (2013)
5. Raspberry Pi: Raspberry pi. Raspberry Pi 1 HDMI 13 Secure Digital 34 Universal Serial Bus 56 Python (programming language), vol. 84, p. 1 (2013)
6. Reynolds, C.: Boids (2006)
7. Zagros Robotics: Building robots with raspberry pi and python (2015)
8. Visinsky, L.M., Cavallaro, J.R., Walker, I.D.: Expert system framework for fault detection and fault tolerance in robotics. Comput. Electr. Eng. **20**(5), 421–435 (1994)
9. Zachariadou, K., Yiasemides, K., Trougkakos, N.: A low-cost computer-controlled arduino-based educational laboratory system for teaching the fundamentals of photovoltaic cells. Eur. J. Phys. **33**(6), 1599 (2012)

Printed in the United States
By Bookmasters